Plasma Deposition of Amorphous Silicon-Based Materials

Plasma–Materials Interactions

A Series Edited by

Orlando Auciello
MCNC
Electron Technology Division
North Carolina State University
Research Triangle Park, North Carolina

Daniel L. Flamm
University of California at Berkeley
Berkeley, California

Advisory Board

A list of the titles in this series appears at the end of this volume.

Plasma Deposition of Amorphous Silicon-Based Materials

Edited by

Giovanni Bruno
Centro di Studio per la Chimica dei Plasmi, CNR
Dipartimento di Chimica
Universita di Bari
Bari, Italy

Pio Capezzuto
Centro di Studio per la Chimica dei Plasmi, CNR
Dipartimento di Chimica
Universita di Bari
Bari, Italy

Arun Madan
MV Systems, Inc.
Golden, Colorado

ACADEMIC PRESS

Boston San Diego New York
Berkeley London Sydney
Tokyo Toronto

This book is printed on acid-free paper. ∞

Academic Press, Inc.
A Division of Harcourt Brace & Company
525 B Street, Suite 1900, San Diego, California 92101-4495

United Kingdom Edition published by
Academic Press Limited
24-28 Oval Road, London NW1 7DX

Library of Congress Cataloging-in-Publication Data

Plasma deposition of amorphous silicon-based materials / by Giovanni Bruno, Pio Capezzuto, Arun Madan.
p. cm. -- (Plasma--materials interactions)
Includes index.
ISBN 0-12-137940-X
1. Amorphous semiconductors--Design and construction. 2. Silicon alloys. 3. Plasma-enhanced chemical vapor deposition. I. Bruno, Giovanni. II. Capezzuto, Pio. III. Madan, A. (Arun) IV. Series.
TK7871.99.A45P55 1995
621.3815'2--dc20 95-12433
CIP

PRINTED IN THE UNITED STATES OF AMERICA
95 96 97 98 99 00 BB 9 8 7 6 5 4 3 2 1

Contents

Contributors

Numbers in parentheses indicate the pages on which the authors' contributions begin.

GIOVANNI BRUNO (1), *Centro di Studio per la Chimica dei Plasmi CNR, Dipartimento di Chimica, Universita di Bari, 70216 Bari, Italy*

PIO CAPEZZUTO (1), *Centro di Studio per la Chimica dei Plasmi CNR, Dipartimento di Chimica, Universita di Bari, 70216 Bari, Italy*

GRAZIA CICALA (1), *Centro di Studio per la Chimica dei Plasmi CNR, Dipartimento di Chimica, Universita di Bari, 70216 Bari, Italy*

BERNARD DRÉVILLON (63), *Laboratoire de Physique des Interfaces et des Couches Minces, CNRS UPR 258, Ecole Polytechnique, F-91128 Palaiseau, France*

C. M. FORTMANN (131), *Electrical Engineering Department, Pennsylvania State University, University Park, Pennsylvania, 16802*

YOSHIHIRO HAMAKAWA (283), *Faculty of Engineering Science, Osaka University, Toyonaka, Osaka 560, Japan*

WEN MA (283), *Faculty of Engineering Science, Osaka University, Toyonaka, Osaka 560, Japan*

ARUN MADAN (243), *MV Systems, Inc., Golden, Colorado, 80401*

DIMITRI S. MATARAS (63), *Laboratory of Plasma Chemistry, Department of Chemical Engineering, University of Patras, GR-26110 Patras, Greece*

HIROAKI OKAMOTO (283), *Faculty of Engineering Science, Osaka University, Toyonaka, Osaka 560, Japan*

JÉRÔME PERRIN (177), *Laboratoire de Physique des Interfaces et des Couches Minces, CNRS UPR A0258, Ecole Polytechnique, F-91128 Palaiseau, France*

DIMITRI E. RAPAKOULIAS (63), *Laboratory of Plasma Chemistry, Department of Chemical Engineering, University of Patras, GR-26110 Patras, Greece*

GUY TURBAN (63), *Laboratoire des Plasmas et des Couches Minces, Institut des Matériaux—UMR110—CNRS, University of Nantes, F-44072 Nantes, France*

Preface

Amorphous semiconductors, specifically amorphous silicon, have become extremely important from a technological point of view. About 40 applications have been identified, some of which have been used commercially, such as TFT's in active-matrix liquid crystal displays, linear image sensors for facsimile machines, solar cells, and electrophotographic drums in photocopying machines. The common underlying theme is the large area amorphous silicon fabrication using a plasma deposition technique.

Plasma Deposition of Amorphous Silicon-Based Materials attempts to link the fundamental growth kinetics involving complex plasma chemistry with the resulting semiconductor film properties and the subsequent effect on the performance of the electronic devices produced.

The intent of this book is to provide active researchers in the field, and also graduate students and engineers working on the production lines, with a tool able to cover details from the fundamentals of deposition kinetics to the salient issues in large area mass production techniques, with the plasma chemistry being the focus.

This volume consists of chapters written by different authors on (1) the chemistry and (2) the diagnostics of the silicon deposition processes, (3) the properties of silicon alloys, (4) the reactor design for amorphous silicon deposition, (5) the relationship between optoelectronic properties and plasma process parameters, and (6) the development and applications of amorphous silicon-based devices.

Finally, the editors wish to dedicate the book to Professor P. G. Le Comber, whose contribution to this book has been missed due to his untimely death.

G. Bruno
P. Capezzuto
A. Madan

1 Chemistry of Amorphous Silicon Deposition Processes: Fundamentals and Controversial Aspects

Giovanni Bruno, Pio Capezzuto, and Grazia Cicala
Centro di Studio per la Chimica dei Plasmi
Department of Chemistry
University of Bari
Bari, Italy

I. Introduction

Hydrogenated amorphous silicon (a-Si : H), although simple in formula, is a highly complex material with unique properties, which have provoked widespread scientific interest and stimulated a variety of technological applications. Consequently, a number of optical, electrical, and structural studies have been carried out on a-Si:H films. These investigations revealed that the material properties depend strongly on the preparation conditions. All the preparation techniques developed thus far can be collected mainly under the umbrella of the chemical vapor deposition (CVD), in which the film is produced *in situ* via chemical reactions. Generally, the deposition process can be expressed as

$$SiX_4 \rightarrow \text{a-Si:X} \qquad (X = H, F, Cl),$$

where the decomposition of silicon volatile compounds (SiH_4, SiF_4, Si_2H_6, $SiCl_4$, SiH_2F_2, ...), affected by different kinds of energy, results in the deposition of X-containing amorphous silicon in the form of thin film. Thermal activation, stimulation by radiation, and electrical excitation have been used to provide energy to the reactive chemical system in low-pressure CVD (LPCVD) [1], photo-CVD [2], and plasma enhanced CVD (PECVD), respectively. Accordingly, the energy is channeled into different freedom degrees of the molecules, as shown in Fig. 1.

In a conventional LPCVD technique, the deposition occurs under thermal activation onto a substrate that is heated at moderate temperature (500–800°C) to promote the deposition reaction and to provide the adatoms with sufficient mobility to obtain the desired structure.

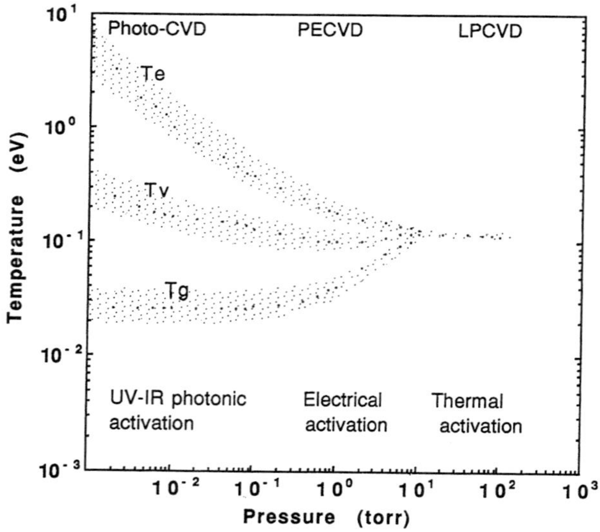

FIGURE 1. Typical dependence of electron, vibration, and gas temperature on pressure in an electrical discharge. The working regions of photo-CVD, PECVD, and LPCVD are also shown.

Photo-CVD utilizes both IR and UV light irradiation to activate the chemical process for film deposition. Specifically, using IR irradiation (laser CVD) the energy is imparted directly to the vibrational levels ($T_v \gg T_g$) up to the dissociation limit, whereas, using UV irradiation, the energy for electron excitation (photolysis) is transmitted to the reactants either directly or through a foreign atomic species (e.g., Hg).

In the PECVD technique, energy is directly imparted to the chemical system by the collision of energetic electrons with the heavy particles. In particular, in a plasma produced at low pressures (glow discharges), the free electrons can gain sufficient kinetic energy to activate processes of excitation (both electronic and vibrational), ionization, and dissociation, while maintaining low gas temperature. The high probability of producing active species (radicals, atoms, ions, and excited species) in the plasma phase and allowing them to interact with a surface at low temperature makes the PECVD technique attractive for producing a-Si based materials and capable of producing unique, high-quality a-Si:H thin films.

Since the work of Chittick (1969) at the Standard Telecommunications Laboratories in Harlow [3] and of Spear and LeComber at Dundee University [4], who demonstrated that amorphous silicon produced in glow discharges could be intentionally doped, the PECVD technique has been intensively studied. The main events generally considered as milestones in the history of the amorphous silicon are depicted in Fig. 2.

At present, from a technological point of view, amorphous silicon solar cells with an efficiency of 11–13% can be routinely produced, and new applications such as electrophotography, contact image sensors for facsimile (FAX) machines, thin-film transistors for display have been commercialized and are discussed more

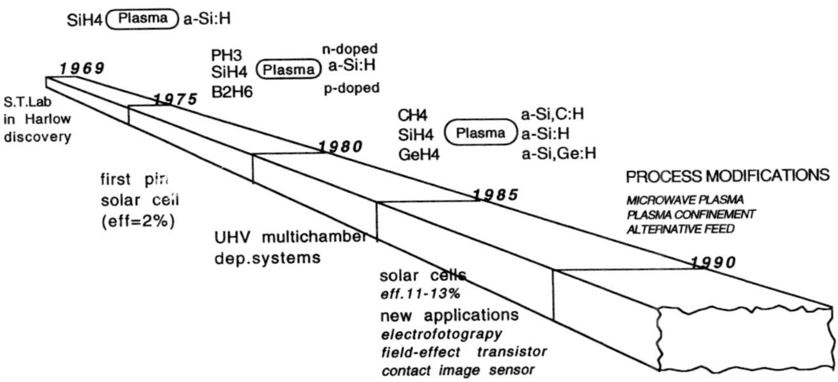

FIGURE 2. History of amorphous silicon: process and technology developments.

fully in Chapter 6. However, from a scientific point of view, intensive research is being carried out to understand the growth kinetics involved in the formation of amorphous silicon, such as process modification involving microwave plasma, plasma confinement, plasma modulation, light irradiation, and use of alternative feedstock gases. As for the future, the research direction will be dictated by answers to some open questions. In particular, from the scientific point of view, arguments still debated are (1) identity of the growth precursors, (2) plasma–surface interaction, (3) a better control of the interfaces between different layers, and (4) microscopic parameters affecting the film quality. From the technological viewpoint, important goals to be reached are (1) the increase of the film deposition rate without affecting the optoelectronic properties, for low cost production; (2) the improvement of the surface homogeneity, for the scaling-up process to large-area devices; and (3) the achievement of the topmost material quality and stability, for high-efficiency and stable devices.

The purpose of this chapter is to review the features of the glow discharge plasma that are of importance in understanding the fundamentals of amorphous silicon PECVD process, and to survey some specific aspects of the amorphous silicon deposition that are used in clarifying the chemistry of the growth and in promoting the technological development.

II. Some Fundamentals on Plasma Deposition

A. Plasma: Fundamental Concepts and Properties

A *plasma* can be simply defined as a partially ionized, quasineutral gas. If the very low number of charged particles (ions and electrons) existing in any gas at any temperature is increased by an external source, the electric field due to charge separation can become strong enough to limit particles' own motion and maintain the macroscopic neutrality; in that case an ionized gas is called a *plasma.* As there is a balance between the densities of negative and positive charges in macroscopic volumes and times, the more appropriate term *quasineutrality* is frequently used.

Among the various kinds of energy that can be applied, the electrical discharge is the simplest and widespread means to sustain a plasma for a long time. Plasmas can be loosely grouped into two generic categories: nonequilibrium (nonisothermal or "cold") and equilibrium (isothermal or "thermal") plasmas. Under conditions of high electrical field [direct current (DC) or radiofrequency (RF)] applied for nonequilibrium plasmas generated at reduced pressure (0.5–500 mtorr), free electrons are accelerated to high energies (1–10 eV). However, because of their large mass, neutrals and ions in the plasma have low energy (a few hundredths of eV). This energy difference results in a high temperature for the electrons (50,000 K) and a low, or "cold" temperature of the neutrals and ions (500 K) (see Fig. 1). Under these nonequilibrium conditions, the initiation of chemical reactions occurs by collisions with the "hot" electrons. This allows the processing temperature to be much lower than in conventional thermal processes using similar chemistry. On the contrary, in a thermal plasma all species have the same temperature, due to the shorter mean free path of particles and higher collision frequency characterizing higher-pressure conditions. The application of an electric field is just a convenient way to supply the energy necessary to achieve a high temperature.

Over the last two decades, a large number of practical applications have been developed in the field of "cold" plasma science. Amongst others, it is worth mentioning that modern very large-scale integrated circuits (VLSIs) would not exist without sophisticated plasma processing techniques, and that many "new" materials have become available with unique chemical, structural, and physical properties, which otherwise would not have been possible (e.g., a-Si-based alloys, diamond-like carbon, Teflon-like polymers, biocompatible materials). As mentioned above, plasma processing allows the opportunity to synthesize materials at a low temperature, while simultaneously allowing chemistry to result from dissociation/ionization of feed gas (by the high-energy electrons), which would have normally occurred in thermal processes operated at a much higher temperature (>900°C).

Although it is possible to utilize thermodynamics for estimating the chemical composition of the plasma and gas kinetics in a thermal plasma condition to estimate the resulting transport properties [5, 6], a large research effort has been focused on understanding plasma under nonequilibrium conditions. These plasmas, utilized mainly for etching and deposition processes and for materials treatment, are characterized by free electrons whose energy distribution function (EEDF) is not Maxwellian, and by molecules in which the distribution among the internal degrees of freedom are non-Boltzmann. Under these conditions, typical values of the density of neutrals are 10^{13}–10^{16} cm^{-3} and charged/neutral species ratios are in the range of 10^{-4}–10^{-7}. The shape of EEDF is the result of two opposite processes: the first is due to their energy loss by collisions and the second, to their energy gain from the electric field of the discharge. For low-pressure plasma containing pure or large amounts of inert gases, the electron energy generally exhibits a Maxwellian distribution. A relevant number of studies have been carried out on the EEDF dependence on plasma parameters and composition, mainly in presence of atomic and diatomic species [7–9], but also when polyatomic molecules are present [10, 11].

In an electrical discharge, electrons tend to dominate the plasma characteristics as they are responsible for inelastic collisions leading to rotational, vibrational, and electronic excitations, dissociations and ionizations, and elastic collisions. In

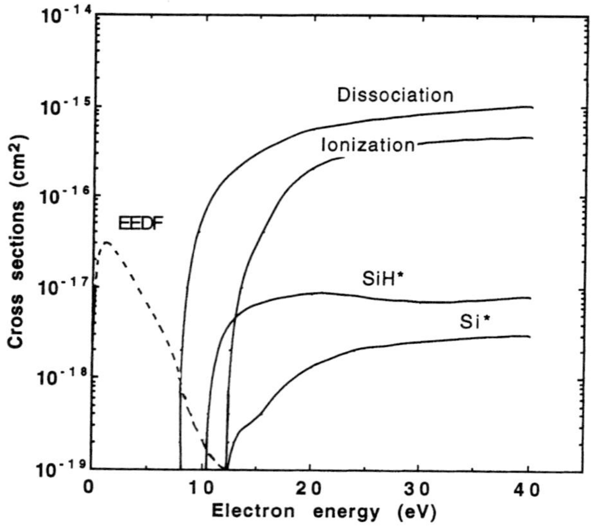

FIGURE 3. Cross sections for electron impact processes on silane. Dashed line represents the typical electron energy distribution function.

the latter case, electrons experience large change in momentum, energy, and direction, whereas the internal energy of neutral species is unchanged by the collisions. During inelastic collisions, the energy transferred from electrons to heavy particles is higher and is channeled into the various degrees of freedom of the target species.

Each collision is characterized by a cross section σ (in the hard-sphere approximation $\sigma = \pi \cdot r^2$) which is a function of energy and measures the probability that the event can occur. In Fig. 3 we show a typical EEDF in a glow discharge and the expected energy dependence of excitation (shown by the asterisks for the species SiH and Si), dissociation, and ionization cross sections, which resemble those reported for the silane molecule [12]. As the rate constants (k_e) of these collisional processes are proportional to the product of the EEDF [$f(E)$] and cross section [$\sigma(E)$] ($k_e = \int E^{1/2} \cdot f(E) \cdot \sigma(E) \cdot dE$), the overlap area of $f(E)$ and $\sigma(E)$ is a measure of the event probability. The main chemical events due to inelastic collisions by electron impact are as follows:

Dissociation.

$$\begin{aligned} &e + AB \rightarrow A + B + e, \\ &e + AB \rightarrow e + AB^* \rightarrow A + B + e. \end{aligned} \tag{1}$$

Ionization.

$$e + AB \rightarrow AB^+ + 2e. \tag{2}$$

Dissociative ionization.

$$e + AB \rightarrow A^+ + B + 2e. \tag{3}$$

Attachment.

$$e + AB \rightarrow AB^-. \tag{4}$$

Dissociative attachment.

$$e + AB \rightarrow AB^- \rightarrow A^- + B. \tag{5}$$

Recombination.

$$e + A^+ \rightarrow A. \tag{6}$$

Excitation processes due to electronic collisions are as follows:

Vibrational excitation.

$$e + AB(v > 0) \rightarrow e + AB(w), \tag{7}$$

where v and w are the vibrational quantum numbers with $w > v$.

Electronic excitation.

$$e + AB(v = 0) \rightarrow AB^* + e. \tag{8}$$

Dissociative excitation.

$$e + AB \rightarrow AB^* \rightarrow A^* + B + e. \tag{9}$$

Besides these elementary processes promoted by free electrons, the collisions among heavy particles in the redistribution of the different kinds of energy introduced by the electrons [13] are noteworthy. In particular, in the presence of molecules, the following processes have to be taken into account:

(a) The redistribution of vibrational energy among the vibrational manifold of the molecule, through *vibration–vibration* (V–V) energy exchange:

V–V energy exchange.

$$AB(v) + AB(w) \rightarrow AB(v - 1) + AB(w + 1). \tag{10}$$

This "vibrational pumping" can bring the molecule AB up to the limit of dissociation.

(b) The dissipation of the vibrational energy through *vibration–translation* (V–T) exchanges in gas phase and in the interaction with metallic walls:

V–T energy exchange.

$$\mathrm{AB}(v) + \mathrm{AB} \rightarrow \mathrm{AB}(v - 1) + \mathrm{AB}. \tag{11}$$

Wall interaction.

$$\mathrm{AB}(w) \xrightarrow{\text{wall}} \mathrm{AB}(v). \tag{12}$$

(c) The transformation of vibrational energy into electronic (V–E) and chemical dissociation (V–D) energy:

V–E energy exchange.

$$\mathrm{AB}(v) + \mathrm{AB}(w) \rightarrow \mathrm{AB}^* + \mathrm{AB}. \tag{13}$$

V–D energy exchange.

$$\mathrm{AB}(v) + \mathrm{AB}(w) \rightarrow \mathrm{AB}_2 + \mathrm{A}. \tag{14}$$

B. Plasma Anatomy

When there is a separation of a large number of electrons from positive ions in a plasma reactor, the neutrality is violated and an electric field, as in a plane condenser, takes place among the positive- and negative-charge layers. In fact, because of their high energy and small mass, electrons usually diffuse toward all the surfaces in contact with the plasma (walls and electrodes), leaving the bulk positively charged. Thus, all the surfaces in contact with the plasma phase exhibit a negative charge and the relative electric field creates a thin sheath. This aspect becomes particularly important in a RF glow discharge plasma produced in a typical parallel-plate reactor (see Fig. 4).

Here, the potential distribution across the electrode gap is shown for an asymmetrical reactor; i.e., the electrode areas are different. The higher potential fall at the cathode with respect to that of the anode is due not only to the difference of the areas but also to the presence of a blocking capacitor between the RF generator and the cathode; hence a self-bias voltage (V_b) is established onto it. This parameter becomes important as it is determined by a variety of factors:

- Electrode asymmetry
- Electron temperature and density
- Density of negative ions
- Deposition pressure

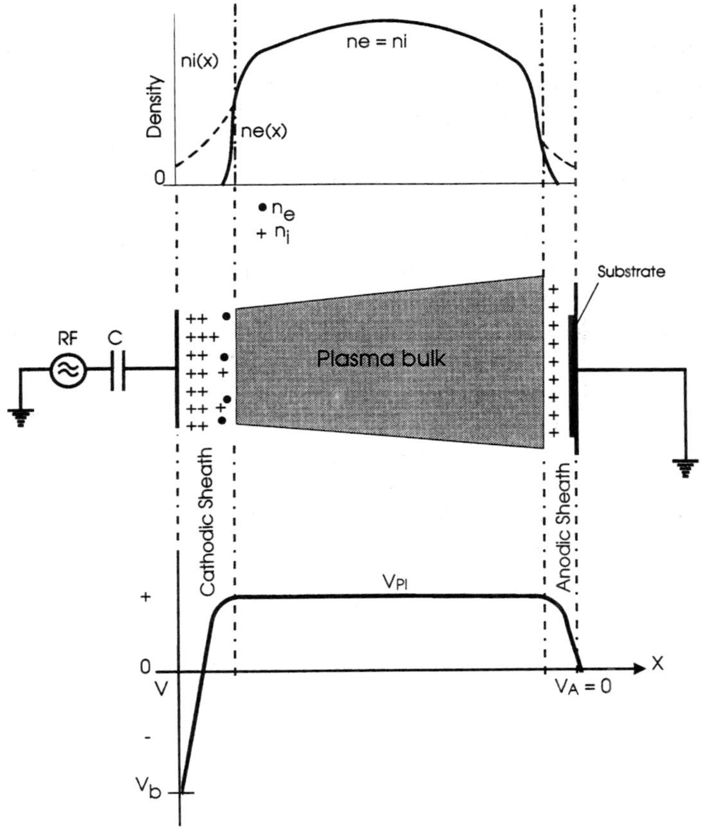

FIGURE 4. Scheme of an RF discharge in a parallel-plate reactor with capacitive coupling of the RF generator; together with the spatial distributions of potential voltage V (plasma potential V_{Pl}, bias voltage V_b, and anode voltage V_A) and electron and ion densities (n_e, n_i).

- RF power
- Electrode distance
- Frequency of the alternating-current (AC) field

The value V_b determines the processes occurring at the RF electrode, mainly physical processes such as ion bombardment, and hence sputtering and secondary-electron emission (α–γ transition; see Chapter 4). All these effects become crucial when the plasma reactor is scaled up to systems for large-area deposition.

The different potential distributions at the cathodic and anodic regions also causes a difference in the thickness of their relative sheaths. Rectification of the

RF field occurs in both sheaths, since the ions are accelerated toward the electrodes and the electrons are reflected to the plasma. This leads to a spatial distribution of n_e and n_i along the interelectrode distance as shown in Fig. 4. Therefore, a low-pressure capacitively coupled RF discharge is well characterized, and many studies [14] have been carried out on the dynamic behavior of the RF sheaths. The study of the sheath properties is of particular interest, because they influence the ion energy distribution and hence the extent of ion bombardment onto the growth surfaces (see Section I.D). This process is more or less effective in dependence on the excitation frequency and pressure. As an example, Matsuda *et al.* [15] have reported that in a silane plasma, the number of impinging ions onto the growing surface strongly increases when the frequency (ω) decreases below the ion plasma frequency ($\omega_i = 1$ MHz). At higher frequency ($\omega > \omega_i$) the ions are unable to follow the AC field and the sheaths can be considered as capacitors [16]. However, on the other hand, at a low frequency regime ($\omega < \omega_i$), the sheath becomes essentially resistive [17]. These considerations apply also to cathode where, in addition, the effect of the frequency variation on the self-bias voltage must be included. As an example, in the range of frequency 13–75 MHz, a strong variation of V_b has been measured for SiH_4–H_2 plasmas as shown in Fig. 5.

Hence, at 13 MHz a very efficient ion bombardment is operative and the secondary emission coefficient γ, relative to electrons coming from electrode be-

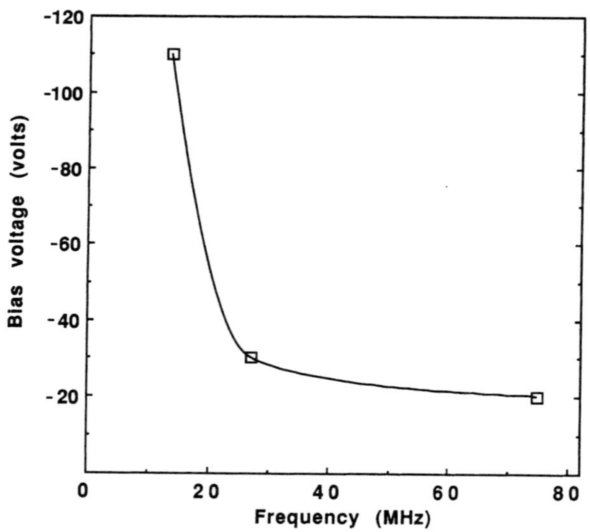

FIGURE 5. Bias voltage of the RF-powered electrode vs. excitation frequency in SiH_4–H_2 plasmas.

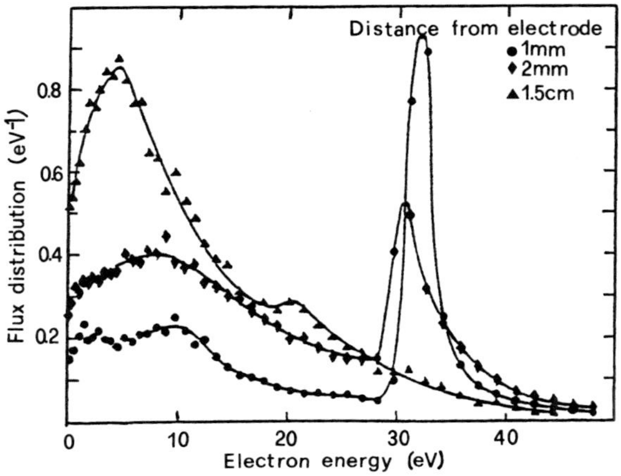

FIGURE 6. Computed electron flux distributions for an RF discharge in CO for $\gamma \gg 1$. The gas pressure is 20 mtorr and electrode separation, 3 cm. The distributions have been normalized separately at each spatial position. (From Kushner [18].)

comes large. Figure 6 shows the spatially resolved electron distribution calculated by Kushner [18] with a Monte Carlo simulation as a function of the distance from the cathode. It is evident that the dominant contribution of energetic secondary electrons is close to the electrode, which strongly increases the electron temperature in the sheath region when γ is greater than 1. At a larger distance from the cathode a typical Maxwellian EEDF is established within the bulk of the plasma. In contrast, when γ is less than 1, the spatial distribution of electron temperature along the interelectrode distance is essentially flat and also close to electrode regions [13]. Experimental measurements of n_e and kT_e spatial profiles, performed by electrical double-probe technique, are shown in Fig. 7 for SiH_4–H_2 plasma under conditions normally employed to produce high-quality amorphous silicon films [19]. The extent of the cathode sheath is very high in comparison with the plasma volume, as deduced by the n_e profile. The trend of electron temperature kT_e is opposite that of the electron density and exhibits a strong increase near the cathode sheath, which is indicative of the fact that the discharge is sustained mainly by the energetic secondary electrons.

C. Gas-Phase Processes in Plasma Deposition

The main plasma reactions, leading to radicals, atoms and ions, and active species in plasma deposition, can be summarized as in the following:

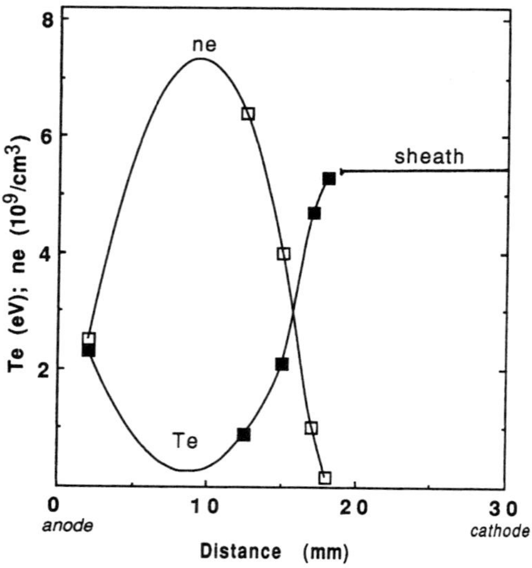

FIGURE 7. Electron density (n_e) and temperature (kT_e) profiles in the electrode gap for SiH_4–H_2 plasma (10 sccm, 0.2 torr, 5 W).

(a) *Electron–molecule reactions.*

$$SiX_4 + e \rightarrow SiX_n + (4 - n)X + e, \tag{15}$$

$$SiX_4 + e \rightarrow SiX_n^+ + (4 - n)X + 2e, \tag{16}$$

$$SiX_4 + e \rightarrow SiX_n^- + (4 - n)X. \tag{17}$$

(b) *Neutral–neutral reactions.*

$$H + SiX_4 \rightarrow SiX_3 + HX, \tag{18}$$

$$SiX_4 + SiX_n \rightarrow Si_2X_{4+n} \rightarrow \rightarrow Si_nX_m. \tag{19}$$

(c) *Ion–molecule reactions.*

$$SiX_n^+ + SiX_4 \rightarrow Si_2X_m^+ + (4 + n - m)X, \tag{20}$$

which can refer to SiH_4, SiF_4, and $SiCl_4$ systems (X = H, F, Cl).

From the literature, it appears that the relative abundance of SiX, SiX_2, and SiX_3 radicals depends on which of the three types of reactions reported in the previous scheme is invoked [20]. This, in turn, is determined by the plasma parameters (pressure, power, frequency, gas composition), mainly those affecting the electron density and energy.

There is some controversy as to the major precursor for growth; for instance, Wagner and Veprek [21] regard SiH_2 as the precursor, whereas Kampas and Griffith [22, 23] invoke both SiH_2 and SiH_3, and Robertson and Gallagher [24] consider only the SiH_3 radical. In addition, the last author invokes reaction of reaction (18) as an additional source of SiH_3 radical.

An exhaustive list of gas-phase processes occurring in SiH_4 plasmas can be found in Kushner [11] where, from the discussed chemical model, SiH_2 and SiH_3 are reported as the main radicals in the discharges involving SiH_4–H_2 and SiH_4–Ar gas mixtures. In the same reference some consideration on the formation of higher silanes are also developed as to their contribute to the film formation. From the experimental viewpoint, silylene (SiH_2) and silyl (SiH_3) radicals, have been detected [25, 26] and their absolute densities of about 10^9 cm^{-3} and 10^{11} cm^{-3}, respectively, have been measured when SiH_4 density ranges between 10^{14} and 10^{15} cm^{-3}.

When the plasma contains halogenated reactants (SiF_4, $SiCl_4$), an additional "chemical activity" is exhibited, in that an etching process can occur. In fact, silicon halides can be active for both etching and deposition processes, depending on the relative abundance of halogen atoms (F, Cl) and silicon radicals (SiF_x, $SiCl_x$), which are etchant species and building blocks for film formation, respectively. The same types of reactions listed in reactions (15)–(20) have been involved in producing halogenated silicon radicals [27–29]. In addition to the reaction process of Eq. (18), the halogen scavenging reaction

$$\mathrm{H} + \mathrm{X} \rightarrow \mathrm{HX} \tag{21}$$

involving H atoms becomes important in controlling the deposition since it prevents the etching process from being effective. This is clearly evidenced in the data of Fig. 8, where the transition between the deposition and etching regimes is seen to occur when the H atom density sharply decreases and, according to reaction (21), Cl atoms strongly increase. The ability of H atoms in controlling the deposition–etching transition has also been confirmed in SiF_4–H_2 system for a-Si:H,F film growth [28].

1. Electron-Impact Excitation Processes: Origin of Emitting Species

We now consider the electron impact excitation processes that determine the light emission through the subsequent radiative decay of electronic excited species. The pioneering work of SiH_4 plasma by Kampas and Griffith [22] revealed that the processes

$$\mathrm{SiH_4} + \mathrm{e} \rightarrow \mathrm{SiH_4^*} \begin{array}{l} \nearrow \mathrm{SiH^*} + \mathrm{H_2} + \mathrm{H} + \mathrm{e} \\ \searrow \mathrm{Si^*} + 2\mathrm{H_2} + \mathrm{e} \end{array} \tag{22}$$

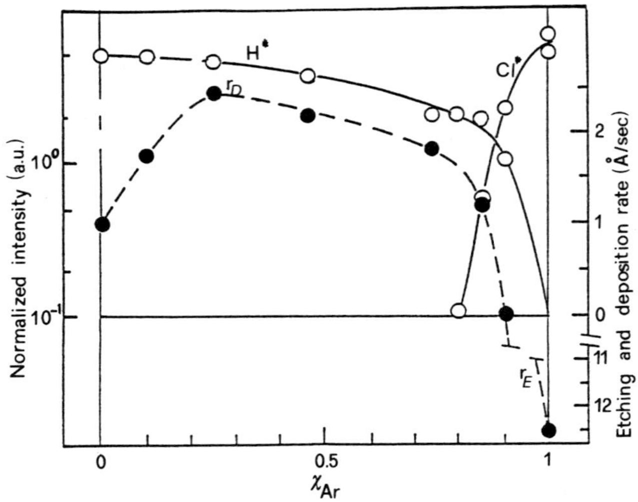

FIGURE 8. Intensity of the emitting species (H* and Cl*), normalized to Ar peak at 750 nm, and of deposition r_D and etching r_E rates vs. H_2–Ar mixture composition χ_{Ar}, for $SiCl_4$–H_2–Ar plasmas.

directly produce SiH* and Si* excited species. Their radiative decays can be expressed by

$$SiH^* \rightarrow SiH + h\nu \quad (3\ eV), \tag{23}$$

$$Si^* \rightarrow Si + h\nu \quad (4.9\text{–}5.1\ eV),$$

which produce typical emissions as listed in Table 1 and were detected by the optical emission spectroscopy (OES) technique. The OES intensities of SiH* and Si* have been used [30, 31] by the same authors to model the deposition of a-Si:H films and, in particular, to correlate their deposition rate with the variation of the electron density and energy when the silane concentration is changed.

In halogenated systems (SiF_4 and $SiCl_4$), the formation of emitting species (see Table 1) SiF_x^* ($x = 3, 2, 1$) and $SiCl_x^*$ ($x = 2, 1$) is, on the contrary, determined by a two-electron process [28]. As an example, for SiF_4 plasmas, whose typical spectrum is shown in Fig. 9, the SiF_x^* formation processes can be described as follows:

$$SiF_4 + e \rightarrow SiF_x + (4 - x)F + e, \quad (x = 1, 2, 3) \tag{24}$$

$$SiF_x + e \rightarrow SiF_x^* + e, \tag{25}$$

$$SiF_x^* \rightarrow SiF_x + h\nu. \tag{26}$$

Table 1 Principal spectral systems observed in SiH_4–H_2, SiF_4–H_2, and $SiCl_4$–H_2 discharges

Species	Emission wavelength (nm)	Transition	Energy of emitting state above ground state (eV)
Si	251–253	UV1 $4s^3P^0$–$3p^{21}D$	4.9
Si	288	UV43 $4s^1P^0$–$3p^{21}D$	5.1
SiH	414.23	$A^2\Delta$–$X^2\Pi$	3.0
SiF_3	240.22–240.73	$^2B_1$–X^2A_1	6.1[a], 5.47[b]
SiF	442.98–443.02	Sistema α $A^2\Sigma^+$–$X^2\Pi$	2.82[b]
SiF_2	395.46	$^3B_1$–$^1A_1$	3.6[a], 3.27[b]
F	703.75	$3p^2P^0$–$3s^2P$	14.75[a]
SiCl	281–282	$B'^2\Delta$–$X^2\Pi$	4.4
SiCl	287–288	$B^2\Sigma^+$–$X^2\Sigma$	4.2
$SiCl_2$	Continuum 310–400	—	—
Cl	755	$4p^4S^0$–$4s^4P$	10.6
H	486.13	H_β $3d^2D$–$2p^2P^0$	12.70[a]
	656.28	H_α $4d^2D$–$2p^2P^0$	12.09[a]
H_2	602.13	$3p\ ^3\Pi$–$2s\ ^3\Sigma$	14.00[a]
Ar	750.38[c]	$4p'(½)$–$s'(½)^0$	13.48[a]

[a] Calculated values.
[b] Experimental values.
[c] Actinometer emission line.

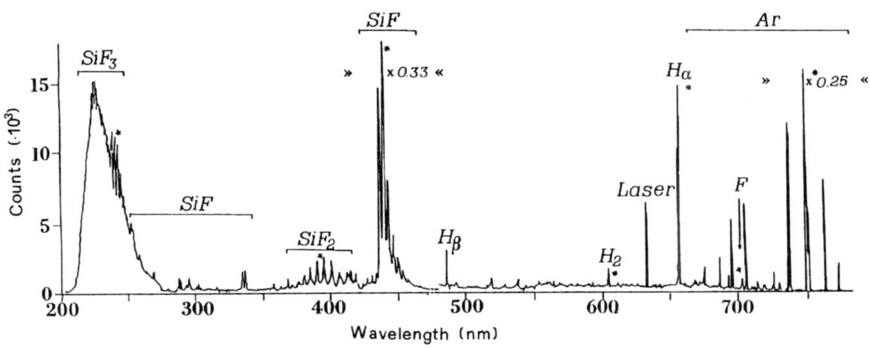

FIGURE 9. Typical optical emission spectrum from RF discharge in SiF_4–H_2–Ar mixture. (From Bruno *et al.* [28].)

Here, the density of SiF_x^*, and hence the emission intensity $I_{h\nu}$, can be related to that of the same species in the ground state, under conditions in which the decay is only radiative, through the equation

$$I_{h\nu} \propto [SiF_x^*] \propto k_e^* \cdot n_e \cdot [SiF_x], \tag{27}$$

where k_e^* is the rate constant of the electron excitation process [reaction (25)] and n_e is the total electron density.

The SiF_x radical density can be evaluated from the emission intensity by utilizing the actinometric approach [28, 32] in the OES technique, in which the actinometry consists of adding a small amount of an inert gas (Ar, He) to the reactive plasma. The emission of the actinometer (Ar, He) is used as a probe for the excitation conditions, i.e., density of electrons (n_e) with energy higher than that of the resonant value. The validity of Ar as a probe for the resonant electron density has been confirmed by OES and Langmuir electrical probe (LEP) joined analysis for both SiH_4–H_2 and SiF_4–H_2 systems [19]. The presence of emitting species can be exploited to study the process kinetics in modulated plasmas coupled with time-resolved OES (TROES) technique. This approach, developed in our laboratory [29] to investigate the SiH_4 and SiF_4 systems, has allowed to confirm the processes of reactions (22)–(26) and to produce data on the formation kinetics of SiF_x radicals [see reaction (24)]. For more details, see Section IV.D.

D. Plasma–Surface Interaction Processes

There is little understanding of the complex and multiple interaction processes that occur at the juncture of the plasma and a surface, at which point there is the breaking and formation of chemical bonds. This knowledge is fundamental for understanding the plasma-assisted etching and deposition processes. The species produced in the plasma react at the solid surface to produce volatile compounds that desorb and/or building blocks for the film growth.

A fundamental study of the bond breaking or molecular dissociation at surfaces in a situation without plasma was made in the 1930s by Lennard-Jones [33], who described the variation in potential energy as a species approaches a surface as shown in Fig. 10. For the sake of simplicity, the energetics of the adsorption/desorption processes are also depicted in Fig. 10. Curve 1 depicts the weak van der Waals interaction of the intact molecule with the surface at large distances; it should be noted that this involves physical adsorption and is not an activated process and, therefore, requires a low adsorption energy and is typically less than 0.5 eV. Curve 2 represents the chemisorption (adsorbed molecules approach the surface by involving covalent bonds) and is an activated process requiring adsorption energies of 1–10 eV; it should be noted that the molecules can be chemisorbed in their molecular state or can be dissociated into atoms. This last event, known as

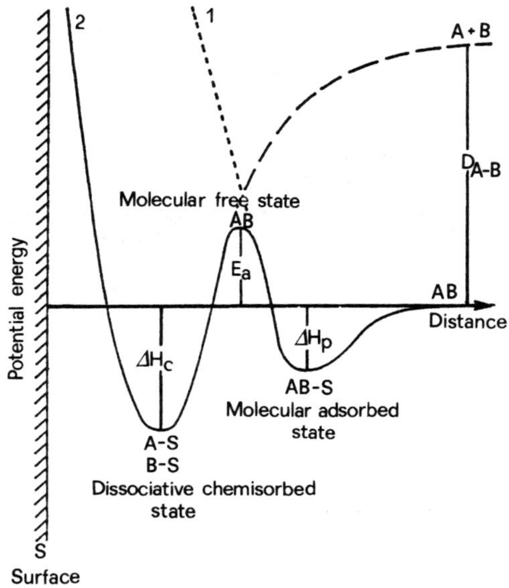

FIGURE 10. Lennard-Jones diagram of the potential energy of AB molecule approaching the surface: ΔH_c = enthalpy of dissociative chemisorption, ΔH_p = enthalpy of physiosorption, E_a = chemisorption activation energy, D_{A-B} = dissociation energy.

dissociative chemisorption, constitutes the essence of the collision-induced dissociation or activation mechanism that causes an unreactive molecule to become reactive on a surface [34, 35]. Hence, the energetics of adsorption/desorption is influenced by the exposure to a plasma and becomes dictated by the extent of the activation energy barriers E_a. The main factors that could change the E_a values in the plasma are a number of different physical processes, such as bombardment with ion, electron, hot species, and photons, which induced excitation of the adsorbed species. The above-cited physical processes can lead to the following:

- The formation of dissociated species with high probability of diffusion into the bulk or interaction with other species at the surface.
- The formation of some surface defect sites leading to a reduction in the barrier activation energy.
- The removal of foreign species from the surface, which can interfere with and inhibit the dissociative chemisorption.

We therefore conclude that the plasma has a dual role: (1) to supply internal energy to the molecules and/or radicals that can potentially dissociate on the surface and (2) to produce energetic species (hot atoms and ions), which, like a hammer, collide with adsorbed molecules giving energy for their dissociation [34], as illustrated in Fig. 11.

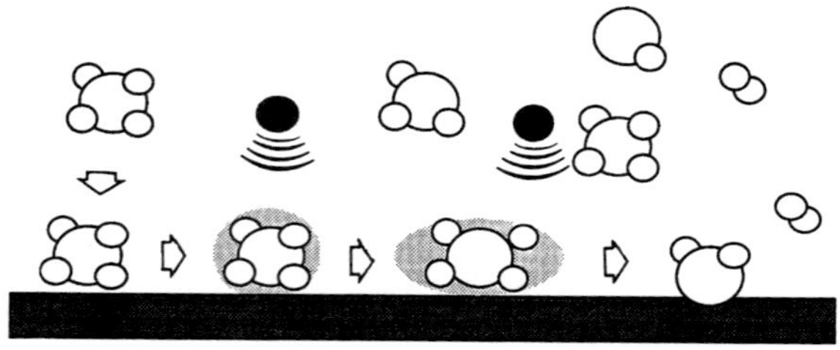

FIGURE 11. Schematic diagram showing the activation of SiX_4 dissociative chemisorption by plasma-produced energetic species.

We now examine in more detail the plasma–surface interaction in a typical system used for the deposition of amorphous silicon. The three main processes are now discussed in turn.

1. Adsorption/desorption process. This is a common hypothesis to consider this process as forced intermediate step of the deposition process. These surface chemical reactions occur along favored routes involving surface species that are in energetic equilibrium with the surface referred to as Langmuir–Hinshelwood reaction [36]. As an example, the fragments SiX_n (with X = H, Cl, F; n = 0-3), emanating from the electronic impact dissociation of SiX_4, undergo adsorption on the hydrogenated silicon surface and can be represented as

$$SiX_n(\text{gas} \rightleftarrows SiX_n(\text{ads}). \tag{28}$$

The same adsorbed species (ads) can originate also from dissociative chemisorption process of SiX_4:

$$SiX_4(\text{gas}) \rightleftarrows SiX_n(\text{ads}) + (4 - n)X(\text{ads}). \tag{29}$$

Since the surface is part of the semiconductor material, the chemisorption process will be strongly affected by its electronic character. In fact, in Section II.D.1 it will be shown that any factor modifying the Fermi level of the semiconductor, will modify the adsorption/desorption equilibrium. Hence, doping and light irradiation play a key role, as will be discussed below. The adsorbed SiX_n species behave as building blocks for the material growth, after subsequent reactive processes, as discussed in the next section.

2. Surface-reactive processes. Once adsorbed, the species may react with each other or with gaseous radicals and after the desorption of stable molecules give rise to free-bond silicon species, active sites for the material growth. These processes also include the reactions of silicon etching. Such a class of surface processes is called Eley–Rideal reaction [37] and is well documented in the kinetics of halogen removal from silicon(100) by hydrogen-atom bombardment. The de-

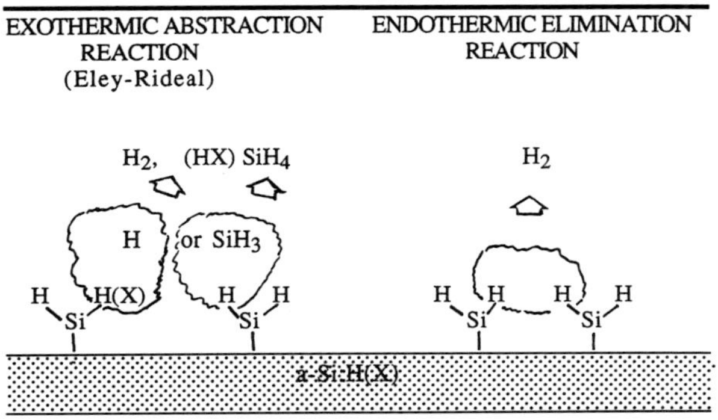

FIGURE 12. Scheme of H-desorption reactions.

sorption reaction of hydrogen halide can be expressed by first-order kinetics with respect to both atomic hydrogen flux and halogen coverage. In addition atomic hydrogen is present in many silicon deposition processes, such as PECVD, and its role to abstract hydrogen [24] and/or halogen [27] atoms may be kinetically important in modelling and controlling the deposition process. This abstraction of hydrogen and/or halogen from the silicon surface is an elementary process of particular interest, since it controls the hydrogen and halogen content in the material, which in turn dictates the opto-electronic properties of the material and is discussed more fully in Chapter 6.

The hydrogen desorption reaction can occur by exothermic abstraction reaction [24] and/or endothermic elimination reaction [38–40] as shown in Fig. 12.

3. Surface bombardment by charged particles. Strongly related to the electrical features of the plasma, the surface bombardment can play a relevant role on the growth kinetics and on the material properties since it acts on both processes (1) and (2) discussed above. In this respect, data reported until now leads to controversial conclusions and will be treated in more detail in Chapters 2 and 4. Generally, the three processes mentioned are contemporaneously present during the deposition of silicon films. However, only one of them, playing the role of rate limiting step, can be evidenced by the kinetic analysis. It should be noted that in some cases the rate of the overall surface process is controlled by reactions involving neutral species [20, 36], while in others it is dictated by ions and electrons bombarding the growth surface [41, 42].

1. Chemisorption on Doped Amorphous Silicon Surfaces

The "electronic factors" related to the bond formation between the chemisorbed species and the surface are important. In particular, semiconductors provide ideal

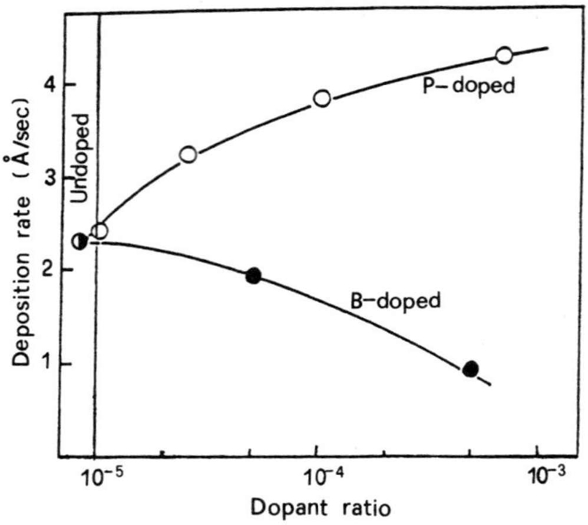

FIGURE 13. Deposition rate r_D of doped (○, ●) and undoped (◑) SiH:Cl samples from $SiCl_4$–H_2 plasmas vs. PH_3 and B_2H_6 addition. (From Bruno *et al.* [27].)

surfaces to study charge transfer during chemisorption. The results on the dopant effect shown in Figs. 13 and 14 provide evidence of the presence of an intermediate chemisorption process in the deposition of a-Si;H,Cl films from $SiCl_4$-H_2 plasmas [27]. In fact, during the deposition of n- and p-doped amorphous silicon, it is well known that the addition of dopant gases (PH_3, AsH_3, B_2H_6) to the feed

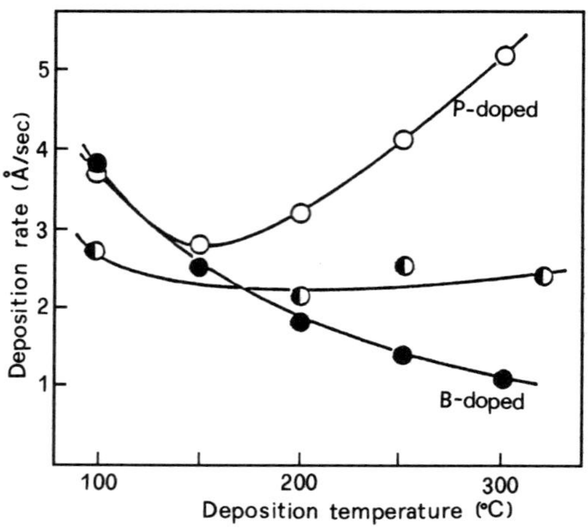

FIGURE 14. Effect of the substrate temperature on deposition rate (r_D) during the growth of undoped (◐) and doped (○, ●) Si:H,Cl films from $SiCl_4$–H_2 plasmas. (After Bruno *et al.* [32].)

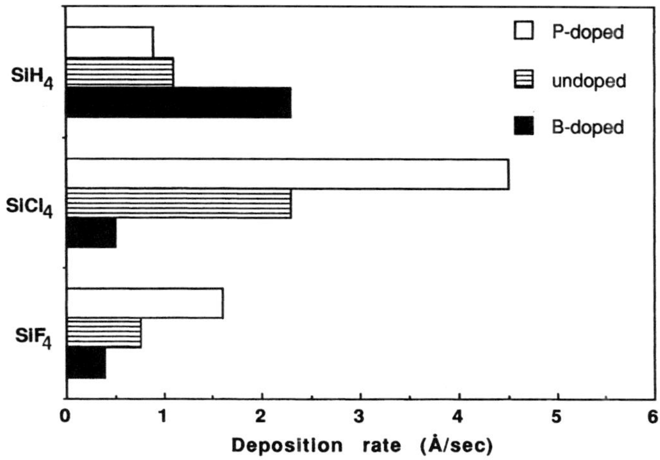

FIGURE 15. Stick diagram of the doping effect on the silicon film deposition rate r_D for SiH_4, $SiCl_4$ and SiF_4 systems. (From Bruno *et al.* [47].)

strongly affects the deposition rate. Figure 15 shows this effect for three different gas systems utilized in the silicon deposition. In particular, when halogenated silicon compounds are used as reactants [32, 43], it has been found that the film deposition rate r_D increases with PH_3 and decreases with B_2H_6 addition. On the contrary, the addition of dopants to SiH_4 exhibits an opposite behavior: r_D increases with B_2H_6 and slightly decreases with PH_3 addition [44]. This apparent discrepancy can be accounted for by referring to the chemisorption model on a semiconductor surface according to the boundary-layer theory [45]. In conformity with this theory, the chemisorption of halogenated compounds ($SiCl_4$, SiF_4) is promoted by the presence of n-doped surfaces, which transfer an electron from their donor level to the adsorbed molecule, thus becoming an anion referred to as *anionic chemisorption.* In contrast, in the presence of hydrogenated reactants (SiH_4), the chemisorption process is favored by the presence of p-doped surfaces, receiving in their acceptor levels an electron from the adsorbed molecule, which transforms to a correspondent cation; this process is known as *cationic chemisorption* [46]. According to the boundary-layer theory, the degree of anionic chemisorption on n-type semiconductors can be proportional to the difference between the adsorbate electron affinity E and the work function Φ (Fermi level) of the semiconductor, i.e., activation energy barrier, $E_a = E - \Phi$ (see Fig. 16). On the other hand, the degree of cationic chemisorption on p-type semiconductors can be proportional to the difference between the semiconductor work function Φ and the ionization potential I of the adsorbate, $E_a = \Phi - I$.

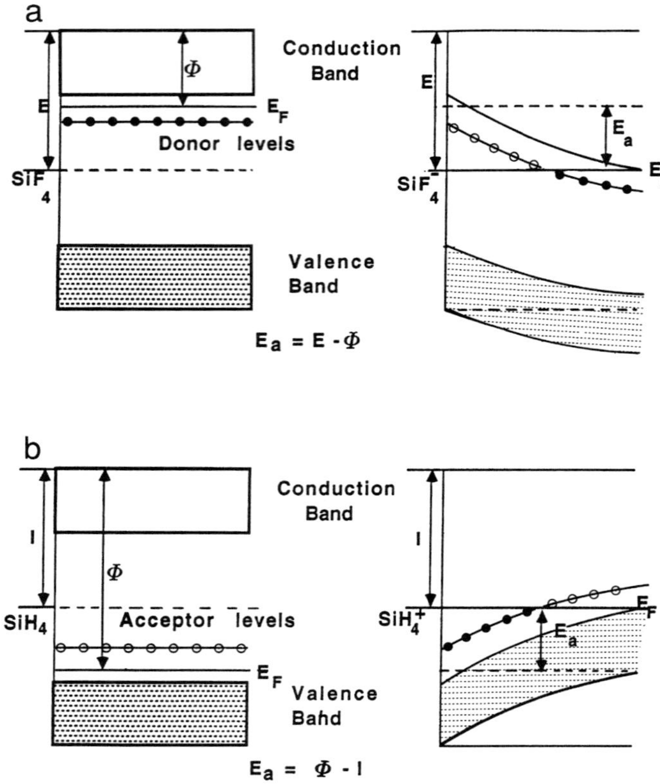

FIGURE 16. Energy-band structure and band bending at the surfaces of a n- and p-doped a-Si:H film in the case of SiF_4 anionic chemisorption (a) and SiH_4 cationic chemisorption (b), respectively.

The behavior of different compounds with respect to chemisorption on doped substrates has been confirmed in a study on the doping effect in the deposition of Si–Ge alloys from SiF_4–GeH_4 mixtures [47]. The experimental data of Fig. 17 are reported in terms of partial Si and Ge deposition rates (r_{Si}, r_{Ge}), besides their sum r_D. The two opposite trends of r_{Si} and r_{Ge} are those expected according to the above-mentioned model, coming Si from halogenated (SiF_4) and Ge from hydrogenated (GeH_4) compounds. Once again, there is clear evidence that hydrogenated species (GeH_4, like SiH_4) preferentially chemisorb on p-type materials, whereas halogenated species (SiF_4, $SiCl_4$) on n-type materials. Perrin *et al.* [48] explain the positive effect of B_2H_6 addition on the deposition rate of a-Si:H from SiH_4 as due to its strong catalytic effect on the growth kinetics. In particular, the hydroboron radical such as BH_3 catalyzes H_2 desorption, by acting as a scavenger,

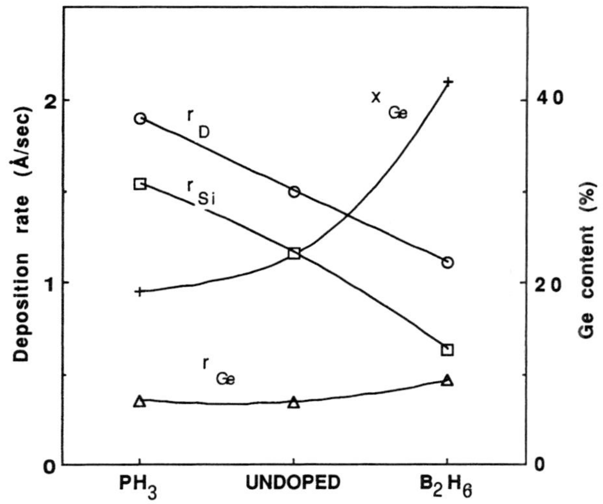

FIGURE 17. Total deposition rate (r_D), Si and Ge partial deposition rate (r_{Si}, r_{Ge}), and Ge content (x_{Ge}) for undoped and doped a-SiGe:H,F films from SiF_4–GeH_4–H_2 plasmas. (From Bruno *et al.* [47].)

through the formation of BH_5 reactive intermediates. The increased number of dangling bonds, due to the presence of B_2H_6, causes an increment in the sticking coefficient of SiH_3 radicals, considered by the authors to be the dominant growth precursors.

2. *Chemisorption under Light Irradiation*

From the preceding considerations, it immediately follows that the progress of the chemical processes occurring at the growing amorphous silicon surface will also depend on the local electronic structure and specifically on the availability of free electrons and/or holes at the surface. Besides dopant addition, another way to modify the local structure and charge distribution at the growing a-Si:H surface is by the irradiation where energy is greater than the optical band gap ($h\nu > E_g = 1.8$ eV), which generates negative (electrons) and positive (holes) carriers. Hence, both anionic and cationic chemisorption can be promoted. This well-established phenomenon [35], referred to as *photoadsorption effect,* can be positive or negative depending on the fact that in some cases the adsorptivity is increased; thus, light irradiation leads to additional adsorption, whereas in other cases the adsorptivity is reduced, i.e., external light favors the desorption of particles from the surface.

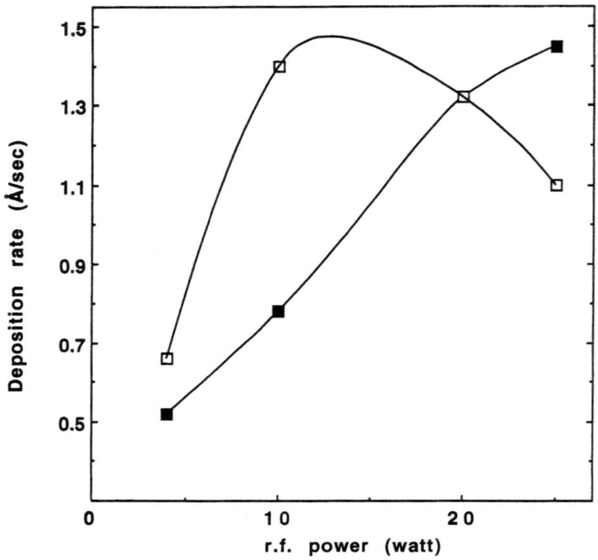

FIGURE 18. Effect of UV light irradiation (Hg lamp 225 nm < λ <500 nm, irradiation density = 0.6 mW/cm^2) on the deposition rate of a-Si:H films from SiH_4–H_2 plasmas (□ illuminated, ■ non-illuminated) as a function of RF power. (SiH_4/H_2 = 1/9; p = 0.3 torr; T_D = 300°C.)

In the deposition of a-Si:H films from SiH_4–H_2 plasmas, the presence of a positive photoadsorption effect has been reported by Bruno *et al.* [49]: the deposition rate of a-Si:H film increases when the film surface is irradiated with UV light produced by a Hg lamp (225 nm < λ < 500 nm, irradiation density = 0.6 mW/cm^2). Under these conditions any effect related to photodissociation can be excluded since the UV photolysis of silane occurs at λ < 160 nm (for more detail, see Chapter 4). Figure 18 shows the dependence of r_D on RF power for samples deposited with and without UV-light irradiation. In addition, other experimentally observed phenomena arise from UV-light irradiation, such as powder formation, and are discussed in Section IV.

In concluding this section, let us restate the importance of the chemisorption in the a-Si deposition chemistries as a process controlling the overall surface kinetics.

3. *Diagnostics of Heterogeneous Processes*

At present, many techniques are used to diagnose the surface processes, some of which are specifically useful to measure the structure and the composition of the surface, and generally are more suitable to perform *ex situ* measurements. *In situ* measurements such as ellipsometry and laser interferometry are more advisable during the PECVD process to study the surface–plasma interactions since they are nonintrusive and inexpensive techniques.

The ellipsometry is based on detecting the change in the polarization of the light reflected from surfaces and is used to determine the deposition rate, the detailed description of the thin-film growth phases (nucleation, coalescence), the film microstructure, and the material density. For a detailed description of this technique, the reader is referred to Chapter 2. Laser reflectance interferometry (LRI) is utilized to measure deposition and etching rates during the process, by monitoring the interference fringes obtained from the rays reflected at the outer surface and those reflected at the substrate interface [50]. This technique utilizes the light from a low-power (1-mW) He–Ne laser (wavelength, $\lambda = 6328$ Å) to impinge onto the growing surface and monitoring the reflected beam using a photodiode detector. Figure 19 shows a typical interferometric output during a deposition and subsequent etching runs of an a-Si:H,Cl film. For a He-Ne laser, the deposition and etching rates ($r_{D/E}$) are given by the following relationship:

$$r_{D/E} = \lambda_{6328}/[2(n^2 - sen^2\alpha)^{1/2} \cdot \Delta t], \tag{30}$$

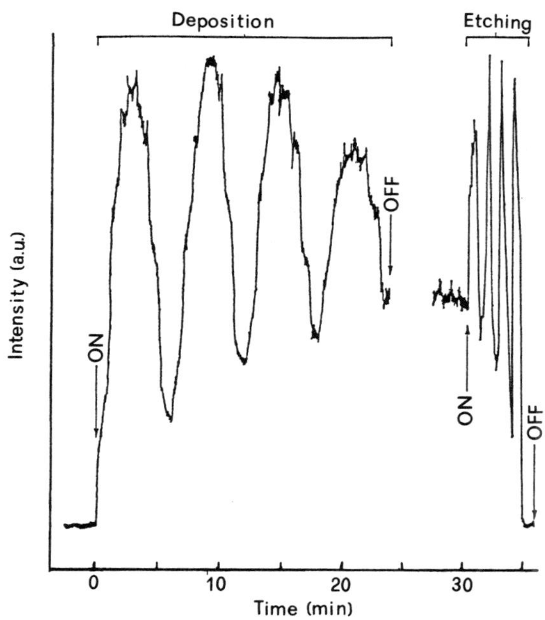

FIGURE 19. Typical laser interferometric traces during the deposition of a-Si:H,Cl in $SiCl_4$–H_2 plasma and subsequent etching run in $SiCl_4$–Ar plasma.

where n is the refractive index of the film, Δt is the oscillation period of the interference fringes, and α is the incidence angle of the beam. The fact that Δt value does not change during the process is an indication of a constant deposition rate or etching rate. The dampening of the amplitude of the oscillations with time during deposition is caused by the absorption of light (6328 Å) by the growing film; obviously the opposite effect is recorded during the etching process.

III. Chemical Systems for Amorphous Silicon and Its Alloys

Since the discovery of the hydrogenated amorphous silicon, various chemical systems have been investigated to deposit a-Si-based materials. Silane (SiH_4), disilane (Si_2H_6), and silicon tetrafluoride (SiF_4), alone or in a mixture with hydrogen (H_2) and noble gases (He, Ar, Xe), have been used to produce hydrogenated (a-Si:H) or hydrogenated and fluorinated (a-Si:H,F) a-Si films. The addition of carbon-source (CH_4, C_2H_6, C_2H_4, C_2H_2, CF_4, C_2F_6), germanium-source (GeH_4, GeF_4), tin-source [$SnCl_4$, $Sn(CH_3)_4$], nitrogen-source (NH_3, N_2, NF_3), and oxygen-source (O_2, N_2O, CO_2) compounds makes the plasma-chemical systems versatile to deposit low-band-gap (a-$Si_{1-x}Ge_x$, a-SiSn), a high-band-gap (a-SiC,), and ultra-high-band-gap (insulating) (a-SiN, a-SiO, a-SiNO) materials. The inherent advantage of this approach provides for altering the optical band gap at will, by varying the content of the alloying element since stoichiometry requirements are relaxed. At present, a-Si-based materials with optimized properties are produced for photovoltaic, electronic, and optical applications. Some properties of typical materials are listed in Table 2. Low values of Fermi level density of states and of Urbach energy, and the absence of SiH_2 stretching in the IR spectrum are indicative of good-quality materials, as discussed in Chapter 5.

Table 2 Some properties of various amorphous-silicon-based materials: optical band gap E_g, Fermi level density of states $g(E_F)$, Urbach energy E_0, and SiH_2 stretching mode at 2080 cm^{-1}, ν_{SiH_2}

Material	E_g (eV)	$g(E_F)$ ($cm^{-3}\,eV^{-1}$)	E_0 (meV)	ν_{SiH_2} (2080 cm^{-1})
a-Si:H	1.74	10^{15}	46	No
a-Si:H, F	1.68	10^{15}	48	No
a-$Si_{1-x}Ge_x$:H	1.50	10^{16}	50	Often
a-$Si_{1-x}Sn_x$:H	1.50	10^{17}	60	Often
a-$Si_{1-x}C_x$:H	1.90	10^{17}	90	Yes

A. Chemical Systems for Hydrogenated and/or Halogenated Silicon Deposition

The enormous number of studies on the deposition of a-Si:H films from SiH_4 plasmas leads to a well-established procedure, whose essential elements can be summarized in the following "recipe":

- Ultra-high-vacuum (UHV) deposition chamber capacitatively coupled to an RF (13.56 MHz) generator (see Fig. 19 in Chapter 4) with interelectrode distance of 2–3 cm.
- Pure SiH_4 or SiH_4 diluted in H_2 or He ($SiH_4 > 10\%$).
- Deposition temperature of about 250°C.
- Power density of about 15 mW/cm^2.

Regarding this last parameter, it is generally accepted that good quality a-Si:H films need these mild conditions. A full discussion of the aforementioned parameters is given in Chapter 5. A general description of this subject can be found in refs. [51–53].

Nevertheless, problems related to process and material properties are still objects of discussion in the scientific community, in order to increase the deposition rate [49] and the material stability [54]. Regarding this last aspect, the common opinion is that the light-induced degradation of the material, Staebler–Wronski effect [55], is strictly related to both chemical bonding and hydrogen content [54, 56]. Both these material characteristics can be quantified by infrared absorption measurements, and a typical spectrum is shown in Fig. 20. The main infrared features deriving from the spectrum for a good-quality material are (1) the dominance of Si–H bonds (2000 cm^{-1}) in the stretching mode region and (2) a SiH/SiH_2 ratio of the bending mode intensities higher than 20. Alternative feeds to silane have been investigated in order to reduce hydrogen or to partially substitute it with deuterium or fluorine, since it has been reported that an increased hydrogen content leads to a larger instability of a Si:H films [57, 58].

The introduction of deuterium in the material network, starting from pure SiD_4 or SiH_4–D_2 plasmas [59], is found to strongly reduce the photodegradation rate. The same effect, as well as the thermal stability improvement, has been found to occur when hydrogen is substituted by fluorine in the material [60, 61].

Fluorinated silicon materials—a-Si:F and a-Si:H,F—are obtained by the reduction of fluorosilanes, SiH_3F [62], SiH_2F_2 [62, 63], $SiHF_3$ [64], SiF_4 [61], and exafluorodisilane, Si_2F_6 [65], with or without H_2, and by the oxidation of SiH_4 with F_2 [66]. These alternative silicon sources cause a modification in both deposition rate and material properties. Films of only fluorinated amorphous silicon, a-Si:F, were initially prepared by sputtering a crystalline silicon target in an atmosphere of SiF_4 [67] and SiF_4/Ar [68], but these materials did not exhibit any

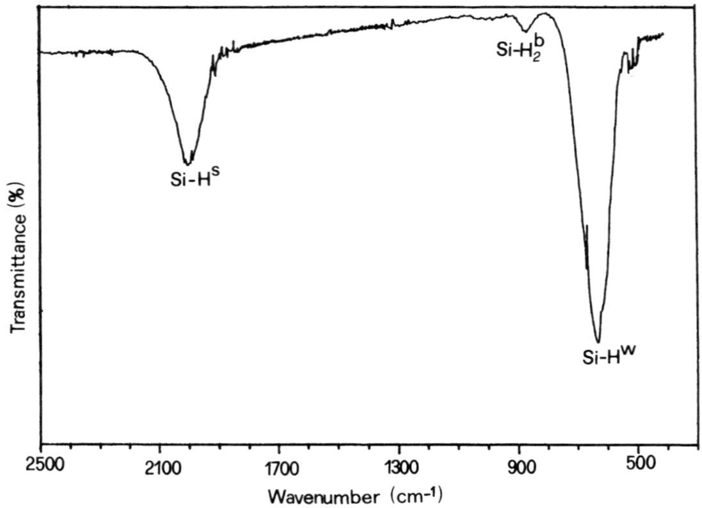

FIGURE 20. Typical infrared spectrum of a-Si:H film obtained from SiH_4–H_2 plasma (SiH_4/H_2 ratio = 1/9; p = 0.3 torr, T_D = 250°C, W = 5 W).

photoconductivity [69]. On the contrary, PECVD films from SiF_4/SiF_2 [70] and Si_2F_6 [65] are found to have a photoconductive response. Si_2F_6 decomposes more easily than SiF_4, since the latter gas requires higher RF power for the dissociation, and the deposition occurs only in a mixture with SiF_2. Significant work has been carried out in the production of hydrofluorinated silicon, a-Si:H,F, which seems attractive in device applications [60, 61] and exhibits a high photo-to-dark conductivity ratio (10^6) [63] and whose value strongly depends on the fluorine content, c_F. It should be noted that the film deposition rate for an equivalent optoelectronic quality material is lower than that of a-Si:H, because of its competition with the etching process [28] (see Section II).

Disilane, Si_2H_6, chemical systems have been investigated mainly to enhance the deposition rate [71, 72]: values as high as 60 Å/s have been reported [73], but a lower value of dark conductivity and photoconductivity [74] is obtained. The dilution of disilane in H_2 can improve the properties of the produced amorphous films. Doyle *et al.* [75] suggest that the improvement in the film quality going from pure Si_2H_6 to diluted Si_2H_6 in H_2 is due to the transition of the γ–α regimes (see Chapter 4). Under H_2 dilution conditions, the discharge operates in the α regime; i.e., the growing films are subjected to low-energy ion bombardment, considered beneficial to the amorphous silicon film quality; in contrast, the pure disilane discharge operates in the γ regime and hence the deposition occurs in the presence of energetic ion bombardment.

B. Chemical Systems for Silicon-Based Alloy Deposition

We now discuss the deposition of the different silicon alloys, by grouping them according to the value of the band gap.

1. Amorphous Silicon Alloys: Low-Energy Band Gap

The band gap is normally decreased by the addition of Ge and/or Sn as these elements replace the strong Si—Si bonds with the weaker bonds, thus resulting in a lower-energy band gap. SiGe alloy has been widely investigated for its application as the active layers in a photovoltaic device, since the reduced band gap provides a better match to the solar spectrum. The optical gap value can be modulated from 1.7 eV (a-Si:H) to 1 eV (a-Ge:H) by varying the Ge-molar fraction in the material (see Fig. 21). It is common opinion that the photoelectronic properties of a-$Si_{1-x}Ge_x$:H, obtained from silane (SiH_4) and germane (GeH_4), are not as good as those of the best PECVD a-Si:H films, when deposited under equivalent experimental conditions. The resulting poor photoelectronic properties are generally attributed to several factors, such as preferential attachment of H to silicon rather than to germanium [76], increase of SiH_2 groups in the network [77], and large heterostructure present [78].

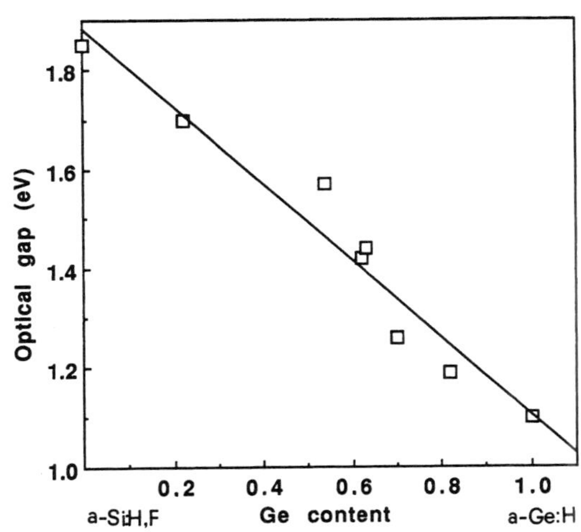

FIGURE 21. Dependence of optical gap on the Ge content for a-$Si_{1-x}Ge_x$:H,F alloys deposited from SiF_4–GeH_4–H_2 plasmas (going from SiF_4–H_2 to GeH_4–H_2 mixture).

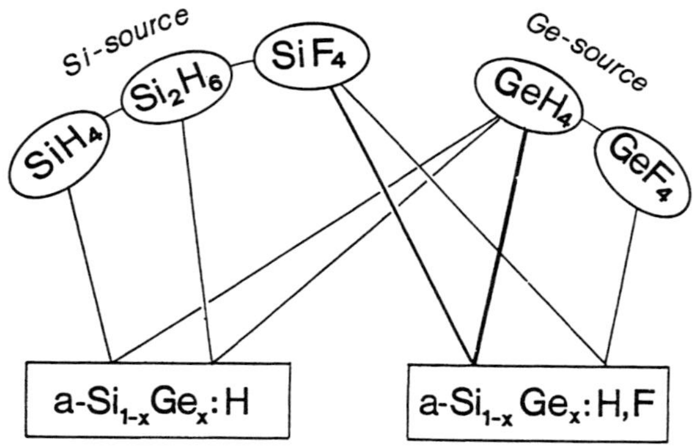

FIGURE 22. Typical mixtures of silicon and germanium volatile compounds utilized in the plasma deposition of silicon–germanium alloys. (From Bruno *et al.* [83].)

In order to improve the properties of a-SiGe alloys, various attempts have been made to synthesize the material in novel ways, such as the use of strong dilution of SiH_4–GeH_4 gas mixtures with H_2 [79, 80] and the use of silicon and germanium fluorinated reactants [81–83] as summarized in Fig. 22. It is also well established that the best-quality SiGe alloys cannot be produced under the "soft" conditions used in obtaining a-Si:H films. After the findings that the deposition of a-Ge:H [84, 85] requires "hard" plasma conditions, i.e., efficient ion bombardment of the growth surface, it has been suggested that [85] good quality SiGe alloys could be deposited under conditions that maximize the ion bombardment process. This is easily obtained when the deposition is carried out on the cathode, with low interelectrode distance, external negative bias voltage, and the use of high RF excitation power. This idea finds support in the improvement of SiGe quality, found when deposited from Si_2H_6–GeH_4 mixture, since according to the findings reported by Doyle *et al.* [86], disilane–germane plasmas operate in the γ regime, and hence under conditions of efficient ion bombardment. The common belief is that hydrogen and fluorine atoms, hot neutral and ion bombardment have the same effect: to break the weak Ge–Ge bonds. This could be extended to all the techniques that restructure the growing layers.

In the production of a-SiGe alloys, it is very important to consider the high cost of germane and, hence, any method leading to an increase of Ge/Si ratio in the material, when compared with that in the gas phase, becomes attractive. This is generally quantified by the so-called Ge enrichment factor, EF (Ge)

$$\mathrm{EF\ (Ge)} = \frac{x_{\mathrm{Ge}}}{[\mathrm{GeH_4}/(\mathrm{GeH_4}+\mathrm{SiX_4})]}, \tag{31}$$

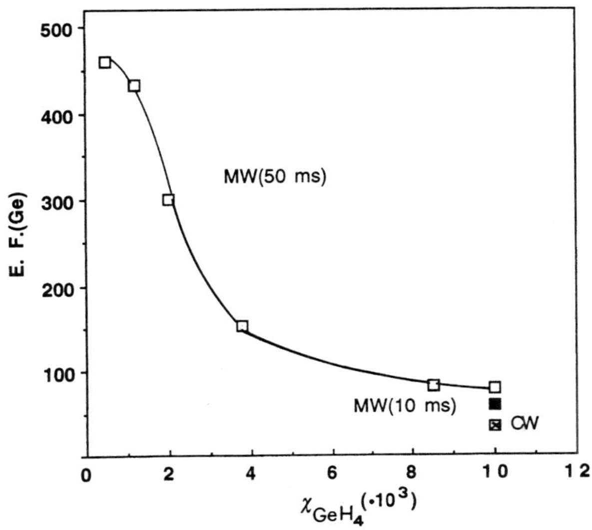

FIGURE 23. Ge enrichment factor EF (Ge) vs. germane amount χ_{GeH_4} in SiF_4–GeH_4–H_2 feeding mixture under plasma modulation conditions (MW) with period $P = 50$ ms. There are also data obtained with $P = 10$ ms and in continuous-wave (CW) conditions ($p = 0.3$ torr; $W = 30$ W; $T_D = 300°C$; $\Phi_{SiF_4} = 10$ sccm; $\Phi_{H_2} = 1$ sccm).

as a large value of EF (Ge) leads to a high Ge content (x_{Ge}) in the material. Bruno *et al.* [83] found that in SiGe alloys deposited from SiF_4–GeH_4–H_2 mixture, x_{Ge} is much higher than the GeH_4 molar fraction (χ_{GeH_4}) in the plasma phase. The same authors reported [87] that a further increase of EF (Ge) can be achieved when the discharge is operated under modulation [modulated wave (MW)] conditions, i.e., when the plasma is periodically switched on and off. This effect is illustrated in Fig. 23, where the EF (Ge) is reported as a function of χ_{GeH_4}. The main features are the increase of Ge in the material from MW plasmas, when compared to [continuous wave (CW)] conditions and its strong increase at lower values of χ_{GeH_4}. The suggested explanations are

(a) A much more efficient electron impact dissociation of GeH_4 than that of SiF_4.
(b) A lower surface mobility of GeH_x radicals than that of SiF_x, due to the larger Ge atomic mass.
(c) A more efficient chemisorption of GeH_4 than SiF_4.

The last point has been claimed in a study [47] on a chemisorption-based deposition model. The prevailing GeH_4 chemisorption is also accounted for by the data

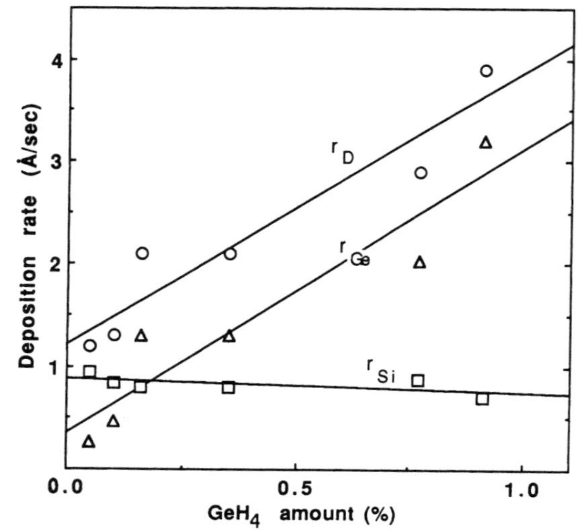

FIGURE 24. Si and Ge partial deposition rates (r_{Si}) and (r_{Ge}) and total deposition rate (r_D) of a-$Si_{1-x}Ge_x$:H,F films vs. germane amount in SiF_4–GeH_4–H_2 mixture under modulation conditions (same experimental conditions as in Fig. 23).

shown in Fig. 24, in that the partial deposition rate of Ge, r_{Ge}, is much higher than that of Si, r_{Si}, and linearly related to χ_{GeH_4}.

The use of SiF_4–GeH_4–H_2 mixtures, besides giving a SiGe of good quality for photovoltaic applications (a-$Si_{0.75}Ge_{0.25}$:H,F with $E_g = 1.5$ eV and a sufficient photo/dark conductivity ratio $\Delta\sigma/\sigma = 10^4$), has the great advantage of requiring χ_{GeH_4} as low as 10^{-3}, instead of 0.1 utilized in SiH_4–GeH_4 mixture [88].

Another low-band-gap silicon-based material is represented by SiSn alloy. It has been found that about 10 atom % Sn is required to decrease the optical band gap from 1.8 to 1.3 eV. In addition, the Sn incorporation causes a transition from n- to p-type conduction mechanism, which can explain the observed loss in photoconductivity and, hence, the poor performance in a solar cell device structure [89, 90]. In the few studies carried out on this subject the silane was mixed mainly with $SnCl_4$ or $Sn(CH_3)_4$. Recently, these poisonous reactants have been substituted by a Sn target in a deposition apparatus combining sputtering and plasma techniques; nonetheless, the quality of the deposited material did not exhibit any significant improvement [91].

2. *Amorphous Silicon Alloys: High-Energy Band Gap*

One of the more intensively studied alloys is the a-Si,C:H alloy as it has importance as an active layer in thin film and Si bipolar transistors, image sensors, photoreceptors, light-emitting diodes (LEDs), photodiodes, and photovoltaic cells (as a p-type window layer). Since 1977, when Anderson and Spear [92] deposited for the first time PECVD a-$Si_{1-x}C_x$:H alloys from SiH_4-CH_4 mixture, many studies have been carried out on this subject. Typical values of refractive index [93], density of states [94], and optical gap [92] for a-$Si_{1-x}C_x$:H material are reported in Fig. 25 as a function of C content. In addition the incorporation of C in a-Si:H matrix, similarly to Ge and Sn, causes the reduction of photoelectronic quality,

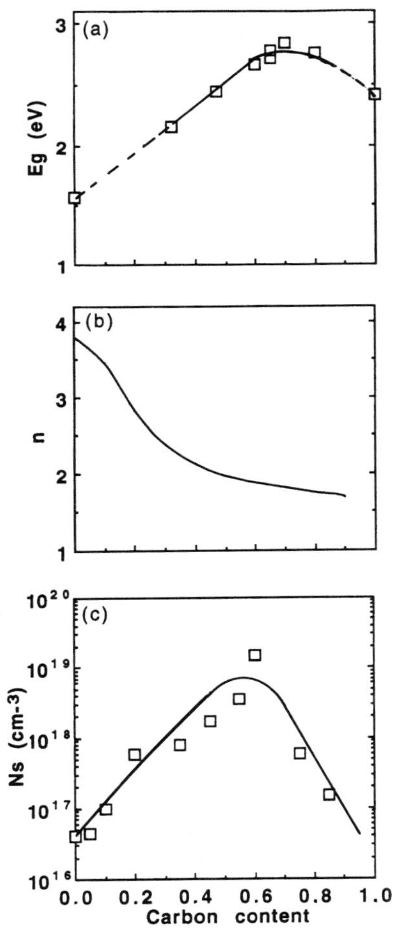

FIGURE 25. Optical gap (E_g), refractive index (n), and spin density (N_s) of a-$Si_{1-x}C_x$:H alloys as a function of carbon content in the material. [(a) From Anderson and Spear [92]; (b) from Sussmann and Ogden [93]; (c) from Liedtke *et al.* [94].]

the increase of Urbach energy, and, hence, of the microstructural disorder [95]. From the structural point of view, four different phases have been identified in the SiC matrix: a-SiC alloy, graphitic carbon, polymeric carbon, and voids [96]. The graphitic C bondings in the alloy are considered the principal cause for the degradation and can be minimized by hydrogen bonded to carbon via CH_n ($n = 2, 3$) formation [97]. As for stability aspects these alloys are unstable under prolonged light irradiation and can undergo two types of photoinduced changes: photobleaching and photodarkening, which become more pronounced at a high C content [98].

Successful attempts to improve the a-SiC film properties have been reported: H_2-dilution effect [79], deposition of μc-SiC with the remote plasma-enhanced chemical vapor deposition (RPECVD) method [99], and growth of fluorinated alloys [100] from SiF_4–C_2F_6–H_2 mixture. In addition, fluorine incorporation inhibits the polymerization and the graphitization of C bonds, through the preferential attachment of F atoms to carbon, rather than to silicon [101].

Tetramethylsilane and tetramethyldisilane have also been used as alternative feeds [102, 103]. In their case the incorporation of CH_3 groups has been observed, giving less connective network and microvoids.

3. *Amorphous Silicon Alloys: Ultra-High-Energy Band Gap*

Silicon oxides SiO, silicon oxinitrides SiON, and silicon nitrides SiN belong to this category. Their main applications are

- As insulating interlayers, for their high values of energy gap.
- As protective antireflection coating on solar cells, for their low value of refractive index.
- As a general protective coating because of their excellent mechanical and chemical properties (strength, hardness, and thermal shock resistance). In fact, in the semiconductor industry (for integrated-circuit technology) they are used as final passivation films against oxidation, corrosion, erosion, and diffusion of moisture or alkaline ions.
- As gate dielectric layer in thin-film transistors.

The transition from silicon oxide to silicon nitride can be carried out gradually through the formation of silicon oxinitride [104].

In the following text, plasma deposited silicon nitride will be described in some detail, because of its superior barrier properties compared to those of silicon oxides, which determine a significant improvement in final device quality. Silicon nitride films produced by PECVD are not stoichiometric and are usually described by the general formula a-$Si_xN_yH_z$. According to y/x values, the silicon nitride can be a quasistoichiometric ($y/x = 1.3$), silicon-rich ($y/x < 1.3$), and nitrogen-rich ($y/x > 1.3$ alloy). The best material quality is obtained from quasistoichiometric composition and low H content. However, some authors [105, 106] have reported that an enrichment in nitrogen content and a shift of H bond from Si–H to N–H are favorable.

The actual efforts in the production of SiN thin films have been directed toward the development of materials exhibiting high dielectric constant, low hydrogen concentration, low leakage current, high breakdown strength, and low interface trap density. When silicon nitride films are used as encapsulation layers for III–V compounds (GaAs, InP) or in metal oxide semiconductor devices, they require low plasma bombardment, low deposition temperature (<300°C), and low H content, in order to achieve a good electrical semiconductor/Si_3N_4 interface. These requirements can be reached by substituting the conventional PECVD process with some modified systems or by replacing the classical feed-gas mixture of SiH_4 and NH_3. The choice of the alternative gas mixtures must be aimed at a material with low etching rate, high density, and H content sufficiently low to reduce its diffusion in the network in order to achieve high stability and low degradation of the final device.

The modified techniques investigated are remote plasma enhanced chemical vapor deposition (RPECVD), electron cyclotron resonance plasma CVD (ECR PCVD), multipolar plasma CVD (MPCVD), photo-CVD, and plasma-enhanced evaporation (PEE).

- *RPECVD* is also known as the *downstream plasma technique* [104], where the plasma chamber is fed with pure N_2 or diluted in He and SiH_4 flows close to the substrate region. The deposition occurs onto the substrate located outside the plasma zone, with the benefit of avoiding bombardment by ions and photons. In addition, the RPECVD technique involves fewer reaction paths and, compared to PECVD apparatus, allows a better control over the film properties [107]. As a consequence, N—H and Si—H configurations are nearly absent in the material structure.
- *ECR PCVD* is another downstream method and is attractive as process of SiN film production, because it works at lower pressures and temperatures (room T). The plasma is generated by microwaves (2.45 GHz) and by resonance conditions obtained by 875-G (gauss) magnetic field in the plasma region. The ECR silicon nitride films [108] have a H content as low as 2% and a very high value of energy gap ($E_g = 5.3$ eV) typical of thermal CVD films, but much larger than that of conventional PECVD samples ($E_g = 2$–3 eV).
- *MPCVD* is characterized by a hot cathode and a magnetic confinement, and produces a plasma, free from energetic species, since it requires modest DC voltage (<75 V). The deposited material [109] contains a very low H concentration compared to typical PECVD ($c_H = 20\%$).

- *Photo-CVD* of SiN films from SiH_4–NH_3 mixtures is carried out by Hg photosensitization at 254 nm [110] or NH_3 photolysis at 185 nm [111]. The absence of ion and electron bombardment of the growing film is highly advantageous for its application as passivant layer in Groups III–V compound devices.
- *PEE* process has been used in the production of SiO_2 and Si_3N_4 at low temperatures (100–400°C) for optical applications. The H content is less than 1% and the dielectrics used in the metal-insulating semiconductor (MIS) structure exhibit very promising electrical properties [112].

With regard to the use of alternative gas-feed mixtures, the substitution of NH_3 with N_2 not only results in a strong decrease in the H content of the material but also decreases the deposition rate and deteriorates in the nitride electrical properties [113]. The dilution of SiH_4–N_2 in He [114] produces silicon nitride, which exhibits physical and electrical properties similar to those of high-quality films prepared at 700°C by the LPCVD technique, and a H content lower than in absence of He. The He dilution causes the reduction of the SiH_x species, and hence inhibits the formation of polymeric silicon species in the gas phase. He dilution also causes an increase in the incorporation of N reaching a quasistoichiometric nitride because the He metastable species can increase, through collisional energy transfer, the density of excited molecular nitrogen and also the amount of N—H bonds in the structure.

The H content in the silicon nitride alloys can also be reduced by utilizing fluorinated gaseous mixtures from SiH_4–NH_3–NF_3 [115], SiH_4–N_2–NF_3 [116], SiH_2F_2–NH_3 [117], and $SiF_4(SiF_2)$–N_2–H_2 [118–120], resulting in materials where some of the H is replaced by F atoms. The fluorinated films have been found to exhibit higher resistivity (10^{14}–10^{16} $\Omega \cdot$ cm), higher breakdown strength (10 MV/cm), and lower trap density [121] than samples without fluorine, in a wide range of y/x ratio (0.6–1.3). Cicala *et al.* [120] have prepared stable and quasistoichiometric fluorinated silicon nitride from SiF_4–N_2–H_2 mixtures under H_2-rich conditions. In fact, among the various gas mixtures utilized and shown in the ternary diagram of Fig. 26, H_2-rich feeds have produced samples with good-quality characteristics.

IV. Effect of Novel Parameters

An investigated problem in the deposition process of a-Si-based materials is, for obvious reasons, the increase of the film deposition rate, while keeping good material properties. The glow discharge pressure, the gas composition, and the RF power have proved to be, in conventional PECVD reactors, the most effective external parameters for the growth kinetics, even though they often cause undesir-

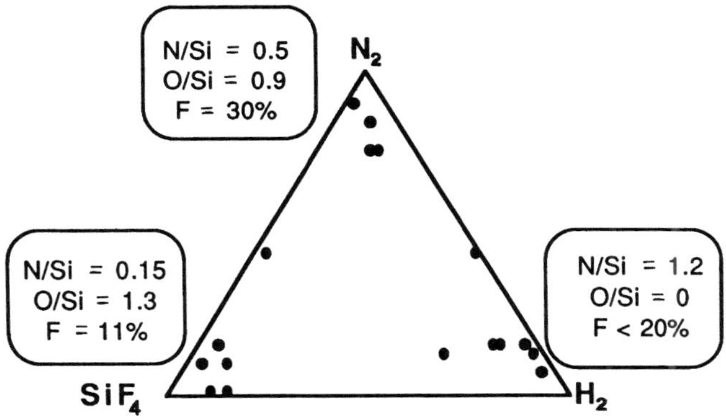

FIGURE 26. SiF_4–N_2–H_2 ternary diagram showing the various gas feed mixtures used for the deposition experiments. Chemical composition data refer to typical films deposited from SiF_4-, N_2- and H_2-rich mixtures ($p = 1$ torr, $W = 50$ W, $T_D = 300°C$).

able effects on material properties. Alternatively, novel growth methods can be used in the attempt to (1) increase the production of active species and thus the deposition rate, (2) reduce powder formation, (3) increase the thickness homogeneity and thus the available deposition area, and (4) better control the material properties. In this respect, novel parameters, such as frequency of RF field [122–124], plasma confinement (both magnetic and electrostatic) [125], light irradiation [126, 127], and plasma modulation [128–130] are experienced, and/or higher silanes (Si_2H_6) and different reactor configurations are used. In addition, some of these new parameters are used as tools to understand the deposition chemistry and to better define some fundamental properties of the material. For instance, Matsuda [131] has used a triode glow discharge method to separate the plasma from the substrate region and to select the specific radicals able to reach the growth surface. He found arguments for discussing the growth mechanism and for suggesting a "guiding principle" to obtain high-quality Si-based alloys.

In the following section, we report the essential features of the effect of (1) frequency of the RF field, (2) gas dilution, (3) light irradiation, and (4) plasma modulation. The reactor configuration subject is treated in Chapter 4.

A. Effect of Plasma Excitation Frequency

The first study on the effect of the excitation frequency in RF glow discharge deposition of a-Si:H films was reported in 1987 by Curtins *et al.* [122]. They describe in detail (see Fig. 27) the variation of the deposition rate in plasma fed

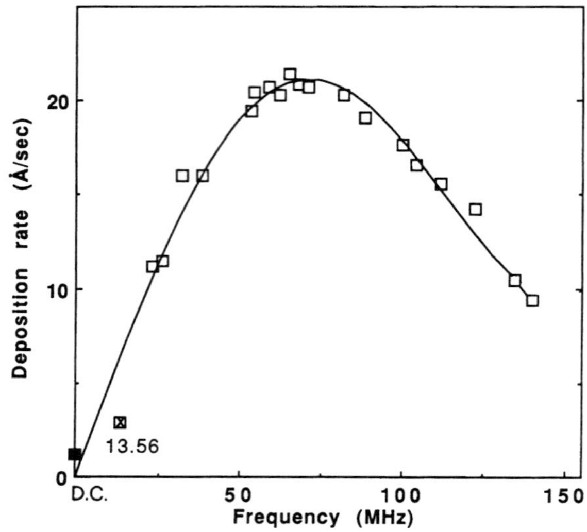

FIGURE 27. Deposition rate of a-Si:H films from pure SiH_4 as a function of plasma excitation frequency. (From Curtins *et al.* [122].)

with pure SiH_4, by varying the frequency in the range 13.56–150 MHz. The trend reported in the figure shows that around 75 MHz the deposition rate reaches a value as high as 21 Å/s. However, an additional frequency effect on the efficiency of the power transmission has also been reported [132] to have a concomitant role in controlling the a-Si:H growth rate. A similar beneficial effect of the frequency on the deposition rate, while keeping satisfactory material properties, has been found by Bhat *et al.* [123], up to 110 MHz of excitation frequency. An increase of the growth rate has also been found in the deposition from SiH_4–H_2 mixture [49]. Figure 28 shows the trends of r_D as a function of RF power at 13.56, 27, and 75 MHz. The plateau at 27 MHz and the decrease at 75 MHz have been explained as due to SiH_4 depletion.

The explanation of this effect is manifold, since the frequency variation causes the parallel variation [133] of

- Shape of the EEDF.
- Spatial distribution of species and electric field (plasma anatomy).
- Time fluctuation of energy and density of species.
- Minimum voltage required to sustain the discharge.

In fact, a theoretical study on SiH_4 and SiH_4-H_2 systems [10, 134] has shown, with frequency variation, both modifications in the shape and fluctuations of EEDF in the time scale of RF cycle; the last phenomenon is seen to increase, going

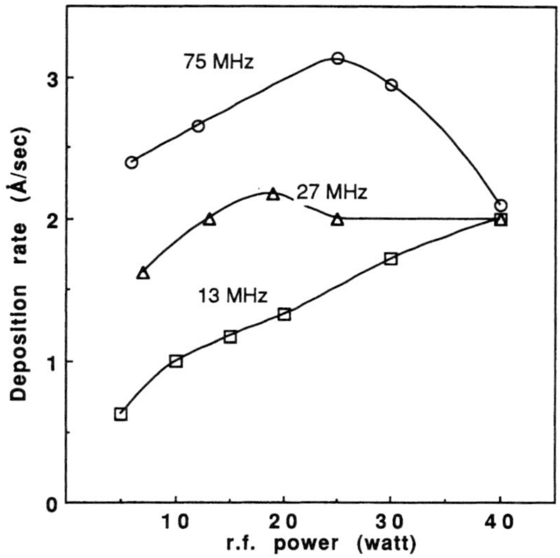

FIGURE 28. Deposition rate of a-Si:H films from SiH_4–H_2 plasmas at three different frequencies as a function of RF power.

toward lower values of excitation frequency. The fact that the shape of EEDF significantly changes with the excitation frequency can be determinant in controlling the plasma process (see Fig. 29) and hence the growth kinetics [135]. This has been invoked by Curtins *et al.* [122] to explain their data of Fig. 27. A similar conclusion has been reported by Moisan et al. [136] to give reasons for the experimental results on surface kinetics at different frequencies in plasma deposition and etching of polymer films.

On the other hand, plasma anatomy, i.e., spatial distribution of the species (both neutral and charged) and electric fields between the electrodes, undergoes strong variation when the excitation frequency is changed. In this respect, Langmuir electrical probe measurements (in double-probe configuration) [49] have evidenced large variation in the electron density n_e and energy $\langle \varepsilon \rangle$ spatial profiles as shown in Fig. 30. A large asymmetrical inhomogeneity in both $\langle \varepsilon \rangle$ and n_e values at 13.56 MHz can be noted, when compared to 27- and 75-MHz profiles. In addition, a strong reduction of the cathodic sheath thickness is observed going from 13.56 MHz to higher frequency values. The evidence that the plasma becomes more energetic near the grounded deposition electrode can account for r_D increasing with frequency.

The fact that the frequency dependence of the deposition rate cannot be discussed without accounting for spatial dishomogeneity of the plasma has been stressed by Beneking *et al.* [137] on the basis of deposition profile and plasma

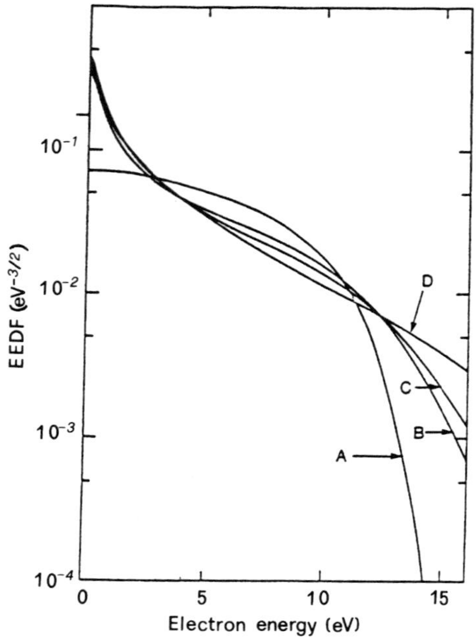

FIGURE 29. Electron energy distribution (EEDF) vs. electron energy ϵ, for argon plasmas of constant $<\varepsilon>$ (3.5 eV), but different values of ν/ω: (curve A) $\nu/\omega \to \infty$ (low-frequency plasma); (curve B) $\nu/\omega = 2$; (curve C) $\nu/\omega = 1.25$; (curve D) $\nu/\omega \to 0$ (microwave plasma). (From Wertheimer and Moisan [135].)

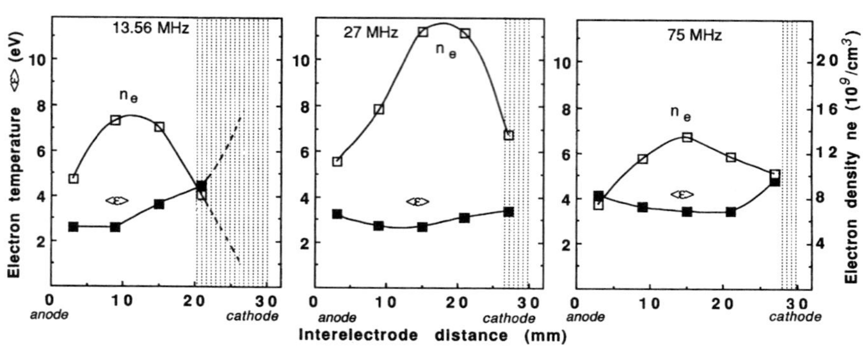

FIGURE 30. Mean electron energy $<\varepsilon>$ and electron density (n_e) profiles measured in Ar plasma at 13.56, 27, and 75 MHz ($p = 0.3$ torr, $W = 10$ W, $\Phi_{Ar} = 10$ sccm).

impedance measurements. They also develop some considerations on the good surface homogeneity at high frequency, but they conclude that the deposition rate averaged on the area does not depend significantly on the frequency at constant power and pressure.

Here, it is useful to recall the data reported in Fig. 5 on the reduction of the self-bias potential with increasing excitation frequency, in that this effect certainly relates to the decrease of the minimum voltage required to sustain the discharge. This has been substantiated by Gottscho and Mandich [138] who found, in BCl_3 system, the peak-to-peak voltage decreasing with increasing frequency. Recently this finding has been confirmed in silane discharge and related to the decrease of ion impact energy on the substrate [136].

A beneficial effect on powder formation [139] and film thickness uniformity [140] have also been reported with increasing frequency.

B. Effect of Gas Dilution

The study on SiH_4 dilution has been carried out with the use of both noble gases (He, Ne, Ar, Kr, and Xe) and H_2. The dilution, obviously, implies a depleting condition of the SiH_4 flow rate [141] and hence the deposition rate is limited. Since Street *et al.* [141] have shown that the gas dilution affects the properties of a-Si:H films, many studies have been reported on this subject. In the last few years, especially, the gas dilution parameter has attracted considerable attention, because of its role in controlling the ion distribution in the gas phase, the material microstructure, and all the parameters affecting the material stability. A parameter related to the microstructure is usually utilized as quality probe of the material and defined by R as the fractional amount of SiH_2 groups to the total SiH_2 + SiH density; this is obtained by infrared absorption measurements of the relative stretching modes (2000 cm^{-1} for SiH and 2090 cm^{-1} for SiH_2). As the SiH_2 configuration is associated mainly with the presence of microvoids in the sample, the R parameter is construed to be correlated to the network microstructure and the material quality improves with decreasing R.

1. Noble-Gas Dilution

Helium dilution has been proved to be useful to increase the deposition rate (>10 Å/s) of a-Si:H films, while maintaining a low defect density in the material [142]. A more thorough study has been carried out by Roca i Cabarrocas *et al.* [143] on a-Si:H films obtained from pure SiH_4 and SiH_4 40% diluted in He, under various experimental conditions and the obtained results allowed the authors to define the causes of material metastability. This analysis revealed that the instability is controlled by the type of the film microstructure and the mobility of H in the film and cannot be correlated to its incorporation or the concentration. In addition, the same authors [144] report that the kink generally present in the conductivity profile of the Arrhenius-like plot (log σ vs. $1/T$) for pure SiH_4 samples and associated with the material instability, was absent under He dilution. Knights *et al.* [142] also found that an important requirement for the stability of the sample is an efficient ion bombardment of the growing surface and that the defect density, the R parameter, and microstructure increase with atomic weight of the inert gas (Ar, Kr) in 5% SiH_4 mixtures at 0.5 W/cm^2 and 0.15–0.4 torr.

Also, a-Si:H films obtained under Xe dilution conditions (SiH_4 16% in Xe) and at moderately high power (0.51 W/cm^2, P = 25–30 mtorr) have been found to be stable [145], since they did not undergo any photodegradation of the photoconductivity after more than 10^4 s of light soaking. The authors had chosen Xe as the diluent noble gas, because the energy of the Xe metastable species is low (8.32–9.45 eV) and the collisions with SiH_4 molecules do not give origin to the formation of SiH and SiH_2 species, which is considered to be the cause of low-quality films [146]. In addition, the steric hindrance effect of the Xe large atomic diameter prevents the microcrystalline nuclei formation. The photoconductivity decay under light irradiation has been found, by the same authors, to be very low for films prepared at low power (0.15 mW/cm^2) provided that a substrate bias between -100 V and -120 V is applied [147], which enhances the ion bombardment process.

The importance of the ion bombardment has also been stressed by other authors [142, 148] who have reported that the type of rare gas and the amount of its dilution affect the ion distribution density of SiH_x^+. In particular, Knights *et al.* [142] reported that in He- or Ne-containing plasmas the density of ions of SiH_x^+ is higher than in the presence of Ar and Kr because the energy of the respective rare-gas metastable species is higher than the SiH_x^+ ionization energy in the first case and lower in the second one. In this last case, the formation of SiH_x radicals is favored (since their production requires a lower energy amount). The role of ion distribution has been emphasized by Hollenstein *et al.* [148]. They reported that with increasing rare-gas dilution the contribution of hydrogen-deficient ions SiH_2^+, SiH^+, and Si^+ becomes comparable to that of SiH_3^+; the latter is predominant in 100% SiH_4 plasmas. The authors suggest that radical composition is responsible for the material quality. They also demonstrate the existence of a relationship between microstructure parameter R and the light-induced degradation of the photoconductivity. Both parameters improve slightly as the silane is diluted down to 10% for Ar and to 20% for Xe; below these values the film properties worsen rapidly.

2. *H_2 Gas Dilution*

The effect of H_2 dilution in silane discharge has been studied from a theoretical and an experimental viewpoint.

Capitelli *et al.* [10] have found, through a theoretical calculation solving the Boltzmann equation, an increase of average electron energy with H_2 addition to SiH_4 plasmas. As an example, the average electron energy (calculated at reduced field strength $E/p = 100$ V cm^{-1} torr^{-1}) increases from 3 to 6 eV passing from pure SiH_4 to a 3% SiH_4 in H_2. In addition, they report, for both systems, the collision frequencies for vibrational and electronic excitation and ionization; the most important result is that in pure SiH_4 plasmas the vibrational impact is 100 times more efficient than the electronic one, whereas in SiH_4-H_2 diluted plasmas the electronic excitation frequency becomes comparable to that of vibrational excitation. Thus, in pure SiH_4 "soft" plasma, the vibrational excitation is the most probable dissociation channel, whereas in "hard" 3% SiH_4 in H_2 plasma the electronic excitation could be predominant. This conclusion should be carefully considered when a detailed study of the a-Si:H deposition kinetics is undertaken.

From the experimental point of view, besides a pronounced variation in the plasma phase, strong changes have been found in the film properties. As an example, the amorphous-to-microcrystalline (or nanocrystalline) silicon transition is favored in H_2 highly diluted SiH_4 discharges at high power density or in sequential alternative discharges of SiH_4 and H_2 gases [149, 150]; in this the H-atom density in plasma phase is very high and hence the interaction with the surface is strongly enhanced. However, the role played by hydrogen during the deposition and on the film properties is still controversial.

Vepřek [151] has found that H atoms act as etchant species and, during deposition, a reversible chemical equilibrium

$$SiH_x \rightleftarrows Si + xH \tag{32}$$

is established. In this respect, Blayo and Drévillon [152] have demonstrated, by *in situ* infrared phase-modulated ellipsometry (IRPME), the etching role of H atoms, by treating a deposited film in a pure H_2 discharge. They found that H atoms preferentially etch the SiH_2 groups, causing the amorphous-to-microcrystalline transition at short discharge duration and the total film etching at longer periods [153].

Nomoto *et al.* [154], by using the above-cited alternative discharges, believe that the role of excess H atoms for the production of μc-Si:H is to increase the hydrogen coverage of the surface, which changes the radical diffusion on the surface.

Shibata *et al.* [155] have invoked another effect played by H atoms, i.e., H induces a "chemical annealing" in the growth zone located below the surface, where the desorption of H,

$$H + SiH_x \rightarrow Si(s) + H_2, \tag{33}$$

releases considerable energy to the surface so that the process of surface nucleation to microcrystallites is favored.

The chemical annealing (CA) preparation technique is also used for producing high-quality a-Si:H with a more stable network [156]. The reactor used in CA technique is composed of an RF diode (13.56 MHz) for the decomposition of SiH_4 and a microwave plasma (2.45 GHz) for the generation of H atoms and excited states of rare gases such as Ar and He. Although the material contains a very low H content ($C_H = 1\%$), the amorphous structure is still maintained; the stability under light irradiation is strongly improved at substrate temperatures higher than 300°C and the hole drift mobility significantly increases up to 0.2 $cm^2\ V^{-1}\ s^{-1}$ at 300 K [157].

Spin density measurements have evidenced that stable a-Si:H is also produced at a high temperature and H_2 dilution in a remote hydrogen plasma (RHP) reactor [158]. The H_2 is decomposed at high microwave power (400 W) in order to produce large amounts of H atoms that not only promote the reaction in the gas phase [see reaction (18)] but also strongly influence the surface reaction. Although the H content remains high ($C_H = 10\%$) at high deposition temperature (400°C), it seems that more stable hydrogen complexes and lower density of weak Si–Si bonds are included. This explains why the RHP samples at 400°C have Urbach energies (50 meV) and defect densities ($3 \cdot 10^{15}$ cm^{-3}) as low as those of GD a-Si:H at 250°C. These results seem contradictory, and it is difficult to establish whether the low H content or the changes in the microstructure are important for the increase of a-Si:H stability. Until now the unresolved principal problem is the identification of the instability causes.

C. Effect of Light Irradiation

The use of visible and UV light irradiation as a parameter affecting the plasma–surface interaction during a-Si:H growth is of recent interest [126, 159–162]. The UV light (Hg lamp) irradiation has been investigated in our laboratory during GD a-Si:H deposition from SiH_4–H_2 mixtures under experimental conditions typical for the deposition of good-quality material [SiH_4:H_2 = 1:9, Φ_t = 20 sccm (cubic centimeters per minute of gas flow at standard conditions of temperature and pressure), $p = 0.3$ torr, $T_n = 300$°C]. Illumination of the plasma with light causes:

(a) An increase of deposition rate at low power (see Fig. 18).
(b) A strong increase of powder formation in the gas phase.

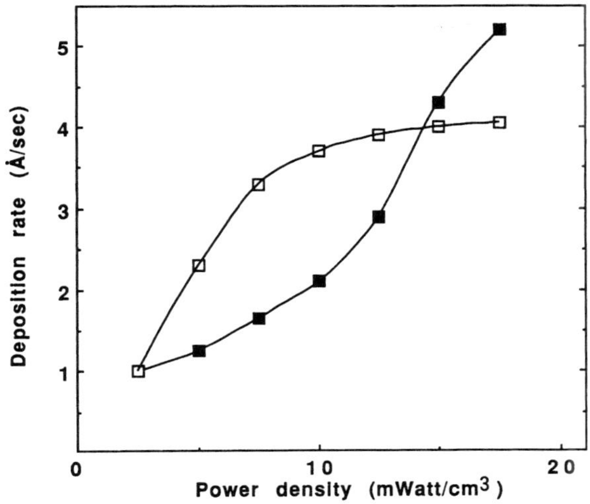

FIGURE 31. Dependence of deposition rate on power density for films produced by plasma UV light assisted (□) or not (■). (From Martins *et al.* [126].)

A similar deposition rate dependence on light irradiation has also been reported by Martins *et al.* [126], whose data are shown in Fig. 31. With regard to this, the opinion of the present authors is that the photoenhancement of the Si-species chemisorption is the main cause for the observed phenomena on deposition rate and powder formation (see Section I). As for the material quality dependence, the study reported by Martins *et al.* [126] can be considered significant: they observe a strong decrease of the photo/dark conductivity ratio for a-Si:H samples deposited under UV light irradiation implying that the material is more defective.

Suzuki *et al.* [159] have studied the effect of UV laser (ArF, KrF, XeF) illumination on a-Si:H growing surface in He-diluted silane plasma. The photoconductivity and the Urbach energy tail improve in illuminated films at deposition temperature below 100°C when compared to nonirradiated film, but the values of energy gap and H content remain unchanged. The effect of UV laser irradiation disappears in films deposited at temperatures above 150°C. Light irradiation with Hg lamp has also been carried out in homogeneous (homo)-CVD systems [160], where the light interacts only with the surface without affecting the gas-phase process, due to the SiH_4 transparency in the utilized UV region (240–310 nm). Under these conditions the photoelectronic properties such as photoconductivity and Urbach tail width are observed to worsen, even though light irradiation drastically reduces the $(SiH_2)_n$ groups, leaving the SiH groups unchanged.

Yamanaka *et al.* [161] have performed the a-Si:H deposition under a visible-light illumination (tungsten halogen lamp) in a plasma reactor equipped with comb-type electrodes. The light-induced degradation of films deposited under illumination is reduced and attributed to a reduction in "latent sites" in the bulk of the material [162].

Silicon nitride films deposited from SiH_4–N_2 plasma under UV irradiation (Xe lamp) exhibit a reduction in the H content of the film, a densification of the material, and a shift from compressive to tensile-type stress [163].

D. Effect of Plasma Modulation

During the last few years, plasma modulation of the PECVD technique systems has been investigated in an attempt to improve the material properties [130] and the deposition process [128, 130]. In most cases, the discharge is 100% modulated by a square-wave audiofrequency variable in the range from 10^{-1} to 10^4 Hz. This, in practice, results in a periodical switching on and off of the plasma at the mentioned frequency, as shown in Fig. 32, in which a schematic diagram of the time sequence of applied modulated RF field is represented together with the behavior of the time resolved optical emission for an Ar plasma. The modulation parameters are the period, viz., the time between two successive discharge ignitions $(t_2 - t_0)$, and the duty cycle, indicating the percentage of the period in which the discharge is on $\{[(t_1 - t_0)/(t_2 - t_0)] \cdot 100\}$.

A detailed study of the modulated plasma morphology, i.e., the electron density profile in the time scale of the modulation, has been carried out by the present authors [29]. In particular, through the analysis of the temporal profile of Ar emission intensity (representative of the fractional density of electrons with energy

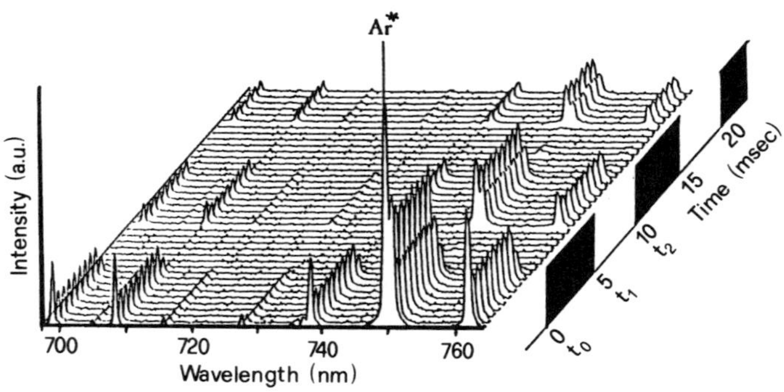

FIGURE 32. Time-resolved optical emission spectrum obtained from Ar-modulated plasma. Modulation parameters: period = 10 ms, duty cycle = 50%.

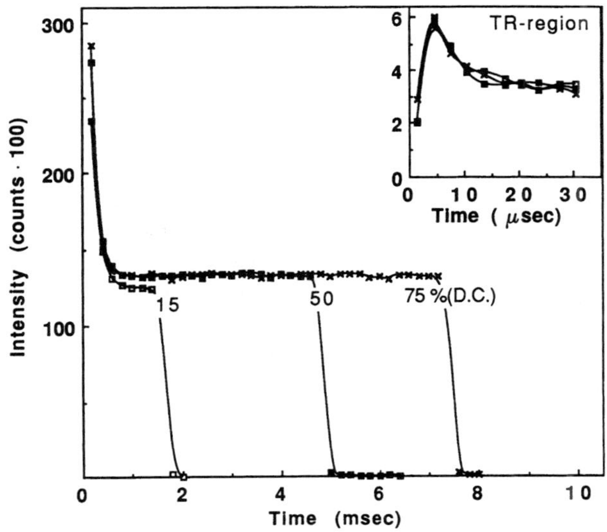

FIGURE 33. Ar* emission intensities from modulated plasmas at a period of 10 ms and duty cycles of 15, 50, and 75%. The inset shows the detailed profile at the plasma ignition (transient region).

>13.5 eV), they described the plasma morphology as characterized by three different regions (see Fig. 33):

(a) A transient region (TR), where the Ar line intensity undergoes a strong variation and reaches a stationary value in a few tenths of milliseconds after a sharp maximum. The first region is characterized by a high transient electron temperature, which decreases toward the steady state value of the continuous-wave (CW) plasma, while the total electron density is increasing [130].
(b) A stationary region (ST), where the electron density and temperature reach the stationary values of the CW conditions, and hence the Ar* emission intensity exhibits a constant value.
(c) An afterglow region (AG), characterized by a decrease to zero of the Ar* emission, after the plasma is switched off. In this region all the excited species present in the plasma phase decay according to the corresponding lifetime value.

The analysis of the temporal profiles of the excited species, as derived from time-resolved OES (TROES) spectra of the type shown in Fig. 34, reveals useful information on the origin as well as on the formation and decay kinetics. As an

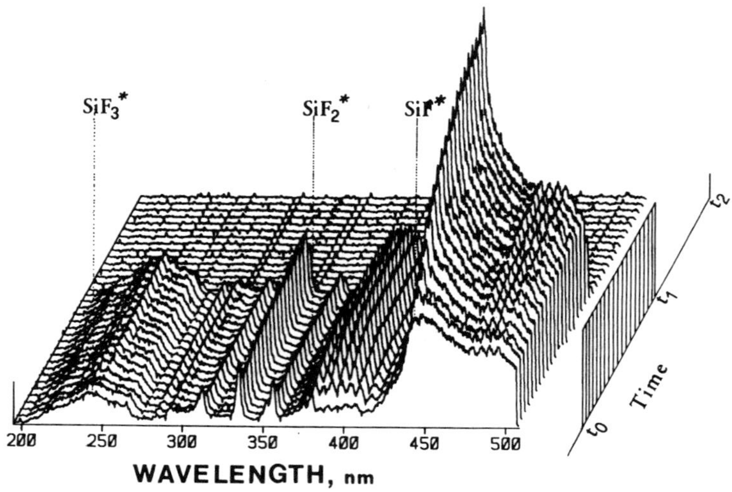

FIGURE 34. Temporal evolution of the optical emission spectrum in $SiF_4:H_2:Ar(10:1:0.1)$ modulated plasma (period = 10 ms, duty cycle = 50%) at 27 MHz, $p = 0.3$ torr, $\Phi = 11$ sccm, $W = 20$ W.

example, Fig. 35 shows the profiles obtained for SiH* species in a SiH_4–H_2 plasma and SiF* in a SiF_4–H_2 plasma [29]. The strong difference between the two trends is accounted for by the fact that SiH* (with a profile similar to that of Ar*) originates from the direct dissociative excitation [reaction (22)] of SiH_4 gas, whereas SiF* is derived from a direct excitation of the same species in the ground state [reaction (25)] and its trend mirrors the SiF_x formation kinetics [reaction (24)]. More details on modulated plasmas examined with TROES diagnostics can be found in Cicala *et al.* [29].

With regard to the effect of plasma modulation on the deposition process, the following phenomena are usually observed:

- An increase in deposition rate.
- A better surface uniformity.
- A decrease in the powder formation.
- An improvement of the material quality.

Contradictory data have been reported on how the plasma modulation affects the deposition rate. Some authors [128, 129] provide evidence of an increase in the deposition rate by varying the period and duty cycle of the modulated plasma fed with a highly diluted silane mixture (0.5% SiH_4 in He). They ascribe this effect to an increase of the electron density under plasma modulation conditions. On the contrary, our investigation carried out on modulated discharge of 10% SiH_4 di-

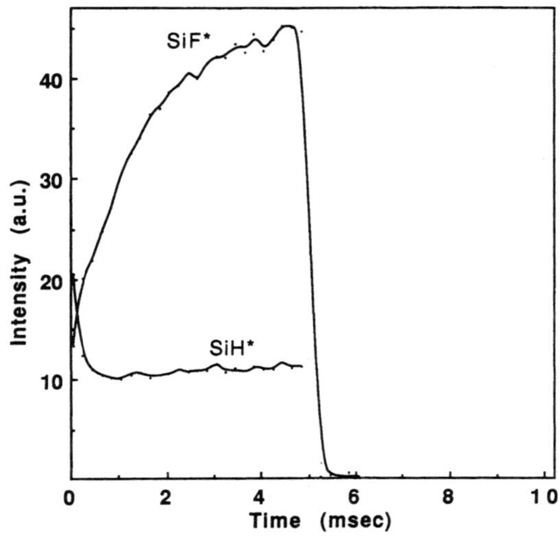

FIGURE 35. Emission intensities of SiH* and SiF* in modulated plasmas (period = 10 ms, duty cycle = 50%) fed with SiH_4–H_2 and SiF_4–H_2 mixtures, respectively.

luted in H_2 did not show evidence of changes in the deposition rate with electron density. However, the theoretical model developed by Park and Economu [164] predicts that the deposition rate can be enhanced under modulation by pulsing the RF signal on a time scale that is comparable to the characteristic time scales of the process. The authors also provide evidence that the modulation dependence of the deposition rate cannot be discussed without accounting for the film thickness inhomogeneity. This is particularly true for plasma systems in which a large depletion of the precursor gas occurs along the flow direction. In fact, SiF_4–H_2 and SiF_4–GeH_4–H_2 systems, which are known to exhibit large depletion of H_2 and GeH_4, respectively, produce film inhomogeneity in both thickness and composition, which disappears under modulation conditions. These effects are well illustrated in Figs. 36 and 37 by the low value of the uniformity index [UI = $(r_{max} - r_{min})/2r$, estimated at a radial distance of 5 cm] and by the constancy of the Ge/Si compositional ratio in a-SiGe:H,F films.

The modulation of the plasma has also been reported as an important tool to avoid the particulate formation resulting from gas-phase nucleation. The contamination of the glow discharge by the particulate produces a deleterious effect in plasma material processing and influences the material quality. Watanabe *et al.* [129, 165], by using the RF modulation method together with the laser Mie scattering technique, have clarified the growth process of particles and have observed a drastic suppression of powder at low values of frequency and duty cycle (see

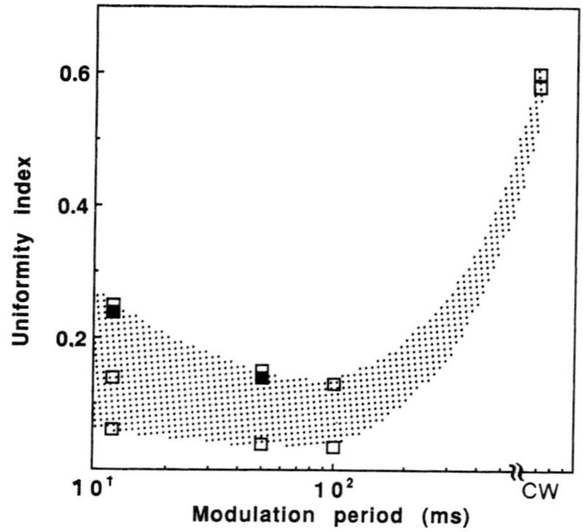

FIGURE 36. Effect of the plasma modulation on the thickness homogeneity of a-Si:H,F (□) and a-SiGe:H,F (■) films deposited from SiF_4–H_2 and SiF_4–GeH_4–H_2 mixtures, respectively. The lower is the uniformity index, the higher is the homogeneity.

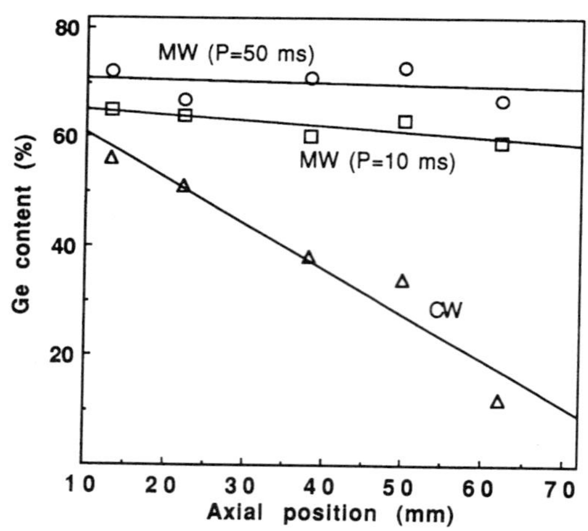

FIGURE 37. Effect of the plasma modulation on the composition homogeneity of a-SiGe:H,F films deposited from SiF_4–GeH_4–H_2 mixtures. The strong variation of Ge content along the gas flow direction in continuous-wave (CW) plasma disappears with modulation. Two samples deposited at the same duty cycle (15%) and different period (*P*) are reported.

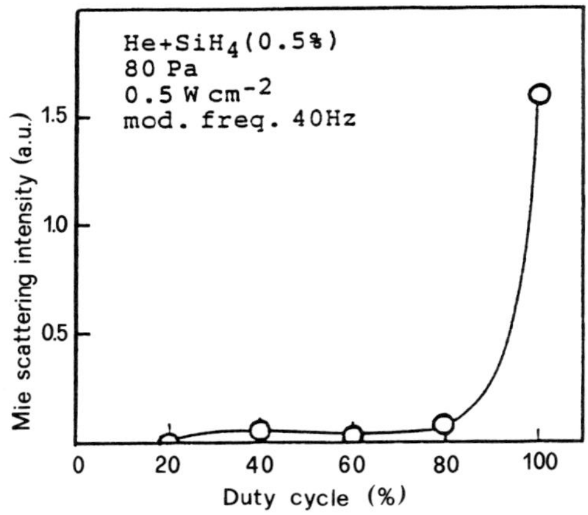

FIGURE 38. Effect of the duty cycle on the powder formation in the He–SiH_4 modulated plasmas. (From Watanabe *et al.* [129].)

Fig. 38). The plasma modulation frequency effect has also been investigated by Howling *et al.* [166], who found at the minimum formation of powders 1 kHz (see Fig. 39). They also illustrated the spatiotemporal distribution of powders by means of video images.

Several studies on dusty plasmas witness an upsurge of experimental and theoretical work focused on silicon clusters. Nevertheless, controversial results exist

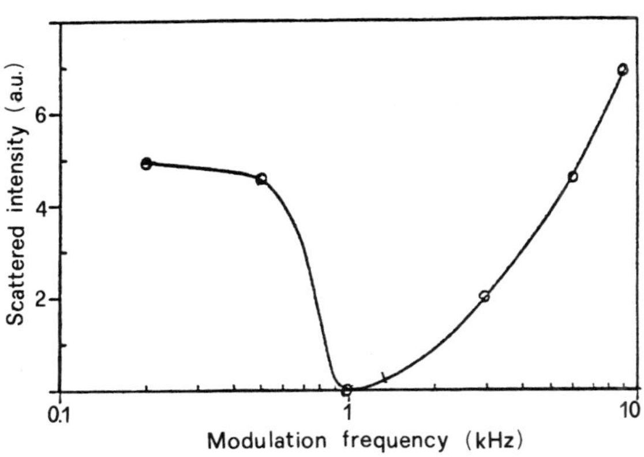

FIGURE 39. Effect of the modulation frquency on the powder formation detected by light scattering in SiH_4 plasmas. (From Howling *et al.* [166].)

about the particle precursors and the growth mechanism in silane plasmas. A fundamental study has been reported by Mandich *et al.* [167], who demonstrated, by a Fourier transform mass spectroscopic method, that the initial silicon clustering reactions require positive ions up to a maximum of three stepwise additions and that the further cluster growth proceeds through a chemisorption of SiD_4 or SiH_4 molecule on silicon clusters, analogously to chemisorption on silicon surfaces (see Section II.D). Other authors [168, 169] ascribed to silicon negative ions ($Si_xH_y^-$) the role of particle formation precursors. Perrin *et al.* [170] believe that negative ions, revealed by a mass-spectroscopic method in multipole discharges, polymerize faster than do positive ions. Additional evidence for this has been given by Howling *et al.* [171] in that they revealed stable negative ions containing up to 16 silicon atoms, whereas positive ions have no more than 6 silicon atoms.

V. Deposition Mechanisms and Controversial Aspects

Since the discovery of the suitability of the PECVD process in producing good-quality amorphous silicon, many theoretical and experimental approaches have been used to model the growth phenomena. The amount and, sometimes, the complexity of the information related to the gas phase and surface processes are so large that one could describe a litany of multiple and, in some cases, contradictory hypotheses concerning the growth mechanism. Here, we have made an attempt to summarize some of them.

One of the main recurring problems in the formulation mechanism has been the identification of the predominant radicals in the gas phase, which are not always detectable under ordinary experimental conditions that give good-quality material. A Si precursor history is presented in Table 3, where a list of different deposition systems and of the relative experimental approaches is given. In our opinion, the diversity in the precursor identity is related to the different experimental apparatus and, to some extent, to the absence of meaningful diagnostic techniques.

It is important to emphasize that, for most of the studies reported in Table 3, the utilized approach was to hypothesize a deposition mechanism, by identifying the most abundant gas-phase radical also as the growth precursor. A wider utilization of diagnostic techniques has stimulated, in these last years, different approaches to the deposition mechanism definition, in that the role of the plasma–surface processes has been emphasized. Thus, the precursor identification has been derived, apart from the gas-phase characterization, from the study of the growth kinetics.

Among the SiH_x radicals, SiH_3 has been the first hypothesized precursor in the deposition of amorphous silicon from silane. This choice was based on various

Table 3 Growth precursors hypothesized in the deposition of silicon films from different feeding mixtures

Gas feed	Precursors	Experimental approach	Refs.
SiH_4–H_2	SiH	Magnetically confined plasma; OES	[172]
Pure SiH_4	SiH	RF plasma; OES, MS	[72]
Pure SiH_4	SiH_2	RF plasma; OES	[22, 23]
Pure SiH_4	SiH_2	DC plasma; MS chemical relaxation technique	[21]
SiH_4–H_2(He)	SiH_2	RF plasma; MS	[173]
SiH_4/SiH_4–Ar	SiH_3	DC plasma; MS	[24, 174]
Pure SiH_4	SiH_3	DC static plasma; MS	[175]
SiH_4–Ar	SiH_3	RF plasma; H content by NMR	[176]
SiH_4	SiH_3	Hg-photosensitized decomposition	[177]
SiH_4–H_2	SiH_3	Hollow cathode tube, diode IRLAS	[26, 178]
Pure SiH_4	SiH_3	Triode RF plasma	[40, 179]
Pure SiH_4	Si_2H_6	Diode RF plasma, MS	[180]
Pure SiH_4	Si_2H_5	Diode RF plasma	[174]
Pure SiH_4	Si_2H_4	Diode RF plasma	[181]
Pure Si_2H_6	Si_2H_y	RF plasma; OES, MS	[72]
Pure Si_2H_6	SiH_3	DC, RF plasma; MS	[75]
Pure SiF_4	SiF	RF plasma; OES	[182]
SiF_4–H_2	SiF_x	RF plasma; OES, LIF, electrical probes	[183, 184]
SiF_4–H_2	SiF_xH_y	RF plasma MS	[185]
SiF_4–H_2	SiF_2	Diode RF plasma; OES	[28]
$SiCl_4$–H_2	$SiCl_2$	RF plasma; MS	[186]
$SiCl_4$–H_2–Ar	$SiCl_2$, $SiCl_4$	RF plasma; MS, OES	[27, 32]

experimental observations. Among these, an estimation of SiH_3 radical density in the plasma phase by threshold-ionization mass-spectrometry measurements has been given by Robertson and Gallagher [24] (see also Section II.C). In addition, on the basis of the measured SiH_3 surface reaction probability, they suggested the growth mechanism as illustrated in Fig. 40 [174]. The main feature of the SiH_3 incorporation mechanism is that this radical, before sticking to the film surface, diffuses on the H-passivated surface until it reaches an appropriate microscopic film valley. This rearrangement causes a smooth and compact film, by avoiding void formation and columnar growth.

The fact that SiH_3 is the most abundant radical in the plasma phase, and is the growth precursor has also been stressed by Matsuda *et al.* [40], and the essential

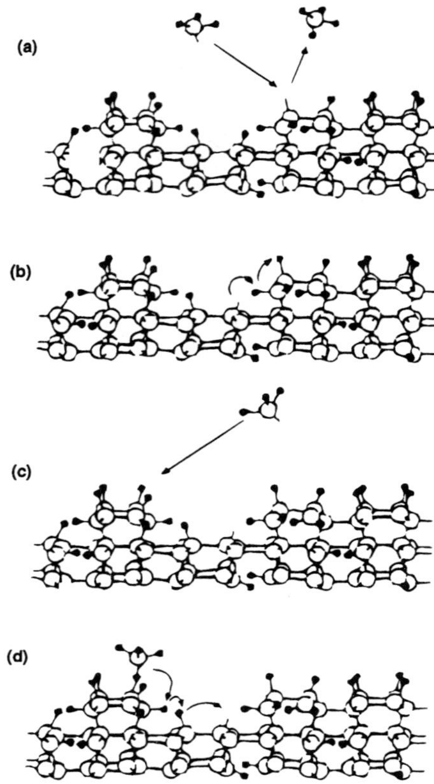

FIGURE 40. Illustration of the incorporation of a gas-phase radical into an a-Si:H: (a) a surface dangling bond is created via an H-abstraction reaction; (b) the dangling bond is shown to diffuse to a lower energy site in a microscopic valley; (c) a free radical adsorbs to the film surface, preferentially at a high point; (d) surface diffusion brings the adsorbed radical into the vicinity of the dangling bond, where incorporation can then occur. (From Doughty *et al.* [174].)

deposition mechanism (in a triode-configuration reactor) is depicted in Fig. 41. The scheme considers that not all of the SiH_3 flux on the surface is effective for the a-Si:H growth: it is in part adsorbed (β), on the surface and in part reflected ($1 - \beta$) in the gas phase. The adsorbed fraction is, in turn, partially incorporated in the film (sticking coefficient s) and partially desorbed as stable volatile compounds (recombination probability γ) formed by surface recombination reactions with atoms or other radicals. The authors state that "the only creation mechanism is the H-abstraction process by SiH_3 itself." A common feature of the above-reported mechanisms is that both are based on SiH_3 radical interacting with the

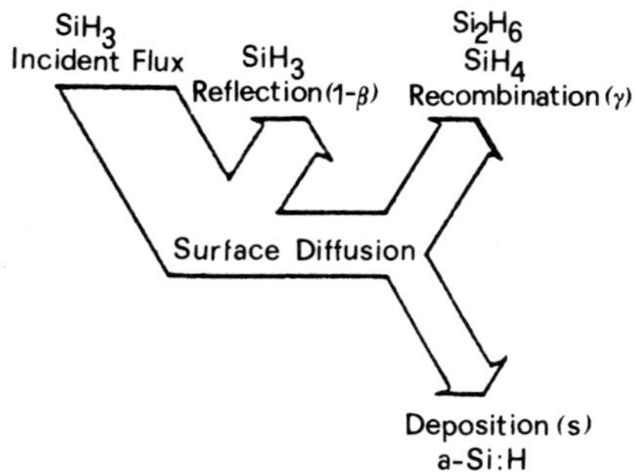

FIGURE 41. Schematic concept of the surface-reaction process of SiH_3 radicals reaching the film-growing surface. (From Matsuda *et al.* [40].)

surface, and on its predominant role in promoting the growth by abstracting the adsorbed hydrogen. Nevertheless, criticism has been expressed [180] on the "direct" measure of SiH_3 density by mass spectrometry [24] and on the radical-separation technique by the mesh electrode [179], which are the fundamental elements supporting the SiH_3-based mechanisms. In this context, the measurement of SiH_3 radical density by infrared (diode) laser absorption spectroscopy (IRLAS), also spatially resolved [178], assumes noticeable importance. The authors make a good estimation of SiH_3 density between the electrodes and they are able to support the fact that "a deposition rate of 1 Å/sec can be sustained by the SiH_3 flux density to the substrate." In contrast, Vepřek states, in many publications, that the estimated deposition rate from SiH_3 density is "negligible as compared to the generally observed deposition rates of > 1 nm/sec of high-quality a-Si:H" [180, 187]. Hence, from waht has been stated above, the observation that SiH_3 is the dominant radical in the plasma does not necessarily mean that the growth proceeds directly from this species.

More recently, it has been claimed of the possibility that higher silanes (Si_2H_6, Si_3H_8) [187, 188] or higher silicon hydride radicals (Si_2H_5, Si_2H_4) [174, 181] contribute to the film growth. Vepřek asserts that SiH_2, formed by direct electron dissociation of SiH_4, quickly inserts into the SiH_4 to produce Si_2H_6, whose decomposition on the surface is responsible for the film growth. The mechanism scheme is illustrated by the following equation:

$$\begin{array}{l} SiH_4 + e \rightarrow SiH_2 + H_2 \\ \qquad\qquad\quad \downarrow +SiH_4 \\ \qquad\qquad\quad Si_2H_6(\text{gas}) \rightleftarrows Si_2H_6(\text{ads}) \xrightarrow{X^+} \text{a-Si:H(film)}. \end{array} \tag{34}$$

The validity of this reaction is quantitatively explained by the experimental results on the correlation of the deposition rate and the gas-phase disilane concentration trends. A similar surface process has been reported by Guizot *et al.* [181], although they consider another kind of radical, Si_2H_4, highly reactive in the diode system. In these last mechanisms, an important step is represented by the precursor adsorption, which is not, however, sufficiently emphasized.

The occurrence of the chemisorption process has been substantiated by the present authors in SiH_4, SiF_4, and $SiCl_4$ systems in order to explain the experimental dependence of the deposition rate on the dopant addition (see Section II.D.1). In addition, they have also reported [27, 28] a chemisorption-based deposition mechanism able to describe the surface growth from the kinetic point of view, i.e., to define the relationship between deposition rate and species concentration. The equation

$$r_D = k^* \cdot [H] \cdot [SiX_n], \tag{35}$$

with X = F, Cl and $n = 2, 4$ allows us to correlate the deposition rate of a-Si:H,X films with the gas densities of H atoms and SiX_n species produced in SiX_4–H_2 plasmas. Basically, equation (35) was derived from the analysis of a simple chemical model, whose concise scheme can be depicted as follows:

$$\begin{array}{l} SiX_4(gas) \overset{K}{\rightleftarrows} SiX_2(ads) \rightleftarrows SiX_2(gas) \\ \qquad\qquad\quad k \downarrow H \\ \qquad\qquad\quad \text{—SiX(FB–Si)} \rightarrow \text{a-Si–film.} \end{array} \tag{36}$$

The most important features are

(a) The dissociative chemisorption of silicon volatile compounds (SiX_4) giving active species (SiX_2) adsorbed at the growing surface.
(b) The surface reaction of H atoms to give free-bond silicon species (FB-Si) that are reactive intermediates for the silicon-to-silicon bond formation on the surface.

The kinetic analysis of the chemical model, evidencing the rate-limiting step (point b) of the whole growth mechanism, leads to the formulation of equation (35). The pseudo–rate constant, $k^* (= k \cdot K)$ includes the true rate constant k and the equilibrium constant K of the dissociative chemisorption as in scheme (36). The value of K takes into account the dopant and the light irradiation effects (see Sections II.D.1 and II.D.2), as they are related to the electronic character of the surface, and the plasma irradiation, as it energetically affects the chemisorption process (see Fig. 11). From the chemical model it is evident that the growth precursors are the starting reactant SiX_4, by far the most abundant species in the gas phase, and SiX_2, produced either in the plasma by electron impact dissociation [see reaction (24)] or on the surface, by dissociative chemisorption. The quantitative evaluation of H and SiX_2 densities have been evaluated by the OES technique with actinometric analysis, according to the procedure reported in Section

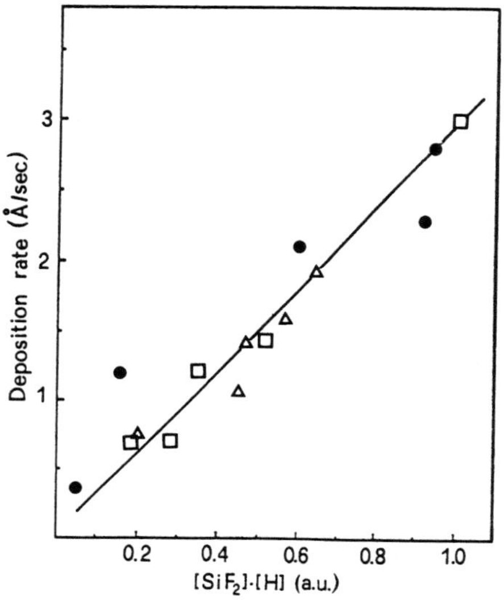

FIGURE 42. Deposition rate (r_D) vs. the product of [SiF_2] and [H] densities for data obtained in SiF_4–H_2 plasmas at different total pressure (□), RF power (△), and gas composition (●). (From Bruno *et al.* [28].)

II.C.1. An example of the validity of the kinetic equation (35) is reported in Fig. 42 for the SiF_4–H_2 system, by examining all the experimental results on species concentrations ([H] and [SiF2]) and deposition rate obtained by varying the power, the pressure, and the feed composition. It is important to stress that the relationship of Eq. (35) is not verified when Si, SiX, and SiX_3 density is considered.

In conclusion, even though the literature has been theater of several diatribes on the plasmachemical processes producing the main radicals and on the growth precursor identity, the actors seem to converge, recently, toward the representation of a final scene with surface chemisorption as protagonist. This aspect, the object of most of our work in the past, has been duly emphasized in this chapter.

References

1. B. S. Mayerson, B. A. Scott, and D. J. Walford, *J. Appl. Phys.* **54,** 1461 (1983).
2. H. Stafast, *Appl. Phys. A.* **45,** 93 (1988).
3. R. C. Chittick, J. H. Alexander, and H. F. Sterling, *J. Electrochem. Soc.* **116,** 77 (1969).
4. W. E. Spear and P. G. LeComber, *J. Non-Cryst. Solids* **8–10,** 727 (1972); W. E. Spear and P. G. LeComber, *Solid State Commun.* **17,** 1193 (1975).

5. M. Capitelli, E. Ficocelli, and E. Molinari, *Z. Naturforsch. A* **26A,** 672 (1971).
6. M. Capitelli, *J. Phys. (Paris), Colloq.* **38,** C3-227 (1977).
7. G. Capriati, G. Colonna, C. Gorse, and M. Capitelli, *Plasma Chem. Plasma Process.* **12,** 237 (1992).
8. C. Gorse, M. Cacciatore, M. Capitelli, S. De Benedictis, and G. Dilecce, *Chem. Phys.* **119,** 63 (1988).
9. C. Gorse, M. Capitelli, R. Celiberto, R. Winkler, and J. Wilhelm, *J. Phys. D* **23,** 1041 (1990).
10. M. Capitelli, C. Gorse, R. Winkler, and J. Wilhelm, *Plasma Chem. Plasma Process.* **8,** 399 (1988).
11. M. J. Kushner, *Mater. Res. Soc. Symp. Proc.* **68,** 293 (1986).
12. J. P. M. Schmitt, *J. Non-Cryst. Solids* **59/60,** 649 (1983).
13. M. Capitelli and E. Molinari, *Top. Plasma Chem.* **90,** 59 (1980).
14. M. A. Lieberman, *IEEE Trans. Plasma Sci.* **PS-17,** 338 (1989).
15. A. Matsuda, T. Kaga, H. Tanaka, and K. Tanaka, *Jpn. J. Appl. Phys.* **23,** L567 (1984).
16. P. M. Vallinga, P. M. Meijer, and F. J. de Hoog, *J. Phys. D* **22,** 1650 (1989).
17. P. M. Vallinga and F. J. de Hoog, *J. Phys. D.* **22,** 925 (1989).
18. M. J. Kushner, *J. Appl. Phys.* **54,** 4958 (1983).
19. G. Cicala, G. Bruno, P. Capezzuto, P. Manodoro, and V. Pische, *Proc. Int. Symp. Plasma Chem., 9th (ISPC–9) (IUPAC), Pugnochiuso, Italy* (R. d'Agostino, ed.), Vol. 3, p. 1405 (1989).
20. G. Turban, *Pure Appl. Chem.* **56,** 215 (1984).
21. J. J. Wagner and S. Vepřek, *Plasma Chem. Plasma Process.* **3,** 219 (1983).
22. F. J. Kampas and R. W. Griffith, *J. Appl. Phys.* **52,** 1285 (1981).
23. F. J. Kampas and R. W. Griffith, *Appl. Phys. Lett.* **39,** 407 (1981).
24. R. Robertson and A. Gallagher, *J. Appl. Phys.* **59,** 3402 (1986).
25. K. Tachibana, T. Shirafuji, and Y. Matsui, *Jpn. J. Appl. Phys.* **31,** 2588 (1992).
26. N. Itabashi, K. Kato, N. Nishiwaki, T. Goto, C. Yamada, and E. Hirota, *Jpn. J. Appl. Phys.* **27,** L1565 (1988).
27. G. Bruno, P. Capezzuto, G. Cicala, and F. Cramarossa, *Plasma Chem. Plasma Process.* **6,** 109 (1986).
28. G. Bruno, P. Capezzuto, and G. Cicala, *J. Appl. Phys.* **69,** 7256 (1991).
29. G. Cicala, M. Losurdo, P. Capezzuto, and G. Bruno, *Plasma Sources Sci. Technol.* **1,** 156 (1992).
30. F. J. Kampas, *J. Appl. Phys.* **54,** 2276 (1983).
31. F. J. Kampas and M. J. Kushner, *IEEE Trans. Plasma Sci.* **PS-14,** 173 (1986).
32. G. Bruno, P. Capezzuto, G. Cicala, and F. Cramarossa, *J. Appl. Phys.* **62,** 2050 (1987).
33. J. E. Lennard-Jones, *Trans. Faraday Soc.* **28,** 333 (1932).
34. S. T. Ceyer, *Science* **249,** 133 (1990).
35. T. Walkenstein, "Electronic Processes on Semiconductor Surfaces During Chemisorption." Consultants Bureau, Plenum, New York, 1991.
36. K. J. Laidler, *in* "Catalysis" (P. H. Emmett, ed.), Vol. 1, p. 128. Reinhold, New York, 1954.
37. J. T. Yates, Jr., *Chem. Eng. News* **30,** 22 (1992).
38. W. A. P. Claassen, J. Bloem, W. G. J. N. Valkenburg, and C. H. J. Van Den Brekel, *J. Non-Cryst. Solids* **57,** 259 (1982).
39. A. Gallagher, *Mater. Res. Soc. Symp. Proc.* **70,** 3 (1986).
40. A. Matsuda, K. Nomoto, Y. Takeuchi, A. Suzuki, A. Yuuki, and J. Perrin, *Surf. Sci.* **277,** 50 (1990).
41. S. Vepřek, K. Ensslen, M. Konuma, and F. A. Sarott, *Proc. Int. Symp. Plasma Chem., 7th (ISPC–7), Eindhoven, Neth.* (C. J. Timmermans, ed.), p. 74 (1985).
42. K. Ando, M. Aozasa, and R. G. Pyon, *Appl. Phys. Lett.* **44,** 413 (1984).
43. G. Bruno, P. Capezzuto, G. Cicala, P. Manodoro, and V. Tassielli, unpublished results (1988).
44. J. C. Knights, *J. Non-Cryst. Solids* **35/36,** 159 (1980).

45. F. S. Stone, *in* "Chemisorption," p. 812 (W. E. Garner, ed.). Butterworth, London, 1957.
46. K. J. Laidler, "Chemical Kinetics," p. 312. McGraw-Hill, New York, 1965.
47. G. Bruno, P. Capezzuto, G. Cicala, P. Manodoro, and V. Tassielli, *IEEE Trans. Plasma Sci.* **PS-18,** 934 (1990).
48. J. Perrin, Y. Takeda, N. Hirano, Y. Takeuchi, and A. Matsuda, *Surf. Sci.* **210,** 114 (1989).
49. G. Bruno, P. Capezzuto, and G. Cicala, *Pure Appl. Chem.* **64,** 725 (1992).
50. G. Bruno, P. Capezzuto, and F. Cramarossa, *Thin Solid Films* **129,** 217 (1985).
51. M. H. Brodsky, ed., *Top. Appl. Phys.* **36** (1979).
52. R. A. Street, J. C. Knights, and D. K. Biegelsen, *Phys. Rev. B* **18,** 1880 (1978).
53. J. D. Joannopoulos and G. Lucovsky, eds., *Top. Appl. Phys.* **55** (1984).
54. B. von Roedern, *Proc. E. C. Photovoltaic Sol. Energy Conf., 10th* (A. Luque, G. Sala, W. Palz, G. Dos Santos, and P. Helm, eds.), p. 1197. Kluwer Academic Publ., Lisbon, 1991.
55. D. L. Staebler and C. R. Wronski, *Appl. Phys. Lett.* **31,** 292 (1977).
56. B. von Roedern and A. Madan, *Philos. Mag., Part B* **63,** 293 (1991).
57. W. B. Jackson, *Phys. Rev. B* **41,** 1059 (1990).
58. M. Nishikuni, H. Haku, M. Isomura, S. Okamoto, K. Sayama, T. Takahama, T. Matsuoka, S. Tsuda, S. Nakano, M. Ohnishi, and Y. Kuwano, *J. Non-Cryst. Solids* **114,** 181 (1989).
59. P. Ganguly, A. Suzuki, S. Yamasaki, K. Nomoto, and A. Matsuda, *J. Appl. Phys.* **68,** 3738 (1990).
60. S. R. Ovshinski and A. Madan, *Nature (London)* **276,** 482 (1978).
61. A. Madan, S. R. Ovshinski, and E. Benn, *Philos. Mag., Part B* **40,** 259 (1979).
62. A. Matsuda, K. Yagii, T. Kaga, and K. Tanaka, *Jpn. J. Appl. Phys.* **23,** L576 (1984).
63. H. Matsumura and H. Ihara, *J. Appl. Phys.* **64,** 6505 (1988).
64. H. Koinuma, H. Natsuaki, K. Fueki, K. Sato, T. Hirano, and M. Isogaya, *Tech. Dig. Int. PVSEC-1, Kobe, Jpn.,* p. 743 (1984).
65. H. Koinuma, T. Manako, H. Natsuaki, H. Fujioka, and K. Fueki, *J. Non-Cryst. Solids* **77/78,** 801 (1985).
66. J. Hanna, S. Oda, H. Shibata, H. Shirai, A. Miyauchi, A. Tanabe, K. Fukuda, T. Ohtoshi, O. Tokuhiro, H. Nguyen, and I. Shimizu, *Mater. Res. Soc. Symp. Proc.* **70,** p. 11 (1986).
67. H. Matsumura, Y. Nakagoma, and S. Furukawa, *Appl. Phys. Lett.* **36,** 439 (1980).
68. T. Shimada, Y. Katayama, and S. Horigoma, *Jpn. J. Appl. Phys.* **19,** L265 (1980).
69. H. Matsumura and S. Furukawa, *Jpn. J. Appl. Phys.* **20,** 275 (1981).
70. M. Janai, L. Frey, R. Weil, and B. Pratt, *Solid State Commun.* **48,** 521 (1983).
71. W. Luft, "High-Rate Deposition of Hydrogenated Amorphous Silicon Films and Devices," SERI/TR211-3052. Solar Energy Res. Inst., Golden, Colorado (1987).
72. D. S. Shen and P. K. Bhat, "Amorphous Silicon Solar Cells Prepared at High Deposition Rate," SERI/TP-211-3949. Solar Energy Res. Inst., Golden, Colorado (1990).
73. K. Ogawa, I. Shimizu, and E. Inoue, *Jpn. J. Appl. Phys.* **20,** L639 (1981).
74. A. Matsuda, T. Kaga, H. Tanaka, L. Malhotra, and K. Tanaka, *Jpn. J. Appl. Phys.* **22,** L115 (1983).
75. J. R. Doyle, D. A. Doughty, and A. Gallagher, *J. Appl. Phys.* **71,** 4771 (1992).
76. W. Paul, D. K. Paul, B. von Roedern, J. Blake, and S. Oguz, *Phys. Rev. Lett.* **46,** 1016 (1981).
77. I. Shimizu, M. Kumeda, A. Morimoto, Y. Tsujimura, and I. Kobayashi, *Mater. Res. Soc. Symp. Proc.* **70,** 313 (1986).
78. K. D. Mackenzie, J. H. Burnett, J. R. Eggert, Y. M. Li, and W. Paul, *Phys. Rev. B* **38,** 6120 (1988).
79. A. Matsuda and K. Tanaka. *J. Non-Cryst. Solids* **97/98,** 1367 (1987).
80. L. Mariucci, F. Ferrazza, D. Della Sala, M. Capizzi, and F. Evangelisti, *J. Non-Cryst. Solids* **97/98,** 1075 (1987).
81. K. Nozawa, Y. Yamaguchi, J. Hanna, and I. Shimizu, *J. Non-Cryst. Solids* **59/60,** 533 (1983).
82. K. D. Mackenzie, J. Hanna, J. R. Eggert, Y. M. Li, Z. L. Sun, and W. Paul, *J. Non-Cryst. Solids* **77/78,** 881 (1985).

83. G. Bruno, P. Capezzuto, G. Cicala, and F. Cramarossa, *J. Mater. Res.* **4,** 366 (1989).
84. W. Kasper, R. Plättner, and J. Eichmeier, *J. Non-Cryst. Solids* **137/138,** 709 (1991).
85. W. Paul, *J. Non-Cryst. Solids* **137/138,** 803 (1991).
86. J. R. Doyle, D. A. Doughty, and A. Gallagher, *J. Appl. Phys.* **71,** 4727 (1992).
87. G. Bruno, P. Capezzuto, M. Losurdo, P. Manodoro, and G. Cicala, *J. Non-Cryst. Solids* **137/138,** 753 (1991).
88. M. Stutzmann, R. A. Street, C. C. Tsai, J. B. Boyce, and S. E. Ready. *J. Appl. Phys.* **66,** 569 (1989).
89. H. Itozaki, N. Fujita, T. Igarashi, and H. Hitotsuyamagi, *J. Non-Cryst. Solids* **59/60,** 589 (1983).
90. A. H. Mahan, D. L. Williamson, and A. Madan, *Appl. Phys. Lett.* **44,** 220 (1984).
91. V. Marakhonov, N. Rogachev, J. Ishkalov, J. Marakhonov, E. Teurkov, and V. Chelndeov, *J. Non-Cryst. Solids* **137/138,** 817 (1991).
92. D. A. Anderson and W. E. Spear, *Philos. Mag.* **35,** 1 (1977).
93. R. S. Sussmann and R. Ogden, *Philos. Mag., Part B* **44,** 137 (1981).
94. S. Liedtke, K. Jahn, F. Finger, and W. Fuhs, *J. Non-Cryst. Solids* **97/98,** 1083 (1987).
95. B. von Roedern, A. H. Mahan, D. L. Williamson, and A. Madan, *Symp. Mater. New Process. Tech. Photovoltaics, 5th, New Orleans,* 1984.
96. K. Mui, D. K. Basa, R. Corderman, and F. W. Smith, *J. Non-Cryst. Solids* **97/98,** 999 (1987).
97. A. H. Mahan, B. von Roedern, D. L. Williamson, and A. Madan, *J. Appl. Phys.* **57,** 2717 (1985).
98. S. Iida and S. Ohki, *Jpn. J. Appl. Phys.* **21,** L62 (1982).
99. C. Wang, G. Lucovsky, and R. J. Nemanich, *J. Non-Cryst. Solids* **137/138,** 741 (1991).
100. N. Shibata, N. Shida, A. Tanabe, J. Hanna, and I. Shimizu, *Proc. Int. Symp. Plasma Chem., 8th (ISPC–8) (IUPAC), Tokyo* (K. Akashi and A. Kinbara, eds.), Vol. 3, p. 1496 (1987).
101. A. H. Mahan, D. L. Williamson, M. Ruth, and P. Rabiosson, *J. Non-Cryst. Solids* **77/78,** 861 (1985).
102. A. H. Mahan, P. Raboisson, and R. Tsu, *Appl. Phys. Lett.* **50,** 335 (1987).
103. H. Koinuma, M. Funabashi, K. Kishio, M. Kawasaki, T. Hirano, and K. Fueki, *Jpn. J. Appl. Phys.* **25,** 1811 (1986).
104. G. Lucovsky, D. V. Tsu, and R. J. Markunas, *Mater. Res. Soc. Symp. Proc.* **68,** 323 (1986).
105. B. L. Jones, *J. Non-Cryst. Solids* **77/78,** 957 (1985).
106. Z. Yin and F. W. Smith, *J. Non-Cryst. Solids* **137/138,** 879 (1991).
107. M. J. Kushner, *J. Appl. Phys.* **71,** 4173 (1992).
108. Y. Manabe and T. Mitsuyu, *J. Appl. Phys.* **66,** 2475 (1989).
109. P. Boher, M. Renaud, L. J. Van Ijzendooru, J. Barrier, and Y. Hily, *J. Appl. Phys.* **63,** 1464 (1988).
110. T. Fuyuki, B. Allain, and J. Perrin, *J. Appl. Phys.* **68,** 3322 (1990).
111. J. L. Guizot, P. Alnot, F. Wyczisk, J. Perrin, and B. Allain, *Semicond. Sci. Technol.* **6,** 582 (1991).
112. H. Lorenz, I. Eisele, J. Ramm, J. Edlinger, and M. Buhler, *J. Vac. Sci. Technol., B* **9,** 208 (1991).
113. H. Dun, P. Pan, F. R. White, and R. E. Hampy, *J. Electrochem. Soc.* **128,** 1555 (1981).
114. G. N. Parsons, J. H. Souk, and J. Batey, *J. Appl. Phys.* **70,** 1553 (1991).
115. R. E. Livengood and D. W. Hess, *Appl. Phys. Lett.* **50,** 560 (1987).
116. C. P. Chang, D. L. Flamm, D. E. Ibbotson, and J. A. Mucha, *J. Appl. Phys.* **62,** 1408 (1987).
117. N. Watanabe, M. Yoshida, Y. C. Jiang, T. Nomoto, and I. Abiko, *Jpn. J. Appl. Phys.* **30,** L619 (1991).
118. S. Fujita, H. Toyoshima, T. Ohishi, and A. Sasaki, *Jpn. J. Appl. Phys.* **23,** L144 (1984).
119. S. Fujita, H. Toyoshima, T. Ohishi, and A. Sasaki, *Jpn. J. Appl. Phys.* **23,** L268 (1984).
120. G. Cicala, G. Bruno, P. Capezzuto, and M. Losurdo, *Mater. Res. Soc. Symp. Proc.* **284,** 27 (1993).
121. S. Fujita, T. Ohishi, H. Toyoshima, and A. Sasaki, *J. Appl. Phys.* **57,** 426 (1985).
122. H. Curtins, N. Wyrsch, M. Favre, and A. V. Shah, *Plasma Chem. Plasma Process.* **7,** 267 (1987); H. Curtins, N. Wyrsch, and A. V. Shah, *Electron. Lett.* **23,** 228 (1987).

123. P. K. Bhat, C. Marshall, J. Sandwich, H. Chatam, R. E. I. Schropp, and A. Madan, *IEEE Proc. Photovoltaics, 20th,* p. 91 (1988).
124. S. Oda, J. Noda, and M. Matsumura, *Mater. Res. Soc. Symp. Proc.* **118,** 117 (1988).
125. T. Hamasaki, M. Ueda, A. Chayahara, M. Hirose, and Y. Osaka, *Appl. Phys. Lett.* **44,** 1049 (1984).
126. R. Martins, N. Corvalho, E. Fortunato, A. Maçarico, M. Santas, I. Baía, M. Viera, and L. Guimarães, *J. Non-Cryst. Solids* **97/98,** 1399 (1987).
127. I. Sakata, M. Yamanaka, and Y. Hayashi, *J. Appl. Phys.* **67,** 3737 (1990).
128. L. J. Overzet and J. T. Verdeyen, *Appl. Phys. Lett.* **48,** 695 (1986).
129. Y. Watanabe, M. Shiratani, Y. Kubo, I. Ogawa, and S. Ogi, *Appl. Phys. Lett.* **53,** 1263 (1988).
130. T. Yoshida, Y. Ichikawa, and H. Sokai, *Proc. E. C. Photovoltaic Sol. Energy Conf., 9th* (W. Palz, G. T. Wrixon, and P. Helm, eds.), p. 1006. Kluwer Academic Publ., Freiburg, Germany, 1989.
131. A. Matsuda, *Pure Appl. Chem.* **60,** 733 (1988).
132. F. Finger, U. Kroll, V. Viret, A. Shah, W. Beyer, X.-M. Tang, J. Weber, A. A. Howling, and C. Hollenstein, *J. Appl. Phys.* **71,** 5665 (1992).
133. D. L. Flamm, *J. Vac. Sci. Technol., A* **4,** 729 (1986).
134. R. Winkler, M. Capitelli, C. Gorse, and J. Wilhelm, *Plasma Chem Plasma Process.* **10,** 419 (1990).
135. M. R. Wertheimer and M. Moisan, *J. Vac. Sci. Technol., A* **3,** 2643 (1985).
136. M. Moisan, C. Barbeau, R. Claude, C. M. Ferreira, J. Margot, J. Paraszczak, A. B. Sá, G. Sauvé, and M. R. Wertheimer, *J. Vac. Sci. Technol., B* **9,** 8 (1991).
137. C. Beneking, F. Finger, and H. Wagner, *Proc. E. C. Photovoltaic Sol. Energy Conf., 11th* (L. Guimaraes, W. Palz, C. De Reyff, H. Kiess, and P. Helm, eds.), p. 586. Harwood Academic Publ., Montreux, Switzerland, 1992.
138. R. A. Gottscho and M. L. Mandich, *J. Vac. Sci. Technol., A* **3,** 617 (1985).
139. J. L. Dorier, C. Hollenstein, A. A. Howling, and U. Kroll, *J. Vac. Sci. Technol., A* **10,** 1048 (1992).
140. A. A. Howling, J.-L. Dorier, C. Hollenstein, U. Kroll, and F. Finger, *J. Vac. Sci. Technol., A* **10,** 1080 (1992).
141. R. A. Street, J. C. Knights, and D. K. Biegelsen, *Phys. Rev. B* **18,** 1880 (1978).
142. J. C. Knights, R. A. Lujan, M. P. Rosenblum, R. A. Street, and D. K. Biegelsen, *Appl. Phys. Lett.* **38,** 331 (1981).
143. P. Roca i Cabarrocas, J. Merten, C. Godet, B. Equer, Y. Bouizem, P. Sladek, M. L. Theye, D. Mencaraglia, Z. Djebbour, J. Sib, J. P. Kleider, C. Longeaud, R. Meaudre, and M. Meaudre, *Proc. E. C. Photovoltaic Sol. Energy Conf., 11th* (L. Guimaraes, W. Palz, C. De Reyff, H. Kiess, and P. Helm, eds.), p. 184. Harwood Academic Publ., Montreux, Switzerland, 1992.
144. R. Meaudre, M. Meaudre, P. Roca i Cabarrocas, S. Tanidi, Y. Bouizem, and M. L. Theye, *J. Non-Cryst. Solids* **137/138,** 171 (1991).
145. A. Matsuda, S. Mashima, K. Hasezaki, A. Suzuki, S. Yamasaki, and P. J. McElheny. *Appl. Phys. Lett.* **58,** 2494 (1991).
146. C. C. Tsai, J. C. Knights, G. Chang, and B. Wacker, *J. Appl. Phys.* **59,** 2998 (1986).
147. J. Dutta, K. Hasezaki, S. Mashima, P. J. McElheny, A. Suzuki, G. Ganguly, and A. Matsuda, *Jpn. J. Appl. Phys.* **31,** L299 (1992).
148. C. Hollenstein, U. Kroll, A. A. Howling, J. Dutta, J.-L. Dorier, J. Meier, R. Tschrner, and A. Shah, *Proc. E. C. Photovoltaic Sol. Energy Conf., 11th* (L. Guimaraes, W. Palz, C. De Reyff, H. Kiess, and P. Helm, eds.), p. 76. Harwood Academic Publ., Montreux, Switzerland, 1992.
149. A. Asano, *Appl. Phys. Lett.* **56,** 533 (1990).
150. M. Otobe and S. Oda, *Jpn. J. Appl. Phys.* **31,** L1443 (1992).
151. S. Vepřek, *Chimia* **34,** 489 (1981).
152. N. Blayo and B. Drévillon, *J. Non-Cryst. Solids* **137/138,** 775 (1991).

153. M. Fang, J. B. Chevrier, and B. Drévillon, *J. Non-Cryst. Solids* **137/138,** 791 (1991).
154. K. Nomoto, Y. Urano, J. L. Guizot, G. Ganguly, and A. Matsuda, *Jpn. J. Appl. Phys.* **29,** L1372 (1990).
155. N. Shibata, K. Fukuda, H. Ohtoshi, J. Hanna, S. Oda, and I. Shimizu, *Mater. Res. Soc. Symp. Proc.* **95,** 225 (1987).
156. H. Shirai, J. Hanna, and I. Shimizu, *Jpn. J. Appl. Phys.* **30,** L881 (1991).
157. H. Shirai, D. Das, J. Hanna, and I. Shimizu, *Appl. Phys. Lett.* **59,** 1096 (1991).
158. N. M. Johnson, P. V. Santos, C. E. Nebel, W. B. Jackson, R. A. Street, K. S. Stevens, and J. Walker, *J. Non-Cryst. Solids* **137/138,** 235 (1991).
159. A. Suzuki, Y. Toyoshima, P. J. McElheny, and A. Matsuda, *Jpn. J. Appl. Phys.* **30,** L790 (1991).
160. S. Koynov, A. Toneva, and K. Kougia, *J. Non-Cryst. Solids* **137/138,** 685 (1991).
161. M. Yamanaka, Y. Hayashi, and I. Sakata, *Jpn. J. Appl. Phys.* **29,** L217 (1990).
162. I. Sakata, M. Yamanaka, and Y. Hayashi, *Jpn. J. Appl. Phys.* **30,** L326 (1991).
163. N. Suzuki, T. Yoshikawa, K. Masu, K. Tsubouchi, and N. Mikoshiba, *Jpn. J. Appl. Phys.* **28,** L2316 (1989).
164. S. K. Park and D. J. Economu, *J. Electrochem. Soc.* **137,** 2103 (1990).
165. M. Shiratani, S. Matsuo, and Y. Watanabe, *Jpn. J. Appl. Phys.* **30,** 1887 (1991).
166. A. A. Howling, C. Hollenstein, J.-L. Dorier, P. Paris, and M. Favre, *Proc. E. C. Photovoltaic Sol. Energy Conf., 10th* (A. Luque, G. Sala, W. Palz, G. Dos Santos, and P. Helm, eds.), p. 169. Kluwer Academic Publ., Lisbon, 1991.
167. M. L. Mandich, W. D. Reents, Jr., and K. D. Kolenbrander, *Pure Appl. Chem.* **62,** 1653 (1990).
168. A. Garscadden, *in* "Non Equilibrium Processes in Partially Ionized Gases" (M. Capitelli and J. N. Bardsley, eds.), p. 541. Plenum, New York, 1990.
169. J. P. Boeuf and P. Belenguer, *J. Appl. Phys.* **71,** 4751 (1992).
170. J. Perrin, A. Lloret, G. de Rosny, and J. P. M. Schmitt, *Int. J. Mass Spectrom. Ion Phys.* **57,** 249 (1984).
171. A. A. Howling, L. Sansonnens, J.-L. Dorier, and C. Hollenstein, *J. Phys. D* **26,** 1003 (1993).
172. T. Hamasaki, M. H. Hirose, H. Kurata, M. Taniguchi, and Y. Osaka, *Jpn. J. Appl. Phys.* **20,** 281 (1981).
173. G. Turban, Y. Catherine, and B. Grolleau, *Thin Solid Films* **77,** 287 (1981).
174. D. A. Doughty, J. R. Doyle, G. H. Lin, and A. Gallagher, *J. Appl. Phys.* **67,** 6220 (1990).
175. P. A. Longeway, R. D. Estes, and H. A. Weakliem, *J. Phys. Chem.* **88,** 73 (1984).
176. G. Turban, Y. Catherine, and B. Grolleau, *Thin Solid Films* **67,** 309 (1980).
177. J. Perrin, T. Broekhuizen, and R. Benfehrat, *Proc. Eur. Mater. Res. Soc., Strasbourg, Fr.* (1986).
178. N. Itabashi, N. Nishiwaki, M. Magane, S. Naito, T. Goto, A. Matsuda, C. Yamada, and E. Hirota, *Jpn. J. Appl. Phys.* **29,** L505 (1990).
179. A. Matsuda and K. Tanaka, *J. Appl. Phys.* **60,** 2351 (1986).
180. S. Vepřek, M. Heintze, F.-A. Sarott, M. Jurcik-Rajman, and P. Willmott, *Mater. Res. Soc. Symp. Proc.* **118,** 3 (1988).
181. J.-L. Guizot, K. Nomoto, and A. Matsuda, *Surf. Sci.* **244,** 22 (1991).
182. H. U. Lee, R. C. Ross, and J. P. Deneufville, *Mater. Res. Soc. Symp. Proc.* **68,** 149 (1986).
183. N. Mutsukura, M. Ohuchi, S. Satoh, and Y. Machi, *Thin Solid Films* **109,** 47 (1983).
184. H. U. Lee and J. P. Deneufville, *J. Non-Cryst. Solids* **66,** 39 (1984).
185. I. Shimizu, *J. Non-Cryst. Solids* **97/98,** 257 (1987).
186. R. Manory, A. Grill, U. Carmi, and R. Avni, *Plasma Chem. Plasma Process.* **3,** 235 (1983).
187. S. Vepřek and M. Heintze, *Plasma Chem. Plasma Process* **10,** 3 (1990).
188. S. Vepřek and M. G. J. Vepřek-Heijman, *Appl. Phys. Lett.* **56,** 1766 (1990).

2 Diagnostics of Amorphous Silicon (a-Si) Plasma Processes

Guy Turban
Laboratoire des Plasmas et des Couches Minces
Institut des Matériaux
University of Nantes
Nantes, France

Bernard Drévillon
Laboratoire de Physique des Interfaces et des Couches Minces
Ecole Polytechnique
Palaiseau, France

and

Dimitri S. Mataras and Dimitri E. Rapakoulias
Department of Chemical Engineering
University of Patras
Patras, Greece

I. Introduction

In this chapter we review the main diagnostic techniques used for the characterization of silane plasma during the growth of a-Si film.

The plasma phase can be studied by means of two different categories of methods: optical and electrical. The optical diagnostic technique has the main advantage of being noninvasive. Both passive and active spectroscopies have been developed and applied to the study of silane discharges. The optical emission spectroscopy (OES) was the first to be used, and several species such as Si, SiH, Si^+, and H^+ have been identified. However, two important radicals—SiH_2 and SiH_3—cannot be detected by OES. The more recent development of active spectroscopy as laser-induced fluorescence (LIF) and infrared laser absorption spectroscopy (RLAS) permit the detection of virtually all types of radicals in silane discharge. Optical probes are well suited for space- and time-resolved analyses, and several examples will illustrate this aspect.

The electrical methods (mass spectrometry and Langmuir probes) are very useful as a complement to optical methods. Mass spectrometry (MS) is unique for study of ion chemistry and is also a simple technique to control the neutral gas chemistry: dissociation of silane, formation of higher silanes, and hydrogen. In some conditions neutral radicals can also be detected with MS. Very useful information on ion–molecule reactions, kinetics, and the neutral chemistry mechanism can be inferred with this technique, as we will demonstrate. The Langmuir probe is certainly more difficult to employ in the silane reactive plasmas, particularly with RF excitation. However, it is the only simple technique to use for getting data on electron concentration and energy.

The knowledge of the deposition mechanism a-Si film and the control of the process need real-time *in situ* surface diagnostics. Ellipsometry is compatible with the environment of silane discharge and permits this real-time study. In this chapter we describe the application of the phase-modulated ellipsometry in the near-ultraviolet (UV) visible range and that of the infrared ellipsometry. Typical examples will illustrate how the microstructure evolution of the a-Si film and the control of the film growth can be followed by this technique. Various aspects of the growth kinetics will be presented as the result of the *in situ* spectroellipsometry analyses.

II. Optical Diagnostics

A. Introduction

The glow discharge of reactive gases is a complex reactive environment in which, besides electron collision processes, a series of secondary gas-phase reactions takes place among neutral and stable molecules, radicals, and positive and negative ions. In addition, the plasma is in continuous interaction with the sustaining electromagnetic field and with the electrodes, the heated substrate, and the containing vessel. The data obtained from the direct experimental observation of such a complex system are still very incomplete, and every piece of information is, therefore, valuable.

Optical diagnostics in silane glow discharge have been widely used, in an effort to understand gas-phase mechanisms and kinetics. This evolved as a natural consequence of the fact that these methods have already been used as plasma diagnostics for many other gases, and also that there is a lack of nonintrusive diagnostic techniques for electric discharges.

The main feature of the discharge itself makes these methods interesting; a spontaneous emission from different electronically excited species is always present. Moreover, the discharge light is not uniformly distributed in what is usually considered the discharge volume; there are several well-distinguishable zones in DC as well as RF discharges. In the latter case we can distinguish two dark regions: the RF or cathode and the grounded or anode sheath, and a nonuniformly illuminated bulk plasma region. Also in RF discharges the intensity at each point of the discharge space changes along with the excitation-source period.

Macroscopically, the oscillating plasma–sheath boundary of both electrodes and the illumination of the plasma change as a function of the discharge conditions. This is due to the spatiotemporal distribution of energetic electrons, which in turn give origin to collisional phenomena, with characteristic topology and frequencies corresponding to different electron groups. The neutral or charged products resulting from these collisions are various, and possess different energetic characteristics, many of which are detectable.

The simplest method to obtain information concerning these species consists in the collection and analysis of the emission light as a function of space and time (although this is not always simple). Moreover, one can force the discharge to give optical information either by measuring the absorption of a light beam passing through it or by stimulating the emission of radiation by a laser.

Concerning silane discharges, several different optical diagnostic approaches have been used over the last 10 years. At an early stage, there have been many studies using optical emission spectroscopy (OES) for the detection of SiH^*, Si^*, H^*, and H_2^*, [1–3]. In particular, different aspects have been examined of the dependence of the emission characteristics of various externally controllable discharge parameters, and the possible relation of these to the discharge and film properties.

More recent experiments have revealed the spatial [4, 5] and temporal [6, 7] dependence of the emission light as a function of various discharge parameters.

In addition, active spectroscopic methods employing lasers have been used. Thus Schmitt *et al.* [8] have reported on a study of the SiH radical by laser-induced fluorescence (LIF). They determined the absolute density, the reaction rate, and the diffusion coefficient of the radical in a multipole discharge. Roth *et al.* [9] have used the same method to detect Si atoms in an RF discharge. In this last case, the spatial concentration was also measured. The absolute density and axial distribution of Si atoms were also determined by Takubo *et al.* in a SiH_4–He–Ar discharge [10]. Later on, LIF was also used to reveal the spatial concentration of SiH [11] as a function of the deposition conditions in pure or diluted silane, for pressures of $\leqslant$500 mtorr [5]. However, this technique failed to detect SiH_2 radical in typical experimental conditions for SiH_4 plasmas [12].

More recently, the systematic detection of the SiH_3 radical, by means of infrared laser absorption spectroscopy (IRLAS), was made possible by Itabashi *et al.* [13]. They have reported data on radical density and its diffusion coefficient, and the reaction rate constant, and spatial distribution in a H_2 diluted RF silane plasma.

Nowadays, it is therefore possible, through optical spectroscopic tools, to detect most of the ground- and excited-state radicals by *in situ,* nonintrusive methods that can be applied in actual deposition conditions. However, there is a need to extend the applicability of these methods to all the possible deposition conditions and to expand their informative content.

Furthermore, a series of other less common spectroscopic techniques were employed in the past. Coherent anti-Stokes Raman scattering (CARS) was used to obtain the spatial and temporal density variation of SiH_4, Si_2H_6, and H_2 [14], and to detect SiH_2 radical [15]. The last radical was also detected by frequency modulation absorption spectroscopy [16].

High-resolution infrared absorption and emission have been used to detect SiH_4 and SiH [17], whereas UV absorption and diode laser absorption were used to detect Si [18] and SiH [19], respectively. Finally, SiH, SiH_2, and SiH_3 were also detected by IR absorption using a matrix isolation technique [20].

However, concerning these last techniques, one must take into account that they are often characterized by a high degree of experimental complexity, and have usually been performed in specially designed "spectroscopic cell" used as discharge chamber.

Significant results have also been obtained in discharges used for the deposition of silicon alloys, such as $SiH_4 + F_2$ [21], $SiF_4 + H_2$ [22, 23], $SiF_4 + GeH_4$ [24], and $SiH_4 + NH_3$ [25]. These systems present advantages for spectroscopic studies, because many radical species are simultaneously detectable, by either LIF or emission, but they are even more complicated systems than the pure silane discharges. Nevertheless, they are a challenge for future applications.

From this list of studies, by no means exhaustive, we will try to present more thoroughly the results concerning mainly the application of OES, LIF, and IRLAS. These are the most widely used *in situ* optical diagnostic techniques in the usual a-Si:H deposition conditions.

In the following sections, we will first discuss some experimental data in detail and then will try to present the most important results obtained by optical diagnostics, and to point out the directions for future research in this field.

B. Experimental Techniques

With the experimental tools available today, the application of optical diagnostics to the deposition systems using RF silane plasmas tends to become simpler. As an example, the use of optical fibers can eliminate the need for small specially designed spectroscopic cells; the intensity of the emitted light as well as its spatial distribution can be measured in the stainless-steel chamber used for the deposition of high-quality a-Si:H alloys, through an ordinary observation window. Moreover, if the chamber is equipped with several windows, more sophisticated experimental procedures employing lasers can be applied.

In the example of Fig. 1, a high-vacuum deposition system is shown equipped with four quartz observation windows perpendicular to each other. They are used for the simultaneous monitoring of the spatially resolved emission of excited spe-

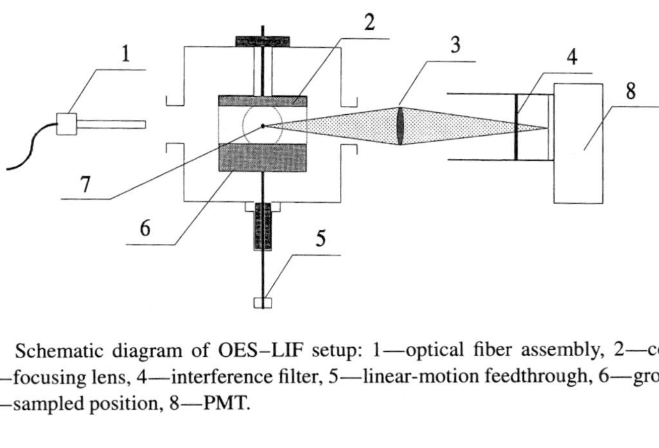

FIGURE 1. Schematic diagram of OES–LIF setup: 1—optical fiber assembly, 2—confined RF electrode, 3—focusing lens, 4—interference filter, 5—linear-motion feedthrough, 6—ground-heated electrode, 7—sampled position, 8—PMT.

cies and spatially resolved LIF of ground-state species. The beam from a tunable N_2 or excimer laser pumped by pulsed dye laser enters the chamber from the upper window. The light emitted, during a few nanoseconds after the laser pulse, from a well-defined small volume of the discharge is collected at the right angle by a lens and focused at the entrance slit of a photomultiplier tube (PMT) through a suitable-wavelength interference filter. On the opposite side of the chamber, the emission light from the same position of the discharge is collected through a collimated field-restrained, optical fiber connected to the entrance slit of a remote monochromator. There are equivalent setups utilizing optical fibers or focusing lenses for both collection paths.

Also, as signal detectors, it is possible to use monochromators instead of interference filters, optical multichannel analyzers (OMA) instead of PMT, and even gated OMA for simultaneous detection of entire LIF bands.

In the simplest case, the sampled emission light gives, in real-time scale, a continuous signal largely distributed in the near-UV and visible range. However, for weak or very weak signals, photon counting equipment must be used. In both cases, if the instrument has a good spectral resolution, the emission band can be resolved, in a more complete form, showing the entire rotational–vibrational spectra of the electronic transition.

These measurements can be furtherly extended by adding spatial and/or temporal resolution. In the first case, a specific narrow area of the discharge is sampled and focused with a lens and/or an optical fiber, or it is monitored by a single element of a diode array. Usually, the value of the required spatial resolution (in the millimeter range) is not crucial, since the light intensity profiles inside the reactor are mostly smooth.

Nevertheless, although simple in principle, these techniques have not been extensively used. In addition, there is space for more sophisticated future applications, such as in obtaining two-dimensional and Abel inverted profiles [26], with temporal resolution. If the obtained information is adequately interpreted and the experimental technique is further simplified, these data could be used as an identity reference to characterize and control the deposition system.

In the case of LIF, the main advantage is that one can get direct quantitative information about ground-state radicals, which are considered the most important species in the glow discharges used for amorphous silicon deposition. The difference between LIF and OES signals lies in the fact that the fluorescence signal lasts for only a few hundreds of nanoseconds after the exciting laser pulse, depending on the lifetime of the excited state, whereas the averaged emission signal is quasi-continuous in time. This permits the extraction of the fluorescence signal with an enhanced signal-to-noise ratio (SNR) from the plasma emission background.

In both LIF and OES, if one repeatedly integrates the signal intensity over the desired time period (larger than the RF half-period), an average intensity is obtained that is proportional to the time-averaged concentration of the sampled species. When in presence of weak signals, we can even know the exact number density of the sampled species, by counting the number of the emitted photons after the calibration of the collection efficiency of the overall light collection system.

In the past, the signal detection, the acquisition and the processing used in these optical diagnostics were complex, but later on they have been simplified by the use of new generations of fast and sensitive PMT, digital processing and personal computers. As an example, the traditional apparatus used for recording LIF signals, which is based on the "boxcar" integrator and avenger, for the integration within an adjustable time gate of a specific part of the analog signal, can be replaced in many cases by a digital storage oscilloscope and a personal computer. The oscilloscope, triggered by the laser pulse, can accumulate and average entire waveforms and, after subtraction of tray signals (emission and reflected laser light), can integrate the signal to give the fluorescence intensity in nanovolt (nV) scale.

The idea of spatial resolution comes naturally in this case, since the laser beam dimensions are usually narrow. To obtain spatial profiles, in the axial or radial direction, one has to scan simultaneously the excitation and collection path through the discharge space. This infers great difficulties and uncertainties concerning the alignment and calibration of the system. However, the use of micropositioners moving the chamber in both directions is by far the simplest and most accurate technique for spatial resolution measurements.

Temporal resolution is more difficult to obtain. The measurement of the emission intensity at a specific point on the RF excitation waveform requires the use of the synchronization output pulse from the RF generator. In the case of LIF, this pulse signal can be used to trigger the laser, whereas in the case of OES it serves as a start signal to trigger a multichannel analyzer. In both cases one has to know exactly the delays in the cables, the electronics and the PMT, in order to calibrate the system.

Finally, infrared laser absorption spectroscopy (IRLAS) is a technique based on the absorption of light from ground-state species; this has recently given the opportunity of detecting SiH_3 radicals. The experimental technique is based on a White-type multipass arrangement (the chopped laser beam passes up to 60 times before exiting the reactor). The absorption signal is then enhanced by a dual-phase lock-in amplifier.

C. Detection of Radicals

Figure 2 presents the emission spectrum obtained from a pure silane glow discharge [27]. The majority of the observed lines comes from the spontaneous emission of the $\nu' = 0$, $\nu'' = 0$ of the $A^2\Delta$–$X^2\Pi$ electronic transition of the silylidine

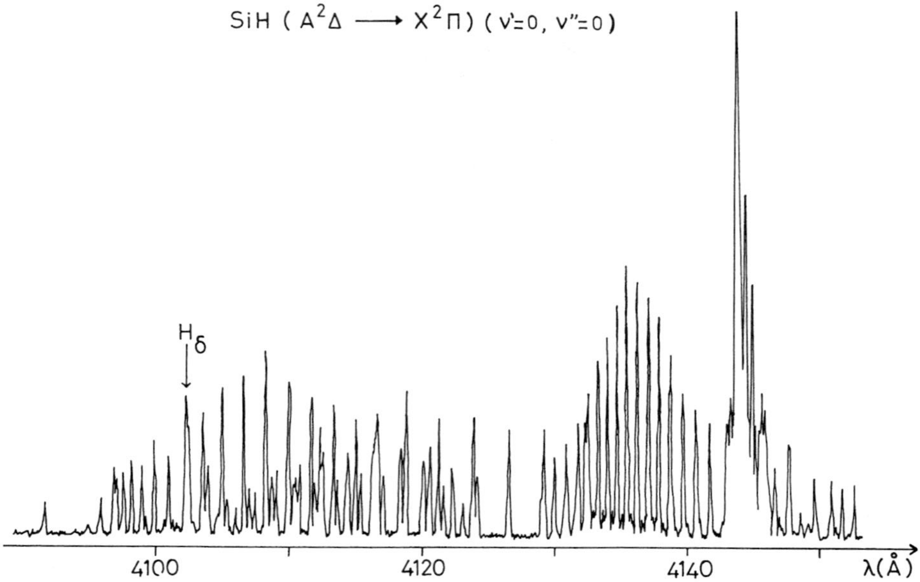

FIGURE 2. Optical emission spectrum from a silane discharge. (From Perrin [27].)

free radical (SiH*), around 414 nm. This gives the characteristic blue color to the discharge light.

The hydrogen Balmer lines are also detectable, together with some lines attributed to Si* and to Si^+ and SiH^+ ions ($A^1\Pi - X^1\Sigma^+$). Unfortunately, the possible emission from the A^1B_1–X^1A_1 transition of the silylene radical (SiH_2^*) is probably below the detection limits, whereas there is no SiH_3^* signal. Also from disilane, which is an important secondary product in this process, there is no signal in the emission spectrum. Molecular hydrogen, which is the most important secondary gas-phase product, exhibits emission lines of weak intensity in pure silane discharges that are usually operating in low-depletion conditions.

Although there is no parametric study of the exact concentrations of the active species described above (SiH, Si, H, . . .), their amount is less than 0.1% of the silane density in typical deposition conditions. Correspondingly, the amount of electronically excited species is less than 10% of the same species in the ground state. Thus, there should be no logical connection between the emission intensity and the deposition rate or the film characteristics, as it has been claimed in the past, unless an indirect relation exists, through a common generation mechanism, with other more abundant species.

Since these species have no significant participation to the film growth and are indirectly, if at all, related to the main film precursors, one can regard this information with skepticism. However, as it will be seen, the pieces of information obtained by emission-based experimental techniques have significantly enriched our knowledge about the process.

A first set of data, obtained by optical emission methods, concerns the production and reactivity of the detected species. The production path of the electronically excited species is quite similar to the production path of the ground-state radicals, making exception of the higher electron energy requirements.

Both types of radicals, excited- and ground-state, are produced by inelastic collisions that lead to the final products through a high-energy excited or superexcited state of silane; As an example, the formation channel for SiH* can be written as follows:

$$e^- + SiH_4 \rightarrow SiH_4^* + e^-,$$

$$SiH_4^* \rightarrow SiH^* + H_2 + H.$$

Here, the excess energy in SiH_4^* is also released in the rotational and vibrational excitation of the products. However, electronically excited species are spontaneously deactivated by emitting a photon without having much time for collisional relaxation with surrounding molecules. As an example, for SiH*, which has a lifetime of $\approx$500 ns, rotational temperatures as high as 2000 K [1] are reached. The ground-state SiH radicals, on the contrary, also produced with sufficiently high initial rotational temperature (950 K measured by LIF [8]), have enough time for collisional relaxation, before reacting with silane or impinging on the growing surface, and to reach an equilibrium temperature close to that of the surrounding gas. The initial temperature difference between SiH and SiH* species is due to the fact that they originate from different energy electron collision processes. In fact, dissociative excitation is energetically closer to the ionization than to the dissociation electron collision process.

These observations have produced information, which have radically changed our point of view concerning the glow discharge dissociation of silane. In particular, after the inelastic collision, with electrons having higher energy than the thermodynamically necessary [28], silane molecules are in a superexcited unstable state. Then, the excess energy is redistributed to the fragments, with the lighter fragments bringing most of it. This results in high initial rotational–vibrational excitation for all the fragments and, especially, for H-atom high translational energy. Hence, it can be concluded that the recombination of two H atoms in this case is difficult. This makes the SiH_4 dissociation reaction toward $SiH_2 + H_2$ less probable as a primary dissociation path, and outlines the differences between thermal and plasma CVD processes.

In addition, the linear relationship of SiH* intensity with RF power has confirmed the hypothesis that these radicals come from a one-electron collision process with silane, and not from a possible excitation of the ground-state SiH radicals. More specifically, actinometric emission measurements [2], using traces of N_2, have shown that the $[SiH^*]/[N_2^*]$ and $[Si^*]/[N_2^*]$ intensity ratios as a function of power are close to unity, whereas $[H^*]/[N_2^*]$ and $H_2^*]/[N_2^*]$ are approximately proportional to the square of the emission intensity of $[N_2^*]$. This means that the first species are produced by the same mechanism as N_2^*, whereas the two last species need two-electron collision processes in order to be produced.

Thus it can be concluded that H is, indeed, a primary silane decomposition product, since the alternative production through H_2 dissociation would require three electrons for H^* formation and, therefore, the $[H^*]/[N_2^*]$ ratio should be proportional to the third power of nitrogen emission $[N_2^*]$.

D. Spatial Distribution

In conjunction with optical emission data, a series of LIF measurements of the resonant fluorescence intensity measured of some ground-state radicals provides additional information. This method allows the detection of Si and SiH radicals. Thus, for SiH radical the formation kinetics and the transport coefficients have been determined under specific deposition conditions; also, the kinetic constant for the reaction of SiH with silane has been measured [8].

Spatially resolved LIF measurements have been used to record axial concentration profiles of Si and SiH radicals in various deposition conditions [11], up to 100 mtorr in pure silane and up to 500 mtorr in silane diluted with He, Ar, or H_2 [5]. A series of new information has emerged by comparison of these profiles to the equivalent emission profiles. Thus, as an example, Fig. 3 shows the variation

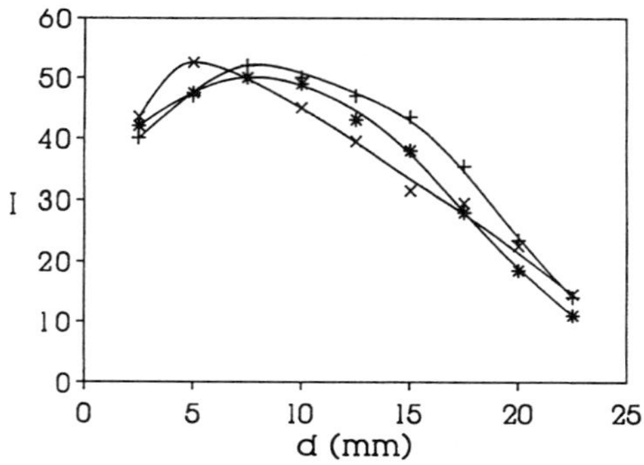

FIGURE 3. OES intensity (I) of SiH* vs. sampled position (d) for pure silane plasmas at 21 (+), 39 (*), and 50 (x) mtorr (From Mataras *et al.* [5].)

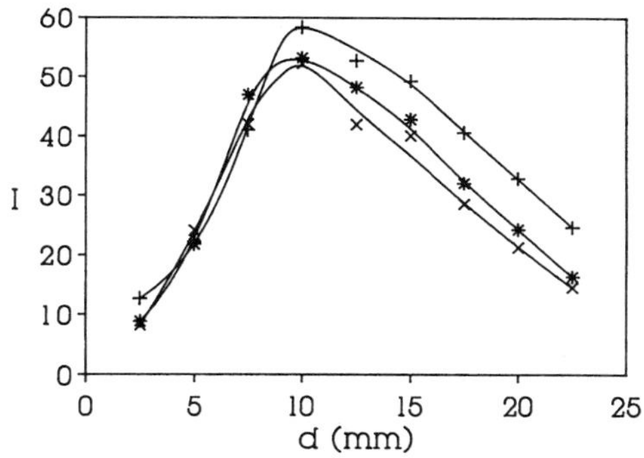

FIGURE 4. LIF intensity (I) of SiH vs. sampled position (d) for the same conditions as in Fig. 3. (From Mataras *et al.* [5].)

of OES and LIF profiles for three different total pressures [5]. It can be observed that the displacement of the emission curve toward the RF electrode as the pressure rises; this is a typical feature of gas discharges [29]. The LIF profile (Fig. 4), on the contrary, does not present a similar behavior; it presents a maximum that is not sensitive to the variation of pressure, at least for silane partial pressures up to 100 mtorr. This indicates that the generation pattern of the two radicals is not only energetically different but also spatially differentiated. This dissimilarity provides information about the spatial distribution of the electrons having sufficiently high energy to produce these radicals and can give idea on the origin and the heating mechanism of these electrons.

Figures 5 and 6 show, respectively, the variation of the OES and LIF profiles as a function of interelectrode distance, at constant RF power [30]. The integrated amount

$$I_{\mathrm{LIF,OES}} = \int_0^d I(x)dx,$$

where d is the interelectrode distance and $I(x)$ is the normalized intensity profile, is representative of the SiH average density present in the reactor at different interelectrode distances. The fact that I(LIF)/I(OES) ratio increases with interelectrode distance d (Fig. 7) means that SiH density is higher than SiH* density and, hence, that the two radical generation processes, dissociation and dissociative excitation, are differently influenced by the modification of d. The process with higher energetic requirements is favored.

This is due mainly to the fact that both cathodic and anodic electrode sheaths extend toward the plasma volume as the two electrodes come closer. Thus, the contribution of sheath-related fast electrons is enhanced and, hence, the dissociative excitation process producing SiH* is favored. In fact, the ground-state SiH generation seems to be influenced mostly by the concentration of relatively fast

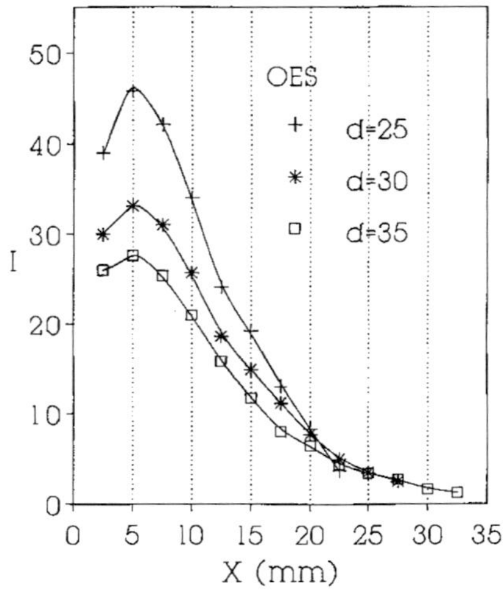

FIGURE 5. OES intensity (I) of SiH* vs. sampled position (X) for three interelectrode distances (P = 50 mtorr). (From Mataras *et al.* [30].)

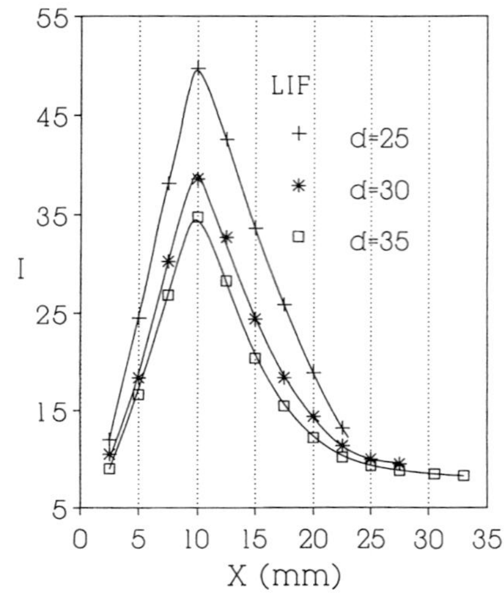

FIGURE 6. LIF intensity (I) of SiH vs. sampled position (X) for three interelectrode distances. Same conditions as in Fig. 5. (From Mataras *et al.* [30].)

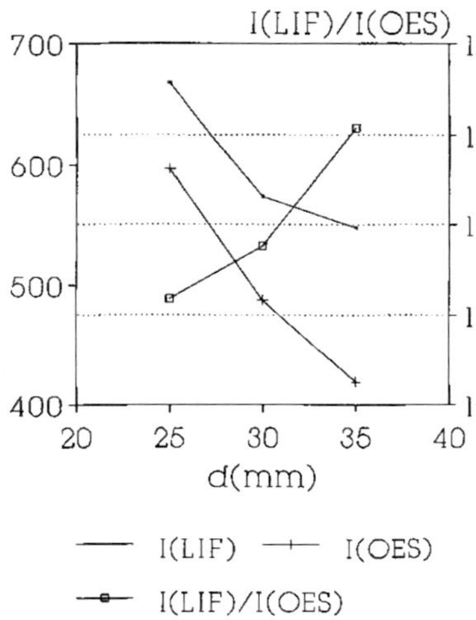

FIGURE 7. Overall intensity of LIF and OES, signal related to SiH and SiH* species, respectively, and intensity ratio from Figs. 5 and 6. (From Mataras *et al.* [30].)

electrons (field-heated or wave-riding electrons) present in the bulk of the plasma.

To furtherly illustrate the relative importance of the sheaths with respect to plasma volume, the diameters of the two electrodes can be modified, thus modifying the powered-to-grounded electrode area ratio $R = A_{RF} \backslash A_g$. This effect has been studied in our laboratory by using two different electrodes having diameter of 10 and 6 cm, respectively.

(a) In the first case the large electrode is connected to the RF generator and the small one is grounded ($R = 2.8$).
(b) In the second case the small electrode is powered and the large one is grounded ($R = 0.36$).

The results obtained (Fig. 8) show that the LIF maximum position is almost the same in both cases, unlike the emission maximum that is significantly influenced. More specifically, SiH* concentration is higher and dislocated toward the anode in case (a), whereas in case (b) it is nearly flat in the center of the discharge.

These observations indicate a close relation of SiH* with the potential and magnitude of both cathode and anode sheaths. At the same time, the SiH concentration seems to be influenced only by the density of energetic electrons in the plasma bulk, which should be higher in case (b) (small-RF electrode) than in (a) (large-RF electrode).

The importance of these results resides in the fact that the generation of other

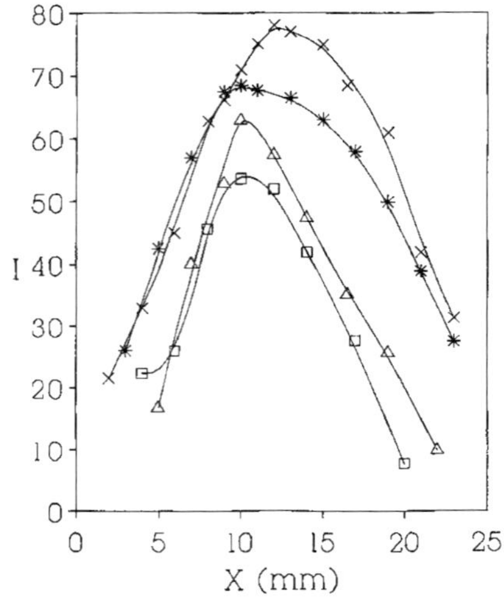

FIGURE 8. LIF and OES intensity for SiH and SiH* species, as a function of sampled position for two different area electrode sets: Δ, LIF, *, OES for 6-cm cathode and 10-cm anode; □, LIF, X, OES for 10-cm cathode and 6-cm anode. (From Mataras *et al.* [30].)

ground-state radicals, such as SiH_2 or SiH_3, follows quite similar and less energetic pathways. Thus, considering the fast reaction of SiH with silane and taking into account the limited diffusivity of the species, the LIF profile of SiH radical can be a good approximation of the generation profile of SiH_2 and SiH_3 radicals, too.

The similarity between these ground-state radical generation pathways is confirmed by the spatial profile of SiH_3 radical obtained by IRLAS (Fig. 9) [13]. As can be seen, the shapes of this profile and of the SiH LIF profile (see Fig. 6) are quite similar, although the first is more blunt because of the higher diffusivity of the species. A SiH_3 density of 5.9×10^{11} at 65 mtorr silane partial pressure, and a reaction rate of 1.5×10^{-10} cm^3 $mol^{-1}s^{-1}$, have been obtained by the same experiment. These data do not clarify which is the primary silane dissociation product but, as far as we know, the SiH_3 radical is the most abundant in silane plasmas.

Moreover, the SiH profile itself is of relevant importance for two reasons:

1. Its high sticking coefficient (close to unity) makes it important for film quality. In fact, the growth of the amorphous network requires many "good" radicals, but, at the same time, only a few "bad" radicals are enough to worsen the film quality. These phenomena can be better illustrated by considering that the impurities in a deposition chamber, which are $< 10^{-7}$ torr, are usually sufficient to create a large number of defects. Our data indicate that the

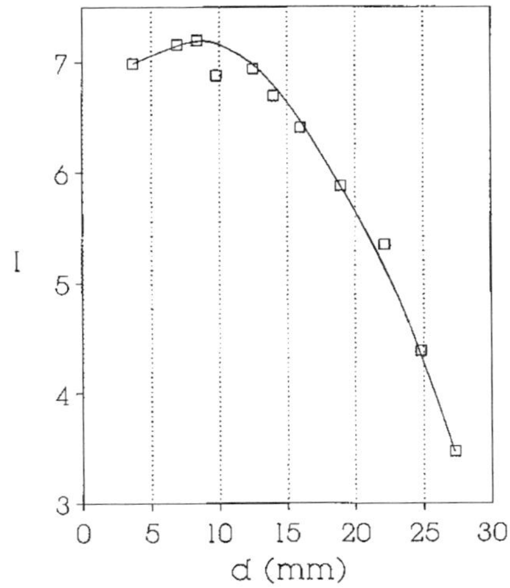

FIGURE 9. SiH_3 density ($\times 10^{11}$ cm^{-3}) as a function of sampled position from IRLAS measurements. (After Itabashi *et al.* [13].)

film can accommodate a limited number of irregularities created by high-sticking-coefficient radicals; an increase beyond these limits leads to a sudden deterioration of the film quality. More specifically, the film quality, as expressed by the density of states (DOS), is systematically worsened for films deposited at small interelectrode distances, and this phenomenon is more pronounced at lower pressures. DOS rapidly increases for electrode distances less than 27 mm, which in our system coincides with the distance where a sharp increase in SiH generation rate begins [30, 31].

2. If the exact generation profile of the radicals is known, it is possible to calculate the exact concentration profile, the space abundance, and the radicals flux to the surface [32].

As explained before, SiH profiles can give a good approximation of the radicals' generation profile. The same concept has been successfully applied in the past by using emission profiles [28, 33]. However, the pressure dependence of these profiles and, moreover, the differences between the dissociation and dissociative excitation processes, contradict this approach, as outlined in previous paragraphs.

E. Detection Sensitivity

The examples of spatially resolved optical diagnostics discussed above illustrate the high sensitivity of these methods to local discharge phenomena. An additional

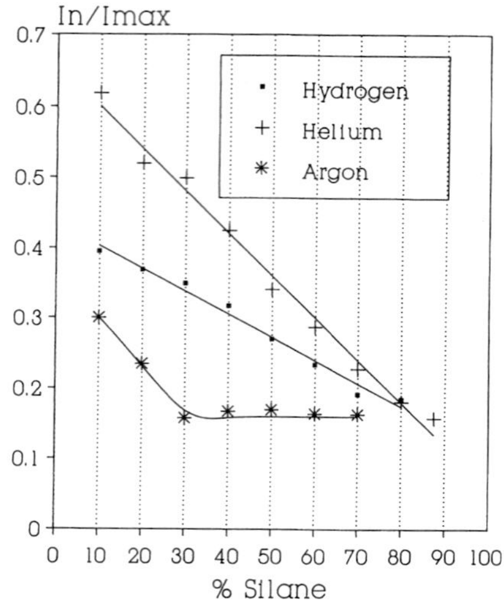

FIGURE 10. Normalized increase of LIF intensity (SiH species) as a function of percent silane (%): I_n—3 mm from anode; I_{max}—10 mm from cathode. (From Mataras *et al.* [5].)

example concerns an effect related to silane dilution. Figure 10 presents the LIF intensity I_n (measured at 3 mm from the anode) divided by the maximum LIF intensity I_{max} (measured at 10 mm from cathode) as a function of dilution, in order to take into account the general increase due to dilution. SiH concentration increases locally near the anode with increasing dilution percentage.

This effect is similar to that observed by Asano *et al.* [34] when heating the substrate. They reported an increase of the LIF and OES intensity observable at 4–10 mm from the anode, with increasing temperature. In both cases the local increase of the signal cannot be explained by the increase of the radical diffusivity or by an improbable SiH desorption from the surface. The most probable cause is the enhancement of the contribution of secondary electrons to the local silane decomposition. This may be due to the increase in the penetration depth of γ electrons in the bulk as well as to the probability of an increase in the secondary electron emission coefficient. Regardless of the cause, which is under investigation by numerical modeling, these conditions lead to higher local concentrations of all the radicals and higher probability for them to reach the growing film surface.

However, although this high sensitivity in small or local discharge variations leads to generally valid conclusions, it does not always permit the transfer of experimental results to other deposition systems. For example, the presented profiles, recorded in asymmetrical silane discharges, are very sensitive to geometric

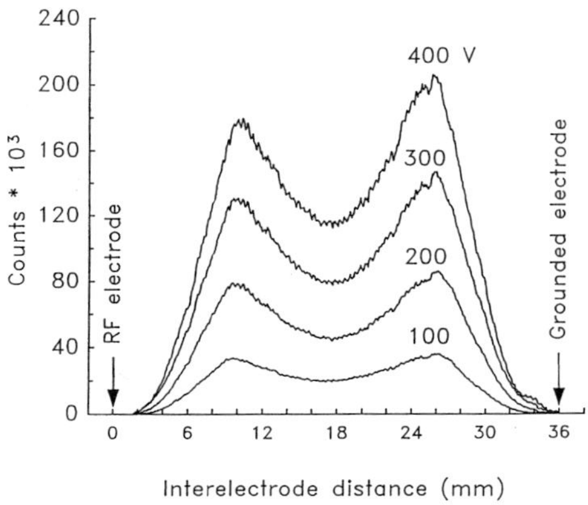

FIGURE 11. SiH emission profiles at 55 mtorr for increasing values of RF voltage [35].

changes (as in Fig. 8). In the case of discharges designed to be as close as possible to symmetry, the situation drastically changes.

Bohm and Perrin [35] have reported spatial OES profiles for a nearly symmetrical RF discharge. The 55-mtorr and the 180-mtorr profiles are presented in Figs. 11 and 12, respectively. In the low-pressure case, these profiles present two maxima, located at the time-averaged sheath boundary. In the high-pressure case, one can observe what the authors initially called "α–γ transition," by analogy to the transition observed by Godyak in He discharges [36]. This transition occurs at relatively high pressure and V_{RF} values and, in silane discharges, is accompanied by an increase in the deposition rate. In the case of He, this transition has been

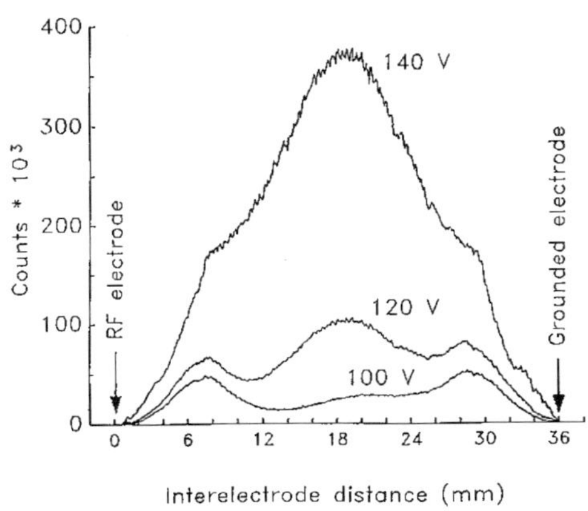

FIGURE 12. SiH emission profiles at 180 mtorr for increasing values of RF voltage [35].

attributed to a rapid change in the electron heating mechanism. Namely, in the α regime the sheath oscillation is responsible for most of the electron heating, whereas in the γ regime secondary electrons play the major role. However, an attempt to analyze by numerical modeling [37], similar phenomena observed in silane discharges have revealed that other factors, such as powder formation, must be included, in order to reproduce the observed transition. The presence of powder, in addition, makes LIF measurements almost impossible, because of the strong light scattering, and it also influences the recording of emission through noise-overriding signals.

F. Temporal Resolution

Temporally resolved diagnostics are the appropriate tool extending the information given by spatially resolved optical diagnostics, in order to elucidate the electron heating mechanisms and their influence on radical generation.

Tochikubo *et al.* have exploited this method to investigate the emission profiles in SiH_4–H_2 as well as in CH_4–H_2 and Ar discharges [7, 38]. In the case of SiH_4, spatiotemporal profiles were obtained in an asymmetrical discharge system at a total pressure of 1 torr (50% in H_2). The profiles were then deconvoluted by considering the influence of the effective lifetime of the excited state, in order to investigate the electron transport in the discharge.

The deconvoluted profiles (Fig. 13), giving, in the case of Si, the relative net excitation rate, present three different peak areas, in the second half-period, which were attributed by the authors to

(a) The volume excitation by electrons reflected at the sheath.
(b) The electrons heating due to the oscillating field in the bulk.
(c) The electrons acceleration by the local anode field, respectively.

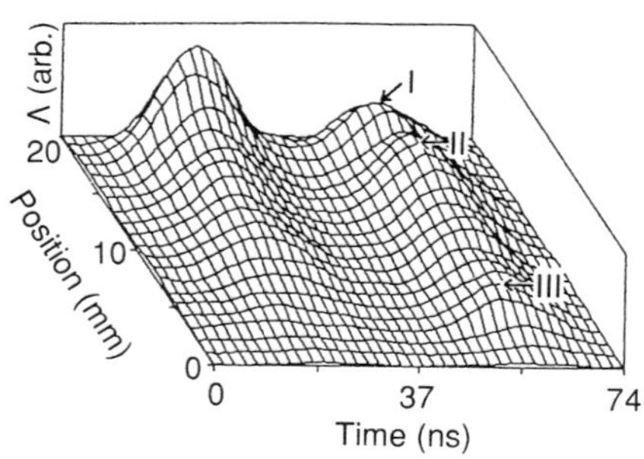

FIGURE 13. Net excitation rate of Si as derived from the deconvolution of spatiotemporal emission profiles at 1 torr (50% in H_2). (From Tochikubo *et al.* [7].)

Since these profiles significantly differ from those obtained in typical electropositive gases, such as CH_4 or Ar, the influence of negative ions is assumed to be important under these conditions. Indeed, in this high-pressure discharge regime, it is expected, as was shown earlier, that particle formation probably induces an increase in the electron loss rate. In this case, also, further experimentation, including the spatiotemporal detection of the ground-state radicals in various deposition conditions, is needed in order to draw clear conclusions.

Time resolution can be used also in another experimental approach. It can give information about the decay of excited species in the afterglow period of pulsed or modulated-wave discharges. In the first case one can determine the reaction and diffusion constants [8], while in the second case one can determine the formation and decay rates of the radicals. This is clearly depicted in Chapter 1, where the time-resolved optical emission spectrum for SiF_4–H_2 plasma is therein also shown (see Fig. 34 in Chapter 1). This type of time profiles, in complex systems such as SiF_4–H_2–Ar or SiF_4–GeH_4–H_2, compared to the same profiles of Ar has given valuable information about the origin and behavior of the emitting species. In fact, it has been shown that SiF_x^* radicals originate from ground-state SiF_x, whereas Si*, SiH*, Ge*, and GeH* originate from dissociative excitation of silane and germane, respectively.

G. Conclusions

Optical diagnostics are among the few, if not the only, nonintrusive sources of data concerning ground- and excited-state radicals in RF silane plasmas. They provide information concerning the presence, the quantity, the behavior, and in many cases the chemistry of these species. In the past few years, there has been substantial progress in both the experimental techniques and their applications. However, there is significant space for improvements and new applications, in order to provide more information for the characterization and the modeling of the discharge. A major goal in this direction remains the correlation of the discharge "microscopic" characteristics to the film quality. Thus far, we have been able to detect most of the neutral species in the gas phase, to obtain number densities and kinetic and transport coefficients in some conditions. In this direction, parametric studies are needed in order to evaluate the influence of externally controllable parameters. For this, the application conditions must be extended, and the techniques must be further simplified and standardized. Furthermore, spatiotemporal resolution has opened new routes in the study of the radical generation processes. There is also a need for more systematic exploitation of these tools, and for the combination of their results with discharge modeling and characterization of films grown under the same conditions. This will perhaps help us to define the "meaningful quantities" that directly influence the film quality. Moreover, in order to obtain a breakthrough from the present situation, we must extend these efforts to other less usual deposition conditions and more complex gas mixtures.

III. Mass Spectrometry and Langmuir Probes

Mass-spectrometric sampling of silane discharges has been employed by several researchers with the objective of getting information on the ionic and the neutral composition of the plasma. Mass spectrometry can be used in different ways:

1. Direct sampling of ions from the discharge.
2. Line-of-sight sampling of neutral radicals and molecules.
3. Analysis of neutral stable products downstream.

Mass spectrometry is the only available method to identify the ion population of a plasma. It is also a simple method to detect neutral products of the silane gas-phase chemistry. In addition the partial pressures can be obtained through an adequate calibration of the instrument.

A. Experimental Aspects

A schematic representation of a mass-spectrometric setup is given in Fig. 14. The basic arrangement may be divided into three parts:

1. The silane discharge section.
2. The sampling system, including the orifice and the mass analyzer.
3. A buffer chamber in front of the apparatus, including an electrostatic analyzer to select the ion energy.

A beam chopper and synchronous differential counting electronics can be used to separate the beam from the background signals.

The quadrupole mass filter is by far the most popular instrument for mass analysis of low-pressure plasmas. Quadrupole mass spectrometry (QMS) accepts ions with kinetic energies in the range 2–20 eV and is insensitive to variations of the ion entrance conditions, but the optimum ion injection energy is 5–10 eV. This sensitivity strongly depends on the resolution setting $m/\Delta m$ and the mass number.

1. *Ion Sampling*

For a collisionless extraction of the positive ions, the sampling orifice dimensions (both diameter and thickness) should be much less than the sheath thickness. But for practical reasons the most usual values are in the 100–200-μm range.

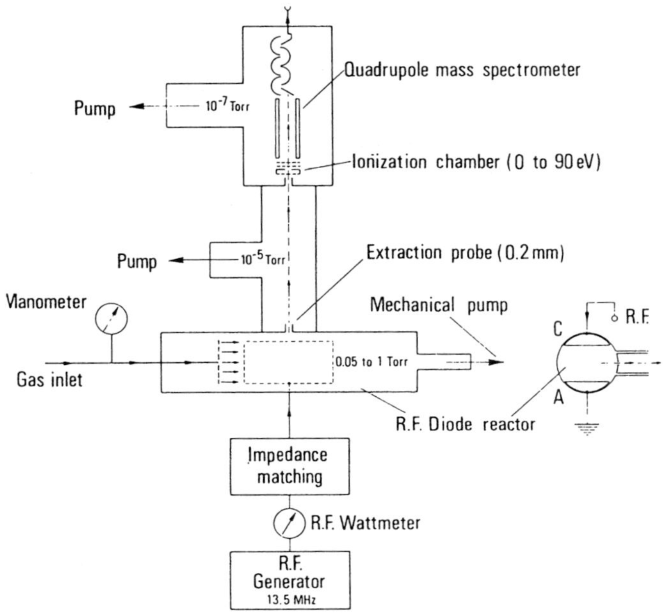

FIGURE 14. The experimental setup for the mass-spectrometric study. (From Turban *et al.* [49].)

The ions arriving at the orifice will not represent exactly the plasma bulk situation if the sheath is collisional, i.e., if total gas pressure exceeds 10–20 mtorr.

2. *Neutral Analysis*

With a line-of-sight sampling, the detection of neutral molecules is straightforward. However, the determination of the real partial pressure in the discharge cell needs calibration.

The pressure is derived from the expression [39]:

$$I_i = p_i \sigma_i T_i,$$

where I_i = measured ion current of species i, with an electron ionizer set at 70 or 90 eV,
p_i = partial pressure of the neutral molecule giving the daughter ion m_i,
σ_i = ionization cross section of the molecule,
T_i = product of the transmission coefficient F_i and the flow coefficient K_i.

The calibration factor T_i is a strong function of the mass m_i species. Forms such as $T_i = A \cdot \exp(-B m_i)$ or $T_i = (A + B m_i)^{-1/2}$ have been reported [40, 41]. The calibration factor depends on the QMS model and the resolution setting.

The constants A and B are calculated by using a standard gas mixture (He, N_2, Ar, Xe, Kr). As the pumping speeds for different gases in a mixture are complicated functions of masses, mean free paths relative to pumping orifices or tube dimensions, and pump type, such a calibration guarantees a correct determination of partial pressures.

The advantage of the use of QMS for neutral analysis is that it allows possible kinetics studies in silane discharge. The detection of reactive neutral radicals, as SiH_n, is a bit more elaborate, as described below. Electron impact, photoionization, for field ionization of radicals have been reported [42, 43].

B. Mass Spectrometry of Ions

The production and the chemistry of ions is a minor phenomenon (<10%) of the observed chemistry in low-power silane discharges. However, ion–molecule reactions provide routes for dissociation of SiH_4 and production of radicals.

Neutral and ion chemistries are not independent in silane plasmas, so ion sampling is needed to obtain a realistic picture of the plasma processes.

1. Silane Ionization Cross Sections

The determination of the ionization cross sections of SiH_4 and Si_2H_6 was one of the first steps in the study of the plasma chemistry of silane glow discharge by means of mass spectrometry. The direct electron impact ionization on SiH_4 involves four dissociative products:

$$e + SiH_4 \rightarrow SiH_3^+ + H + 2e, \tag{1}$$

$$e + SiH_4 \rightarrow SiH_2^+ + H_2 + 2e, \tag{2}$$

$$e + SiH_4 \rightarrow SiH^+ + H_2 + H + 2e, \tag{3}$$

$$e + SiH_4 \rightarrow Si^+ + 2H_e + 2e. \tag{4}$$

The appearance potentials for ions from SiH_4 as measured by Potzinger and Lampe [44], Morrison and Traeger [45], and Chatham *et al.* [46] are compared as follows:

Ion	Appearance potentials (eV)		
	[44]	[45]	[46]
SiH_3^+	12.3	12.2	12.2
SiH_2^+	11.9	11.8	11.6
SiH^+	15.3	14.7	15.0 ± 1
Si^+	13.6	13.3	13.5 ± 2

The total and partial dissociative ionization cross sections have been measured with QMS using an electron ionizer with an energy bandwidth of ± 1 eV [47] or 0.5 eV [46]. The absolute value of cross sections was obtained by normalization with argon or ethane data.

The absence of the parent ion SiH_4^+ is attributed to its predissociation into $SiH_2^+ + H_2$. This would explain the enhanced SiH_2^+ cross section relative to the SiH_3^+ cross section as shown in Fig. 15.

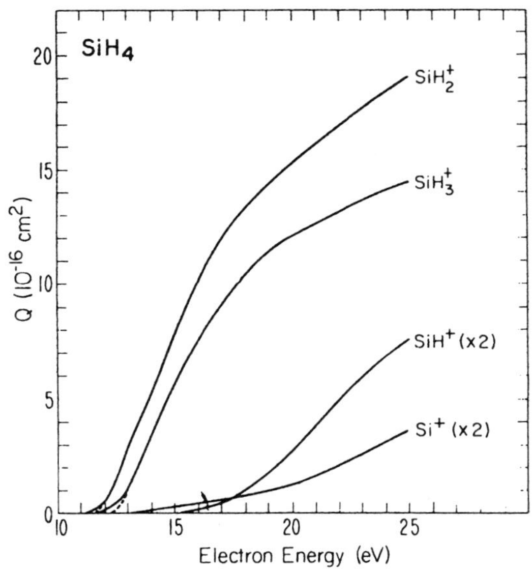

FIGURE 15. Measured SiH_4 threshold regime partial ionization cross sections. (From Chatham *et al.* [46].)

2. *Positive Ions in RF SiH_4 Discharges*

As shown in Fig. 16, SiH_3^+ ($m/e = 31$) is usually the dominant monomeric ion found in silane RF diode discharge, but significant quantities of clusters containing Si with higher molecular weight ($Si_xH_y^+$, $x = 1-9$) have been observed [49–51].

In the case of the monosilane ions, the predominance of the SiH_3^+ ions over the SiH_2^+ ions has been shown to be the result of the ion–molecule reaction:

$$SiH_2^+ + SiH_4 \rightarrow SiH_3^+ + SiH_3. \tag{5}$$

Turban *et al.* [48] have calculated the steady-state concentrations of the monosilane ions, for various pressures and flows in a tubular RF glow discharge reactor. By calculating the amount of ions produced by electron impact ionization, the ion–molecule reaction and the losses by diffusion and ion–molecule reaction, they showed that the SiH_3^+ is the most abundant ion for pressures greater than 0.1 torr, whereas at low pressure SiH_2^+ is dominant in agreement with the MS ion sampling results [49].

The order of distribution of positive ions sampled through a substrate biased at -50 or -100 V is the reverse of that obtained at the anode [52]. SiH^+ and Si^+ are the dominant ions, suggesting a stripping of hydrogen during the crossing of the large ionic sheath.

Ion clusters like $Si_2H_y^+$ and $Si_3H_z^+$ are particularly abundant in pure silane RF discharges [51, 52]. Clustering reactions of SiH_3^+ with SiH_4 form $Si_2H_7^+$ and $Si_2H_5^+$.

Higher-order clustering reactions involving SiH_3^+ are believed to explain large ions detected in such RF silane discharges [53, 54].

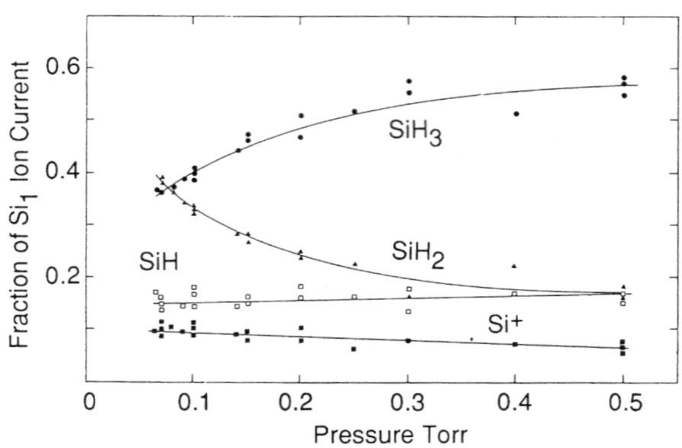

FIGURE 16. Relative monosilicon ion intensities observed in an SiH_4–He RF discharge as a function of total pressure (16 sccm, 10 W). (From Turban *et al.* [49].)

The generalized reaction may be written

$$Si_xH_y^+ + SiH_4 \rightarrow Si_{x+1}H_z^+ + mH_2 + nH, \tag{6}$$

where $4 + y = 2m + n + z$. But these reaction pathways do not appear to explain the formation of hydrogenated silicon dust, as discussed by Mandich *et al.* [55].

A complicating factor in the origin of $Si_xH_y^+$ ions is the simultaneous formation of neutral Si_2H_6 and Si_3H_8 in the silane glow discharge, in reasonable quantities (see Section III.C.2). Direct ionization of disilane and trisilane is believed to account for a noticeable part of the heavy ions [48].

Hydrogen ions (H^+, H_2^+, and H_3^+) are also detected in silane RF plasmas [49, 52]. The reaction of hydrogen ions with SiH_4 to produce SiH_3^+ has a large cross section:

$$H_n + SiH_4 \rightarrow SiH_3^+ + H_2 + H_{n-1}. \tag{7}$$

This reaction reinforces the production of SiH_3^+ in the glow discharge.

3. *Positive Ions in DC Silane Discharge*

In general, the a-SiH films produced in a DC discharge are of a poorer quality than those deposited in a RF discharge. However, the DC discharge study was important to understand many fundamental aspects of a silane plasma.

With a hot-cathode DC discharge in a partially confining multipolar magnetic structure, a silane plasma can be sustained at very low pressure [56].

Below 0.2 mtorr the positive-ion spectrum is identical to the ion fragmentation pattern of SiH_4 by 60–70-eV electron impact [57]. As the pressure is increased beyond 0.2 mtorr, the SiH_3^+, $Si_2H_4^+$, and $Si_3H_7^+$ become dominant as a result of the ion–molecule reactions (5) and (6). At 10 mtorr, $Si_3H_n^+$ ions dominate the ion spectrum of the DC multipole discharge.

In DC silane discharges operating at 0.05–0.5 torr a large part of the ionization process occurs in the DC sheath. This problem is less severe in RF discharges because sheath fields are smaller.

Ions sampled at the cathode have a distribution similar to that obtained in RF with a negative biasing. Because of the large electric field in the DC sheath, the degree of clustering is less pronounced in DC discharge than in RF discharge at the same silane pressure and at equivalent power.

The energy of the ions arriving at the cathode is quite large. A significant fraction of the ions arrives with the full energy of the cathode fall, which is a few hundreds of volts [51, 58]. With the so-called DC-proximity configuration, a third electrode, the substrate, is placed behind a screen cathode. The relative abundance of ions, containing different numbers of silicon atoms that arrive at the substrate,

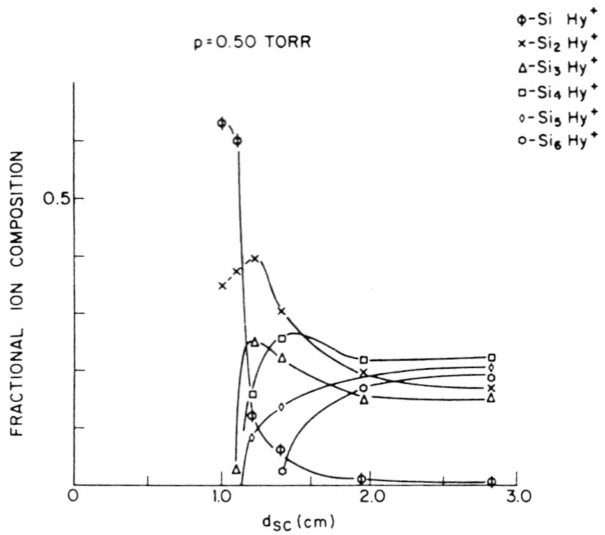

FIGURE 17. Fractional abundances of silane ions containing 1–6 silicon atoms vs. substrate to cathode dsc in a DC silane glow discharge. (From Weakliem *et al.* [59].)

depends primarily on three parameters: the pressure, the distance d_{SC} between the substrate and the cathode, and the substrate temperature [59].

As d_{SC} increases the monosilicon ions are no longer dominant, the ions containing four and five silicon atoms become the most abundant. Cluster ions result from a chain-reaction sequence that is initiated by the primary SiH_y^+ ions. As the distance between the substrate and the cathode increases, more ion–molecule reactions occur during the travel toward the extraction orifice (Fig. 17).

This evolution has been modeled by Koch and Hitchon [60] using 19 ion–molecule reactions. The fractional ion fluxes at the anode was found to be in good agreement with experimental measurements. The two main sources of error in the model are the electron temperature and the values of the ion–molecule reaction rates.

The ion composition also depends on the substrate temperature as shown in Fig. 18. The relative abundances of the higher-silicon-containing ions increase with the increasing substrate temperature. At the same time the abundance of SiH_3^+ relative to SiH_2^+ increases. The variation of the ion composition in the gas phase is probably due to the crucial role of the surface reactions on the hot growing a-Si:H film.

It is believed that the increase in the surface temperature enhances the flux of hydrogen released by the surface. This hydrogen enrichment of the gas near the surface affects the ion–molecule reactions and in this, the yield of product [59].

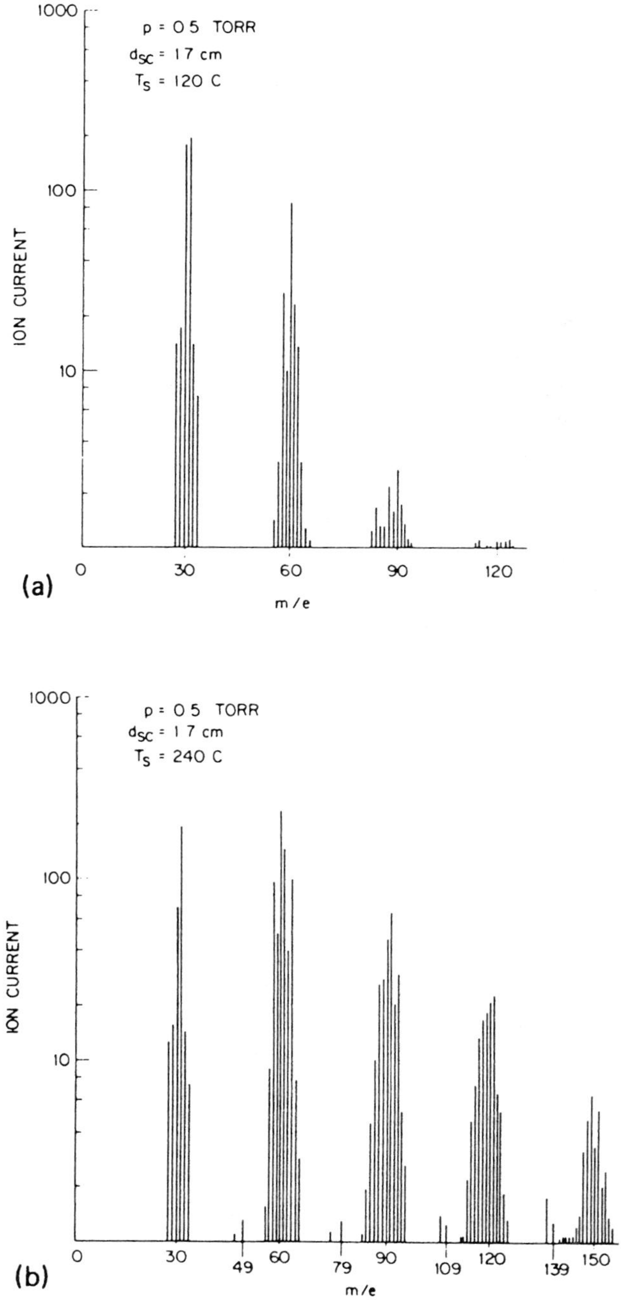

FIGURE 18. Relative abundances of silane ions in a DC silane glow discharge for two substrate temperatures (a) 120°C, (b) 240°C. (From Weakliem *et al.* [59].)

Silanol ions $Si_x H_{2x+1} OH_2^+$ ($x = 1-3$) were identified in the ion spectrum with a substrate temperature of 240°C (m/e = 49, 79, 109). Their abundance relative to the silane ions depends on the history of the deposition reactor.

4. *Positive Ions in Gas Mixtures*

a. SiH_4–rare gas mixtures. Silane is sometimes diluted in rare gas He, Ne, Ar, or Xe in order to increase the dissociation of silane or to reduce the gas-phase nucleation that can produce silicon hydride dust [61].

Some ion spectra have been reported in SiH_4–He and in SiH_4–Ar [47]. The charge-transfer reactions or the Penning ionization via the rare-gas metastables could explain the different ion chemistries:

$$Rg^+ + SiH_4 \rightarrow Rg + SiH_x^+ + \text{hydrogen products}, \tag{8}$$

$$Rg^* + SiH_4 \rightarrow Rg + SiH_x^+ + e^- + \text{hydrogen products}. \tag{9}$$

The cross sections for the reactions of He^+, Ne^+, Ar^+, Kr^+, and Xe^+ have been measured using ion-beam or drift-tube techniques [62, 63]. The products corresponding to reaction (8) are SiH_x^+ for $x = 0-3$ in the case of He^+, Ar^+, and Kr^+, whereas for Ne^+, formation of SiH_x^+ ($x = 0-2$) is observed, and for Xe^+ only SiH_3^+ and SiH_2^+ are noticed.

In all five rare-gas systems, formation of SiH_4^+ is not detected. This observation is consistent with the ion spectra of silane–rare-gas discharges and with the electron impact ionization studies of silane that indicate no SiH_4^+.

In addition to SiH_x^+ ions, ArH^+ or HeH^+ are found in the ion spectra of silane–argon or silane–helium discharges, resulting from reactions between rare-gas ions Rg^+ or metastables Rg^* and molecular hydrogen formed in the dissociation of silane [47].

b. SiH_4–H_2 and SiH_4–D_2 mixtures. The addition of hydrogen to a silane glow discharge tends to increase the degree of hydrogenation of the silane ions and the abundance of SiH_3^+ via ion–molecule reactions involving H_n^+ and SiH_4. The net result is an increase of the relative abundance of SiH_3^+ among monosilane ions and $Si_2H_7^+$ among disilane ions [49]. Knowledge of the reaction mechanisms can be obtained by using the isotopic labeling technique coupled with mass spectrometry [64].

Silane–deuterium discharge also produces SiH_3^+ through the two major ion–molecule reactions:

$$D_2^+ + SiH_4 \rightarrow SiH_3^+ + D_2 + H, \tag{10}$$

$$D_2^+ + SiH_4 \rightarrow SiH_3^+ + D + HD. \tag{11}$$

Moreover, in the SiH_4–D_2 mixture, the formation of partially deuterated neutral silanes is observed, from SiH_3D up to SiD_4. The SiH_2D^+ and $SiHD_2^+$ and SiD_3^+ ions are believed to be the result of the ion–molecule reaction (5) between SiH_2^+, $SiHD^+$, and SiD_2^+ and the neutral deuterated silanes.

The hydrogen atoms produced by the dissociation of H_2 and SiH_4 have been found to etch the a-Si:H film deposited on the reactor walls [64]. The ion spectrum of a hydrogen discharge sustained in a reactor with freshly deposited amorphous silicon on the walls contains the ions SiH_3^+ and $Si_2H_y^+$, as well as the expected H_2^+ and H_3^+ ions.

c. SiH_4–CH_4 and SiH_4–C_2H_4 mixtures. Glow discharges in CH_4–SiH_4–He and C_2H_4–SiH_4–He have been analyzed by mass spectrometry [65, 66]. Apart from the carbon and silicon ions that are characteristic of RF discharges in either methane or silane, the formation of crossed ions at *m*/*e* values of 43 ($SiCH_3^+$), 44 ($SiCH_4^+$), and 45 ($SiCH_5^+$) have been observed in the CH_4–SiH_4 mixture. An increase of the total pressure favors the formation of $SiCH_5^+$ at the expense of SiH_4^+.

The pressure dependence observed in the relative composition of $SiCH_x^+$ follows the same trend as that obtained in a high-pressure mass-spectrometric study of a CH_4–SiH_4 mixture [67].

So it is believed that the $SiCH_x^+$ ions are the result of ion–molecule reactions:

$$SiH_2^+ + CH_4 \rightarrow SiCH_4^+ + H_2, \quad (12)$$

$$'' \quad \rightarrow SiCH_5^+ + H, \quad (13)$$

$$CH_3^+ + SiCH_4 \rightarrow SiCH_3^+ + 2H_2, \quad (14)$$

$$'' \quad \rightarrow SiCH_5^+ + H_2, \quad (15)$$

$$CH_4^+ + SiCH_4 \rightarrow SiCH_4^+ + 2H_2, \quad (16)$$

$$'' \quad \rightarrow SiCH_5^+ + H_2 + H. \quad (17)$$

The work of Cheng *et al.* [67] shows that all primary methane ions react with monosilane, whereas the primary silane ions are much less reactive with methane.

The mixed ions $Si_xC_yH_z^+$ for x, y = 1–3 have also been detected by sampling through the anode of a DC discharge excited in a mixture of CH_4–SiH_4 with no helium dilution [54].

The ion chemistry of the C_2H_4–SiH_4–He discharge shows three groups of positive ions [28]:

Hydrocarbon ions	C_1, C_2, C_3, and C_4
Silane ions	Si_1 and Si_2
Crossed ions	SiC_1 and SiC_2

Except for the three previous SiC_1 ions ($SiCH_3^+$, $SiCH_4^+$, and $SiCH_5^+$), the SiC_2 ions appear to result from ion–molecule reactions:

$$SiC_2H_3^+ \ (m/e = 55), \qquad SiC_2H_5^+ \ (m/e = 57),$$
$$SiC_2H_6^+ \ (m/e = 58), \qquad SiC_2H_7^+ \ (m/e = 59).$$

Among the numerous identified reactions, the most important (i.e., with high reaction rates) concern SiH_2^+, $C_2H_4^+$, and $C_2H_2^+$ ions encountering either C_2H_4, C_2H_2, or SiH_4 molecules.

The previous studies dealt with the plasma chemistry involved in the PECVD preparation of amorphous SiC:H thin films. Silicon carbide films were also deposited from the plasma of monomethylsilane CH_3 SiH_3 [68]. The positive-ion species detected in this plasma included the following:

$$CH_x^+ \quad (2 \le x \le 4),$$
$$SiH_y^+ \quad (2 \le y \le 4),$$
$$SiCH_z^+ \quad (3 \le z \le 5).$$

The composition of the ion flux depends strongly on the electrode examined (anode or RF cathode). This difference is attributed to the influence of the large electrode field resulting from the self-biasing of the RF electrode.

d. SiH_4–B_2H_6 and SiH_4–PH_3 mixtures. The ionic spectrum of a 5% B_2H_6–95% SiH_4 DC discharge exhibits three groups of positive ions [69]:

$$BH_n^+ \quad \text{and} \quad B_2H_n^+,$$
$$SiH_n^+ \quad \text{and} \quad Si_2H_n^+,$$
$$SiBH_n^+ \quad \text{and} \quad Si\ B_2H_n^+.$$

In a similar manner, the ionic spectrum of PH_3–SiH_4 DC plasma includes the expected three categories of positive ions:

$$SiH_n^+, \ Si_2H_n^+;$$
$$PH_n^+;$$
$$SiPH_n^+.$$

Striking differences between the ion composition of DC and RF discharges excited in B_2H_6–SiH_4 and PH_3–SiH_4 mixtures were reported [51].

An enhancement of the $B_2H_n^+$ and PH_n^+ ions compared with the neutral gas composition was observed in RF discharges but not in DC discharges sampled through the anode. In this case the $B_2H_n^+/SiH_n^+$ and PH_n^+/SiH_n^+ ion ratios were found to be nearly equal to the corresponding neutral gas composition. It is believed that ion molecule is produced predominantly in RF discharges [54].

5. *Negative Ions in Silane and Disilane Discharges*

Negative ions were extracted from a DC multipole discharge maintained at very low pressure in pure silane [57]. The extraction orifice was biased with respect to the chamber wall at a positive voltage of 50–100 V.

Among the monosilicon negative ions, SiH_3^- is dominant in either silane or disilane plasma mass spectra. This is attributed to reactions of H^- with SiH_4 or Si_2H_6 yielding SiH_3^-.

Various oligomeric ions were detected:

$$Si_2^-,\ Si_2H_-;$$

$$Si_3H_2^-;$$

$$Si_4H_6^-;$$

$$Si_5H_8^-,\ Si_6H_n^-.$$

Associative attachment reactions of SiH_n^- or SiH_4 and polymerization involving molecular elimination are believed to account for the formation of large oligomeric negative ions in silane plasmas. Formation of dust could have its origin in such negative cluster ions.

C. Mass Spectrometry of Neutrals

Mass spectrometry is an invaluable technique to analyze the neutral composition of a glow discharge plasma. By using a line-of-sight sampling it can measure the species actually impinging on the substrate. A large effort has been made to identify the neutral precursor species involved in the deposition process of a-Si:H film for a decade [70].

1. *Radicals*

The direct detection of neutral radicals produced in a silane discharge can be done by mass spectrometry. The gaseous free radicals are estimated to have low concentration ratios (10^{-4} or 10^{-5}) compared with stable neutral molecules. To discriminate between radicals and molecules two methods can be used: ionization by

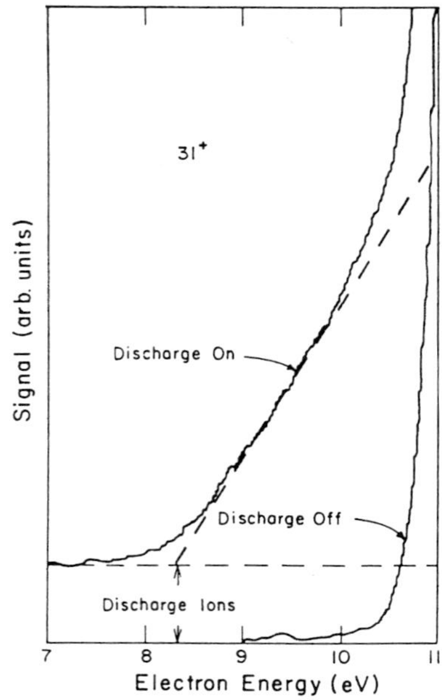

FIGURE 19. Quadrupole mass-spectrometric output current for $m/e = 31$ as a function of electron-beam energy (see text). (From Robertson and Gallagher [71].)

low-energy electrons [42] or field ionization [43]. The first method has been shown to be successful in detecting the SiH_n and Si_2H_n radicals at the cathode of a DC discharge in silane and in silane–argon [71]. The neutral species effuse from the discharge through a 800-μm-diameter orifice. Figure 19 shows the mass spectrometric signal at $m/e = 31$ due to SiH_3^+ as a function of the ionizer electron energy. The signal with the discharge off is due to dissociative ionization of SiH_4 forming SiH_3^+ [reaction (1)] with a threshold at 12.2 eV.

The signal with the discharge on is due to ionization of SiH_3, discharge, and dissociative ionization of the undepleted silane. The ionization threshold of SiH_3 radical is 8.4 eV.

The detection of SiH_2 is most difficult since the threshold energy difference is only ~2 eV (11.9 vs. 9.7 eV). The key point of this detection is to use an ionizer electron with a very narrow energy spread. In the work of Robertson and Gallagher [71] a magnetically confined electron beam with a LaB_6 filament provided a 0.17-eV energy spread that permits a successful detection of SiH_n radicals. The SiH_3 radical was found to be the dominant monosilicon radical under the conditions of a DC-proximity discharge in silane and silane–argon mixtures.

To calibrate the radical signals it is necessary to know the ionization cross sections of radicals and the relative mass spectrometer sensitivity versus mass. The mass sensitivity factor of the QMS can be determined with accuracy, but the ionization cross section of radicals is usually obtained through estimations and not through direct measurements. A typical sensitivity for radical detection is about 10^9 cm^{-3} among 10^{15} silane molecules per cubic centimeter.

Until now there is no report of SiH_n radicals detection in RF silane discharge using mass spectrometry.

2. *Stable Neutrals in Flowing Silane Discharge*

Mass-spectrometric analysis of the neutral gas composition under flowing discharge conditions is a good way to obtain useful information on the kinetics and the mechanisms involved in silane dissociation.

The gas from the plasma chamber is sampled through a small aperture and analyzed mass spectroscopically after electron impact ionization. The partial pressures of H_2, SiH_4, Si_2H_6, and Si_3H_8 in the reactor are determined by measurements of the ion currents $I\ (m/e)$, at values of $m/e = 2$, 32, 62, and 92, respectively. The electron energy of the ionizer is usually set at 70 eV, thus making it possible to use the standard fragmentation patterns. However, a lower electron energy (2–30 eV) is sometimes recommended to reduce the multiple ionization occurring at 70 eV, but with the handicap of a much lower detection sensitivity.

The kinetics of the silane decomposition in a DC discharge has been studied in the conditions of a plug-flow tubular reactor by Nolet [72] and then by Wagner and Veprek [73]. The reaction rate for the depletion of silane is a linear function of the DC current in the discharge. It is possible to determine a first-order rate constant proportional to the electron density and dependent on the electron temperature. Similar results were obtained by considering a backmixed diffusion reactor with an RF discharge [48, 70]. Comparison with CH_4 discharge shows that the decomposition of SiH_4 is about twice that of CH_4 in the same other conditions.

The silane dissociation gives rise to the formation of H_2, Si_2H_6, Si_3H_8, and a solid film. The rate for the depletion of SiH_4 and the formation of products (H_2, Si_2H_6, film) was found to depend on the substrate temperature, in the range 100–250°C [74].

A change in the mass-spectrometer ion current at $m/e = 62$ as a function of the substrate temperature shows that the concentration of Si_2H_6 follows an Arrhenius law with an activation energy of -0.84 kcal/mol. The result is consistent with second-order reaction kinetics of SiH_3 combination on a hot surface.

The concentration of SiH_4, Si_2H_6, and Si_3H_8, as measured by mass spectrometry, versus residence time for several gas temperatures (25–335°C) exhibits two regimes (cf. Fig. 20). At short dwell times (<0.5 s) an increase of the Si_2H_6 and

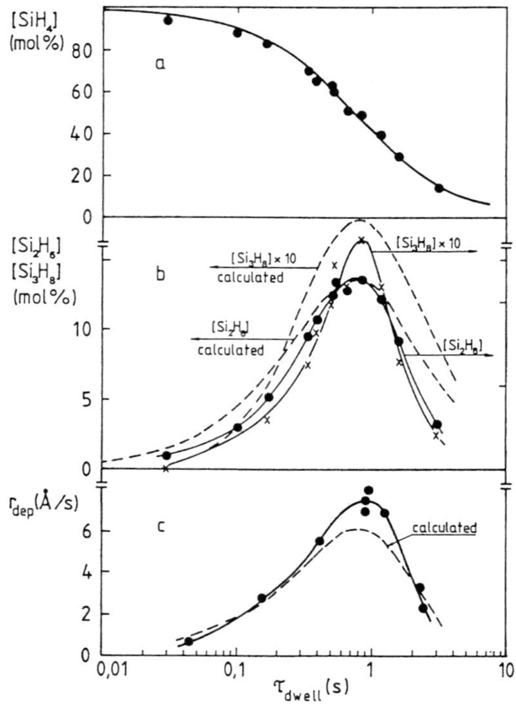

FIGURE 20. Dependence of the silane concentration (a), of the di- and trisaline concentration (b) and of the deposition rate (c) on the residence time for a DC silane glow discharge. (From Veprek and Heintze [75].)

Si_3H_8 concentration corresponds to a decrease of the undissociated SiH_4, the fast pumping away of di- and trisilane is the dominant loss term. At long dwell times (>1 s) the decomposition of Si_2H_6 and Si_3H_8 is dominant and produces a strong decrease in concentration [75]. A good correlation has been found between this disilane (and trisilane) concentration and the silicon deposition rate. These mass-spectrometric measurements have supported an analysis of the mechanism of plasma-induced deposition of a-Si:H and of the neutral precursors involved in it [76–78].

Veprek favors the electron impact dissociation of SiH_4 through the reaction

$$e + SiH_4 \rightarrow SiH_2 + H_2 + e \qquad (18)$$

as the main primary dissociation channel.

By itself, the mass-spectrometric analysis accounts for only the final products of the plasma reactions, not on the detailed mechanism of the chemistry.

Another approach is to study the kinetics of the silane glow discharge with a time-resolved mass spectrometer [79]. A small change of the DC discharge current produces time variations of the concentrations of various species present in the reactive plasma. The relaxation response of silane and molecular hydrogen is analyzed with respect to time. The dominant reaction channels of the silane decomposition are inferred from this time evolution.

3. *Static Discharge Measurements*

Plasma deposition of a-Si:H from silane is normally performed in flowing gas at power/flow ratios that decompose 10–50% of the input silane. The higher silanes (di- and trisilane) produced in the discharge are partially redecomposed into radical products that could play a major role in film growth as proposed by Heintze and Veprek [76].

The SiH_n radicals that have been measured by mass spectrometry at the cathode surface of a DC glow discharge results from the combination of the initial dissociation and the subsequent gas reactions. So these measurements provide only indirect evidence of initial dissociation pathways.

The basic objective of the study of static silane is to try to isolate the characteristics of the initial silane dissociation [42]. After evacuation of the tube reactor, and filling with typically 125 mtorr of silane at 250°C, the discharge is initiated. The response time of the mass-spectrometer signal is ~0.9 s. The mass signals from silane ($m/e = 30$), disilane ($m/e = 62$), trisilane ($m/e = 92$), and tetrasilane ($m/e = 120-128$) are followed as a function of time as shown in Fig. 21.

The silane partial pressure p_1 follows exponential decay up to at least 40% depletion, according to the relationship

$$\frac{dp_1}{dt} = -R_1 p_1,$$

where R_1 is the silane depletion rate.

The gas composition in the reactor changes in time as the result of the silane depletion. The formation of H_2, Si_2H_6, Si_3H_8, and Si_4H_{10} is observed (Fig. 21). The formation of pentasilane, higher silanes, or dust is found to be negligible in these conditions.

Production rates of Si_2H_6 and Si_3H_8 follow the relations

$$\frac{dp_2}{dt} = \beta_{12} R_1 p_1 - R_2 p_2,$$

$$\frac{dp_3}{dt} = \beta_{13} R_1 p_1 - \beta_{23} R_2 p_2 - R_3 p_3,$$

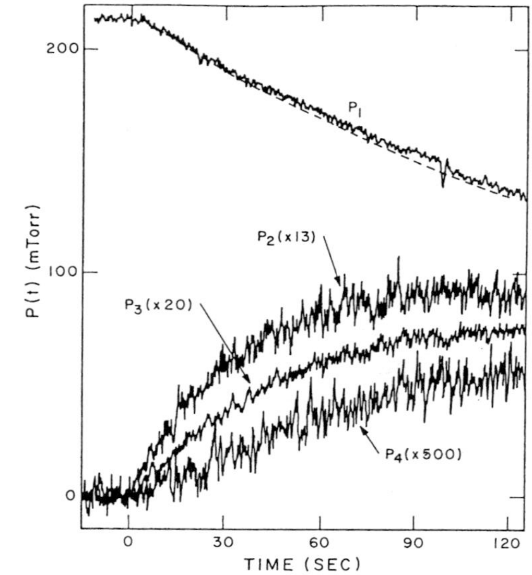

FIGURE 21. Silane (P_1), disilane (P_2), trisilane (P_3), and tetrasilane (P_4) pressures versus discharge time for a DC static silane discharge at 250°C and 210 mtorr. (From Doyle *et al.* [80].)

where β_{12} = fraction of silane depletion producing disilane,
β_{13} = fraction of silane decomposition that produces trisilane by a direct reaction,
β_{23} = fraction of disilane depletion that yields trisilane by a secondary channel.

From these mass-spectrometric measurements Doyle *et al.* [80] conclude that the main dissociation channel of discharge–electron collisions with silane is

$$e + SiH_4 \rightarrow SiH_2 + 2\,H + e, \tag{19}$$

in agreement with the earlier statements of Turban *et al.* [64, 81] and Perrin *et al.* [82].

The discrepancy between reactions (18) and (19) concerns the hydrogen products (H_2 or H). The direct formation of atomic hydrogen resulting from (19) could explain, through the abstraction reaction

$$H + SiH_4 \rightarrow SiH_3 + SiH_3, \tag{20}$$

the large production of SiH_3 radicals.

The experimental results of mass-spectrometric sampling of silane discharge must be considered with caution if a microscopic mechanism is inferred, as illustrated in the work of Longeway *et al.* [83, 84]. The decomposition kinetics of a static DC silane discharge were studied in both absence and presence of nitric oxide (NO), a free-radical scavenger.

The introduction of NO in the silane discharge reduced the Si_3H_8 and Si_4H_{10} production yields to 70 and 21%, respectively, of their values in the absence of NO. The formation of a a-Si:H film was totally suppressed by NO introduction. These mass-spectrometric results were considered as a proof that films must grow largely or entirely from SiH_3 since NO was known to scavenge silyl (SiH_3) radicals. Recent measurements of rate constants for SiH_n (viz., SiH, SiH_2, SiH_3) radical reactions with NO are as follows [85]:

Reaction	Silane discharge rate ($cm^3\ s^{-1}$)
SiH + NO	2.5×10^{-10}
SiH_2 + NO	1.7×10^{-11}
SiH_3 + NO	1.0×10^{-12}
	2.5×10^{-12}

As shown by these data, NO reacts rapidly with all three monosilicon hydride radicals and, in fact, reacts less rapidly with SiH_3. So the Longeway *et al.* [83, 84] mass-spectrometry results require new analysis.

By itself, mass spectrometry is a powerful instrument for understanding the neutral gas composition of the silane discharge and analyzing the dissociation kinetics. The controversy regarding the film growth precursor and the electron impact dissociation channel cannot be solved on the basis solely of the MS data. Both the gas phase and the surface reactions are involved in the plasma chemistry of silane deposition; hence the interpretation of MS results requires the knowledge of constant rates of individual reactions. Other similar studies on static discharge, using mass spectrometry have been performed on Si_2H_6, GeH_4, and mixtures [86].

4. *Interaction with Surface of a-Si:H Film*

The interaction of argon plasma or hydrogen plasma with the surface of freshly deposited a-Si:H films has been studied by means of mass spectrometry [87–90].

In the case of a film submitted to argon sputtering, H_2, SiH_4, and Si_2H_6 have been detected as the result of the interaction. The H-rich top layer of the film, estimated to be five monolayers thick, is converted to H_2, SiH_4, and Si_2H_6 via the ion desorption of H, SiH_3, and Si_2H_5 [91].

Rare-gas ion desorption was also used in order to determine the surface hydrogen content in freshly deposited a-Si:H [92].

Finally the chemical etching of a-SiH film in D_2 discharge has been demonstrated using MS [64]. Silane resulting from silicon hydrogenated film etching was found to be completely deuterated: SiD_4 and Si_2D_6. The hydrogen contained in the film is released as an HD molecule according to the reaction

$$D + \text{a-Si:H (solid)} \rightarrow \text{SiD}_4 + \text{HD}.$$

D. Langmuir Probes

The Langmuir electrostatic probe is used largely in the study of gas discharges. The technique consists of inserting into the plasma a small bare metallic electrode. The current–voltage (*I–V*) characteristics can be related to the plasma parameters: the electron and ion densities, the electron temperature, and the plasma potential.

In its simplest form a single Langmuir probe is a wire or disk used in conjunction with a much larger reference electrode. The greatest interest of this type of diagnostics is to permit a spatially resolved measurement of the charge particle concentrations and energies.

However, the introduction of a small electrode into a reactive plasma that produces deposited films may be a serious problem regarding reliability and reproducibility. Moreover, the silane plasmas are usually excited with radiofrequency fields, and the RF potential fluctuations in the plasma may interfere with the sheath and the probe circuit.

1. *DC Discharge*

The use of electric probes is relatively simple in a DC discharge. The main source of error comes from the contamination of the probe by the deposited a-Si:H film, which introduces a resistive layer in series in the circuit. Probe contamination is often apparent in hysteresis of the *I–V* characteristics. A clean probe is essential for good-quality data. The most current method is sputtering of the probe surface by ion bombardment before recording of the characteristics.

Perrin *et al.* [82] used a single probe in a DC multipole plasma at 1–10 mtorr. They showed that the plasma contains a high-energy group of electrons close to EV_c, where V_c is the potential of the electron-emitted cathode (60 V). A low-energy group of electrons resulting from the numerous electron–molecule collisions was also detected, having an electron temperature KT_e of ~1 eV and an electron density of $\sim 10^8$ cm^{-3}.

Weakliem [51] used a screened probe to measure kT_e and n_e at the edge of the negative glow in a DC silane discharge. In another work performed with a hot-cathode DC discharge in silane [93] the distribution of electron energy was obtained from the second derivative current according to the method of Druyvesteyn.

2. *RF Discharge*

A general problem encountered with probe diagnostics of a RF discharge plasma is the occurrence of a RF voltage component across the probe sheath. This voltage is due to oscillations in the plasma potential with respect to ground. It leads to distorsion in the probe characteristic because of rectification in the nonlinear probe sheath.

The use of electric probes in silane RF discharges has been extensively studied by Mosburg *et al.* [94]. In order to minimize potential fluctuation in the central plane of the discharge, they used a balanced floating push–pull voltage on the two electrodes. As no grounded electrode was available in this arrangement, they introduced a reference electrode placed near the discharge center where the fluctuation in potential is minimum.

As the probe surface was continuously renewed by sufficiently conducting amorphous silicon, Mosburg *et al.* [94] showed that probe current was not disturbed by the a-Si layer. Values of $kT_e = 2–2.5$ eV and $n_e = 1–1.5\ 10^9\ cm^{-3}$ were found for typical conditions of a-Si:H film deposition. In B_2H_6–SiH_4 mixtures, probe measurements became unstable and irreproducible.

Bruno *et al.* [95, 96] noted no difficulty in operating an electrostatic probe in SiH_4–H_2 and SiF_4–GeH_4–H_2 RF plasmas. This was explained by the high photoconductivity values of the deposited film, which behaves as a sufficiently conducting layer in typical glow discharge conditions. Spatially resolved measurements have been performed by these authors [96] in conjunction with OES in order to check the validity of the actinometry hypotheses in plasma systems for a-Si deposition.

Despite the experimental difficulties, due to surface contamination and RF perturbation, it appears that the electric probe technique provides valuable information, in particular for the spatially dependent analysis of the electron population.

E. Conclusion

Applications of mass spectrometry and Langmuir probes to the study of silane plasmas have been reviewed. During the last 15 years, the ion chemistry has been extensively studied by means of mass spectrometry in DC and in RF silane discharges. Cluster ions as $Si_xH_y^+$, with x varying from 1 up to 9, have been shown to be the result of chain reactions sequence involving SiH_y^+ ions and SiH_4.

Mass spectrometric sampling of neutrals gives useful information for partial pressure of silane and by-products. The kinetic study of the silane glow discharge has been performed with both flowing and static discharges. A correct interpretation of these M S results requires a better knowledge of individual reactions rates involving radicals. Further developments can be expected with the use of modulated-beam mass spectrometers and of energy analyzers.

Concerning Langmuir probes, RF-driven technique and time-resolved analysis could improve the determination of the electron population. Reliable measurements of the electron energy distribution function are strongly needed in view of a more complete modeling of the silane discharge kinetics.

IV. *In Situ* Studies of Growth of a-Si:H by Spectroellipsometry

A. Introduction

An important issue in preparing hydrogenated-amorphous-silicon (a-Si:H)-based devices is the monitoring of the films properties from the plasma conditions. a-Si:H films are extensively deposited by plasma-enhanced chemical vapor deposition (PECVD) at moderate temperatures (<300°C). More precisely, most of the a-Si:H films used in practical applications are prepared from an RF discharge (13.56 MHz) of pure silane (SiH_4). The electronic properties of a-Si:H and related materials are generally found to be affected by the film morphology (see Knights [97] and refs. cited therein). Photoelectronic quality a-Si:H is generally a high-density homogeneous material. Besides, the film morphology is related to preparation conditions such as substrate temperature and ion bombardment. More generally, the presence of microstructures in plasma deposited a-Si:H has largely been established [97–100]. In particular, the film morphology near the substrate and the free surface can differ from that of the bulk. This behavior can influence the formation of interfaces, which can also affect the electronic properties of the devices.

Furthermore, the electronic properties of a-Si:H are known to be controlled by the hydrogen incorporation into the silicon network. The hydrogen content of the photoelectronic quality a-Si:H is generally estimated to be about 15%. However, it is generally agreed that a hydrogen-rich radical, SiH_3, is the dominant precursor of these films. Then the hydrogen incorporation at the plasma–film surface can be key to understanding the a-Si:H behavior. The physicochemical nature of the a-Si:H surface can also influence interface formations. In particular, some interfaces such as a-Si:H/a-SiN_x (silicon nitride) are found to depend on the deposition sequence ("bottom" or "top" a-Si:H) [101–103]. In the same way the a-Si:H surface is found weakly reactive as compared to bare substrates.

A detailed analysis of the film deposition necessitates the use of a real-time probe to identify the various possible stages of the growth: nucleation, coalescence, surface roughness evolution, and so on. Likewise, an interface formation can result from various competitive mechanisms with different kinetics [104–107]. However, the *in situ* electronic probes are generally incompatible with the reactive environment of the PECVD processes. Likewise, electron-microscopic techniques are more adapted to *ex situ* measurements. On the contrary, optical diagnostics can be used in any transparent ambient. Besides, optical techniques are noninvasive and do not perturb the reactive plasma. Among the various optical techniques, ellipsometry takes advantage of the simultaneous measurement of two quantities (amplitude and phase). Thus spectroscopic ellipsometry (SE) allows measurements of the complex dielectric functions of thin films [108]. The submonolayer sensitivity of SE has extensively been demonstrated. Furthermore, as a consequence of recent advances in optical instrumentation and numerical signal processing, real-time ellipsometry can be compatible with the kinetics involved in a-Si:H preparations. Thus continuous data recording at a ≥1-kHz frequency can now be achieved with spectroscopic phase-modulated ellipsometry (SPME) [109, 110]. In the case of a semiconductor such as a-Si:H, SE in the near-ultraviolet (UV)–visible range provides a very sensitive probe for surface and interface studies because in the UV the absorption depth can be reduced below 100 Å. In this range, SE measurements are very sensitive to the film morphology. The wavelength range of SPME has recently been extended toward infrared (IR) with comparable sensitivity. In the latter case, SPME allows a direct identification of the vibrational properties, such as the hydrogen bonding, of a-Si:H [111, 112].

Selected examples of applications of *in situ* SPME to the growth of plasma deposited a-Si:H are presented in this review. The ellipsometric techniques and the data analysis methods are briefly described in Section IV.B. Then the plasma–surface interaction during growth is investigated in Section IV.C by IR ellipsometry (IRPME). In particular, it is shown that a-Si:H grows beneath a hydrogen-rich thin overlayer. Other applications of UV–visible ellipsometry are presented in Sections IV.D and IV.E. The microstructure evolution during growth of a-Si:H on a smooth substrate is presented first. Finally the influence of the preparation conditions are described. In particular, the importance of the plasma conditions such as ion bombardment, plasma power, and gas pressure are emphasized.

B. Experimental Details

1. *Ellipsometry and Photometry*

Let us consider the reflection of a plane wave

$$\mathbf{E}(\mathbf{r},t) = \mathbf{E}_0 \exp(i\mathbf{k}\mathbf{r}\text{-}i\omega t)$$

on the surface of an isotropic material. For nonnormal incidence, plane waves are typically referenced to a local coordinate system (x,y,z), where z is the propagation direction and x and y are directions parallel (p) and perpendicular (s), respectively, to the plane of incidence. The complex field amplitudes E_p and E_s represent the projections of the plane wave along x and y.

To realize how different optical instruments take advantage of the information encoded into plane waves, consider the path of $\mathbf{E}(\mathbf{r},t)$ in the (x,y) plane. Since the x and y components oscillate in simple harmonic motion about the origin, in general the path is an ellipse. This polarization ellipse can be characterized by its size and shape. The size is clearly related to the intensity. Instruments that deal with intensity are photometers. For example, reflectometers determine the ratio of outgoing to incoming intensities, generally for a definite state of polarization. The shape of the polarization ellipse is an intensity-independent quantity that must be defined by two parameters. The simplest quantity that meets these requirements is the ratio

$$\chi = E_p/E_s.$$

Optical instruments that deal with the polarization state χ are ellipsometers. In a typical ellipsometric measurement, the sample is irradiated with light in a known state of polarization χ^i, after reflection the state of polarization is transformed into χ^o according to

$$\rho = \chi^o/\chi^i = (E_p/E_s)^o/(E_p/E_s)^i = r_p/r_s,$$

where r_p and r_s are complex reflectances for p- and s-polarized light, respectively. The complex reflectance ratio is often represented in the literature by the angles Ψ and Δ defined by

$$\rho = \tan\Psi \exp(i\Delta). \tag{21}$$

In the case of a reflection at an angle of incidence ϕ on the surface of a semi-infinite medium, the ratio ρ is directly related to the dielectric function ε of this medium according to [108]:

$$\varepsilon = \sin^2\phi + \sin^2\phi \; tg^2\phi \; [(1 - \rho)/(1 + \rho)]^2 \tag{22}$$

This relation shows that ellipsometry allows a direct measurement of the dielectric function of an absorbing material displaying a sharp interface with ambient.

The applications of ellipsometry to multilayers structures lead to more complex analytical relations. A case of particular importance is a thin film of thickness d_f and dielectric function ε_f on a substrate with dielectric function ε_s. If the film thickness is such that $d_f \ll 4\pi/\lambda$ (thin-film approximation), where λ is the wavelength of the light, it can be shown that

$$\langle\varepsilon\rangle = \varepsilon_s + \frac{4i\pi d_f}{\lambda}(\varepsilon_s - \sin^2\phi)\frac{\varepsilon_s(\varepsilon_s - \varepsilon_f)(\varepsilon_f - 1)}{\varepsilon_f(\varepsilon_s - 1)}, \tag{23}$$

where $<\varepsilon>$ is the pseudodielectric defined as the simple transformation of the ellipsometric ratio according to equation (22). The multilayer formalism can be extended to an arbitrary number of layers [108, 113]. In the IR range, instead of equation (23), the ellipsometric measurements are often presented in term of the optical density D defined by [114]

$$D = \log (\overline{\rho}/\rho), \tag{24}$$

where $\overline{\rho}$ refers to the substrate before film deposition. For thin-film approximation, it can be shown that D is directly related to the film vibrational properties according to [112, 115]

$$D = (\varepsilon_f - \varepsilon_s)\ K(\phi,\lambda)A(\varepsilon_s,\phi), \tag{25}$$

where

$$K = (4\pi i d_f/\lambda)\ \sin\phi\ \tan\phi. \tag{26}$$

The contribution of one vibrational mode to the dielectric function can be estimated from the Lorentzian expression, using cgs (centimeter-gram-second) units (see Cardona [116] and refs. cited therein):

$$\begin{aligned}\Delta\varepsilon(\omega) &= (4\pi N\, e_s^{*2}\, m^{-1})/(\omega_0^2 - \omega^2 - i\Gamma\omega)\\ &= F\omega_0/(\omega_0^2 - \omega^2 - i\Gamma\omega),\end{aligned} \tag{27}$$

where ω_0 represents the vibration frequency, Γ the damping constant, e_s^* an effective dynamical charge of the bond, m the effective mass, and N the density of bonds.

2. *Phase-Modulated Ellipsometry*

Two different experimental techniques, based on polarization modulation, are commonly used to perform SE measurements. Rotating-element ellipsometers (REEs) take advantage of a conceptual simplicity and a wavelength insensitivity (for review see Collins [117]). Then they allow high-precision measurements. IR ellipsometers using rotating polarizers have been also used [114, 118, 119]. However, the main limitation of REE is related to the low-frequency modulation provided by the mechanical rotation. In particular, this first technique is often limited to static measurements or to *in situ* applications involving slow kinetics.

In contrast, SPMEs use photoelastic devices to perform the polarization modulation [109, 110, 120, 121]. The main advantage of this technique is the use of a modulation system about three orders of magnitude faster (30–50 kHz) than the mechanical rotation of a polarizer. Moreover, photoelastic modulators can be easily combined with modern numerical data-processing systems [109, 110]. Thus

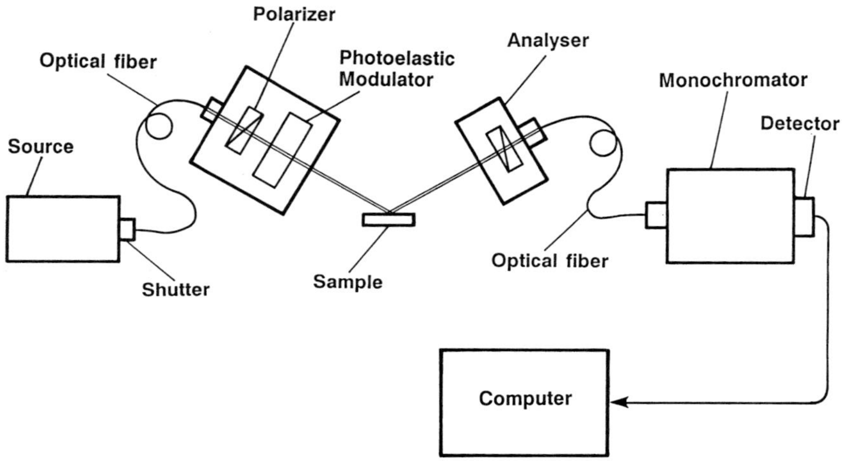

FIGURE 22. Schematic diagram of the phase-modulated ellipsometer in the UV–visible range.

the SPME technique is particularly well adapted to real-time applications. In particular a time resolution of 1 ms can be achieved.

The phase-modulation ellipsometry (PME) technique has been presented in detail elsewhere [109, 110, 121]. The optical setup in the UV–visible range is shown in Fig. 22. Let us briefly recall that two linear polarizers are used to polarize the incident beam and to analyze the reflected beams. The modulation polarization is provided by a photoelastic material, consisting in the UV–visible range of a fused-silica block sandwiched between piezoelectric quartz crystals oscillating at the frequency ω (50 kHz), which generates a periodical phase shift $\delta(t)$ between two orthogonal components of the transmitted electric field. At first-order approximation

$$\delta(\mathrm{t}) = A_{\mathrm{m}} \sin \omega \mathrm{t}, \tag{28}$$

where A_{m} is the modulation amplitude, which is proportional to (V_{m}/λ), where V_{m} is the excitation voltage. In the UV–visible range optical fibers can be adapted in both arms in order to increase the compactness of the ellipsometer (see Fig. 22). Because linear polarizers are used in both beams, SPME is insensitive to any polarization effect due to the optical fibers, in contrast with REE. In the present case, the source is a high-pressure xenon arc lamp. A mechanical shutter is used in order to evaluate the continuous background. Finally the energy of the light is analyzed by a grating monochromator. In the simplest setup a photomultiplier is used as detector. In this case the available wavelength range is 230–830 nm. However,

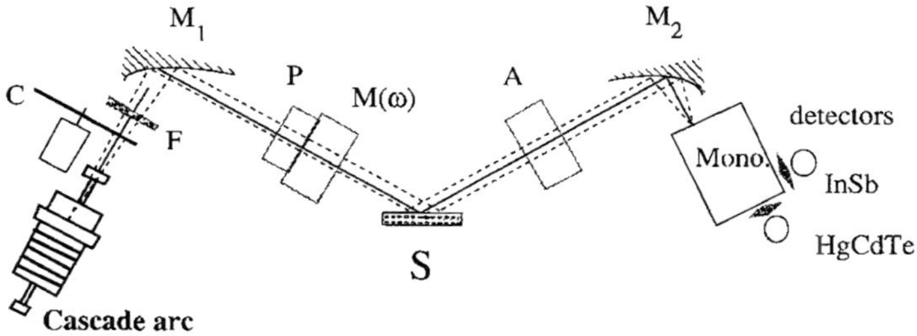

FIGURE 23. Optical setup of the IRPME in the incidence plane: A (analyzer); C (chopper); F (Ge filter); M_1 and M_2 (off-axis parabolic mirror); M(ω) (photoelastic modulator); Mono (monochromator); P (polarizer); S (sample).

this range can be easily extended toward near-IR by using low-band-gap semiconductor detectors.

More generally, the SPME technique has been extended in the IR above 10 μm [112]. The corresponding optical setup is shown in Fig. 23. In the IR range, ZnSe polarizers and modulator are used. In order to increase the signal-to-noise ratio (SNR), the conventional globar source was superseded by a cascade arc corresponding to a blackbody temperature >10,000 K [122]. A low-frequency chopper ($\approx$ 600 Hz) is used to evaluate the DC component of the signal. Ellipsometric measurements can be recorded from 700 up to 4000 cm^{-1} combining photovoltaic InSb and HgCdTe detectors. The spectral resolution of the IR PME varies from 2 to 5 cm^{-1}, depending on the energy wavelength domain.

Whatever the wavelength, the detected intensity takes the general form [109, 110, 120, 121]:

$$I(t) = I[I_0 + I_s \sin \delta(t) + I_c \cos \delta(t)], \tag{29}$$

where I_o, I_s, and I_c are known functions of Ψ and Δ. An ideal photoelastic modulator corresponds to equation (28), then

$$\sin \delta(t) = 2J_1(\mathbf{A}) \sin \omega t + 2J_3(\mathbf{A}) \sin 3\omega t + \cdots, \tag{30}$$

$$\cos \delta(t) = J_0(\mathbf{A}) + 2J_2(\mathbf{A}) \cos 2\omega t + \cdots, \tag{31}$$

where $J_n(\mathbf{A})$ is the nth Bessel function of **A.**

Then, combining equations (29)–(31) leads to a simple linear relation between the DC component S_0 and the first harmonics of S_ω and $S_{2\omega}$ determined from the numerical signal processing and the useful quantities I_o, I_s, and I_c [110, 121]. A more general approach taking into account the presence of higher harmonics in the expression is described elsewhere [121].

3. *Data Interpretation: Effective-Medium Theories*

Effective-medium theories (EMTs) allow the dielectric response ε of a heterogeneous material to be described from the dielectric functions of its constituents ε_i and a few wavelength-independent parameters [123]. Thus EMTs appear as a basic tool in material characterization by optical means. More precisely, EMTs can be used if the separate regions are small compared to the wavelength of the light but large enough so that the component dielectric functions are not distorted by size effects. All EMTs in the quasistatic approximation can be represented by [123]

$$\frac{\varepsilon - \varepsilon_{\mathrm{h}}}{\varepsilon + K\varepsilon_{\mathrm{h}}} = \sum_i f_i \frac{\varepsilon_i - \varepsilon_{\mathrm{h}}}{\varepsilon_i + K\varepsilon_{\mathrm{h}}}, \tag{32}$$

where f_i are the relative volumnic fractions of the various constituents, K is a screening parameter (for spherical inclusions $K = 2$), and ε_{h} the host dielectric function. In the case of Maxwell–Garnett theory [124], the host material is the main component, whereas the Bruggeman approach is self-consistent with $\varepsilon = \varepsilon_h$ [125]. The Bruggeman effective-medium approximation (EMA), with $K = 2$, is generally used to describe the microstructure of a-Si:H.

One application of EMA is to calculate the effect of voids in the dielectric function of a-Si:H (ε_{b}). Taking for voids $\varepsilon_{\mathrm{a}} = 1$ and assuming $f_{\mathrm{a}} \ll 1$, because $|\varepsilon_{\mathrm{b}}| \gg 1$, equation (32) leads to

$$\varepsilon = \varepsilon_{\mathrm{b}}\,(1 - 3/2\, f_{\mathrm{b}}), \tag{33}$$

which shows that voids reduce ε_{b} everywhere uniformly. Moreover, equation (33) reveals how screening decreases the importance of the more polarizable species (silicon) with respect to the less polarizable voids. The previous example can be extended to a film growth model, as illustrated in Fig. 24. A nucleation phase is described as an increase of the film density. The variation of the film density can be deduced from geometric assumptions (e.g., hemispheres development as shown in Fig. 14). During the subsequent growth, the appearance of a surface roughness can be related to the incomplete coalescence of the initial nuclei (Fig. 24).

Fits to the ellipsometric measurements are then performed using the standard least-square method. The free parameters of the fit are obtained by minimization of the error function

$$\delta^2 = \frac{1}{N}\sum_{i=1}^{N} \left| \rho_{i,\exp} - \rho_{i,\mathrm{th}} \right|^2, \tag{34}$$

where N is the number of experimental points.

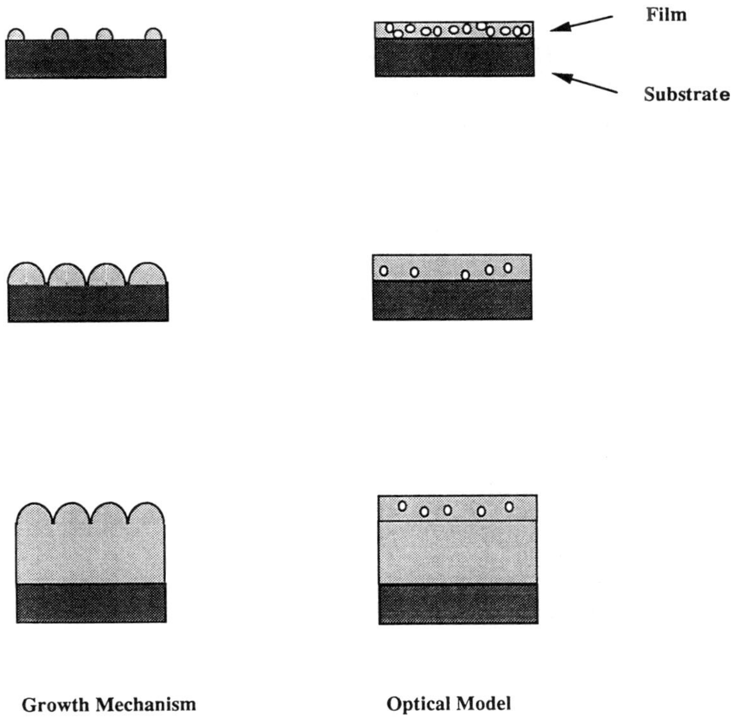

FIGURE 24. Schematic representation of a film growth model based on effective medium theories.

C. Hydrogen Incorporation at the Growing a-Si:H Surface

The first examples of applications of ellipsometry to the growth of plasma deposited a-Si:H consist of recent experiments performed with IR phase-modulated ellipsometry (IRPME).

Detailed *in situ* investigations of plasma–surface interactions during deposition are needed to provide a clear picture of the hydrogen incorporation in a-Si:H. In particular, it has been shown, using various *ex situ* experimental techniques, that the a-Si:H surface is covered by hydrogen [126, 127]. In order to be sensitive to the hydrogen incorporation, such a study requires the identification of vibrational properties by IR measurements. These measurements must be performed in preparation conditions compatible with practical applications. However, the main experimental problem encountered with the IR experimental techniques is their weak sensitivity to SiH_n bonding identification in ultrathin films deposited on conventional substrates like glass. In contrast, the presence of the SiH and SiH_2 vibrations has recently been revealed, for the first time, in 5–10-Å-thick a-Si:H samples deposited on glass at various substrate temperatures, using IRPME [111].

An *in situ* study of the vibrational properties of a-Si:H thin films (4–250 Å) deposited using the conventional PECVD method is summarized here [128]. In the present study, a-Si:H films are deposited by RF glow discharge (13.56 MHz) decomposition of SiH_4 (50 mtorr) at low power (0.05 W cm^{-2}) on glass substrates (Corning 7059) at 250°C, as described elsewhere [129]. The deposition rate is $v_d \approx 1.0$ Å s^{-1}.

Typical IRPME measurements, recorded *in situ* in the SiH_n stretching mode region, on the thinner samples are shown in Fig. 25. The results are presented in terms of optical density defined by equations (24)–(26). The spectra display near 1990 cm^{-1} (SiH) or 2100 cm^{-1} (SiH_2) a Lorentzian behavior corresponding to equation (27). Figure 25 reveals the extreme sensitivity of IRPME, with SiH_n bonds being clearly identified in the 4–8-Å-thick samples. A quantitative analysis of the data, recorded during the early stage of the growth of a-Si:H, can be performed by fitting IRPME measurements using equation (27) and assuming the presence of two layers, as shown in the inset of Fig. 26. The values of the quantities ε and F are assimilated to these of bulk a-Si:H for the Si-H stretching mode,

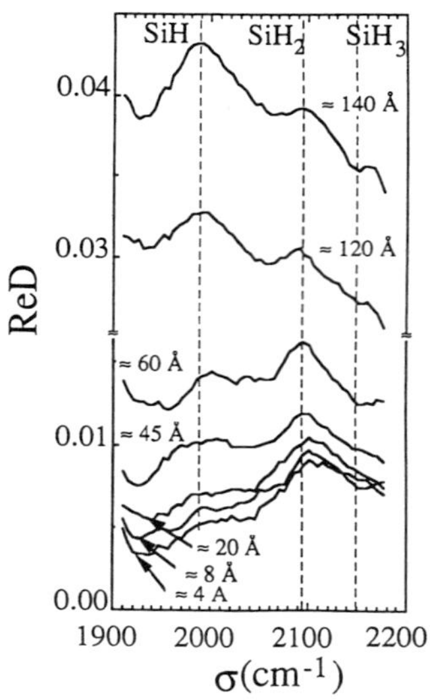

FIGURE 25. Evolution of the real part of the optical density during the early stages of the growth of a-Si:H on glass substrates.

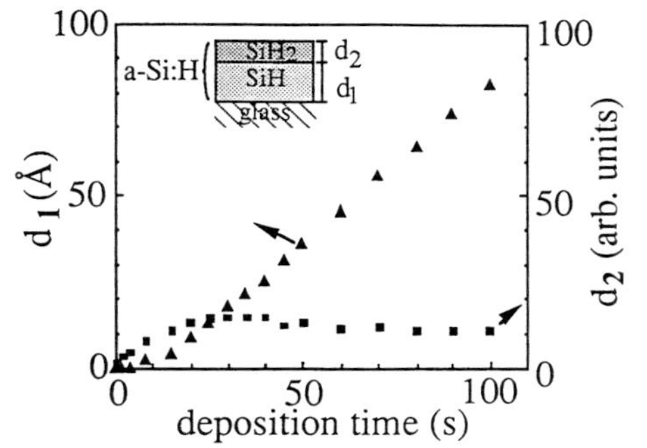

FIGURE 26. Variations of the thickness of the bulk a-Si:H (triangles) and the overlayer (squares).

which determines the sublayer thickness d_1. Such a procedure cannot be used for the overlayer since the environment of surface species is probably very different from this of SiH_2 studied in bulk materials (such as a-Si:H deposited at room temperature); as a consequence, the thickness d_2 cannot unambiguously be determined. The results of the fits are shown in Fig. 26. After the deposition of the first 5 Å, the amount of SiH bonds is found to increase linearly with the deposition thickness. On the contrary, the SiH_2 contribution displays a different behavior. The amount of SiH_2 first slowly increases during the deposition of the first 20–30 Å and then saturates to a constant value. During the subsequent film growth, the thickness d_2 can be estimated to a few hydrogen saturated monolayers [111], in agreement with other analyses [126, 127, 130, 131]. Plasma treatments can be used to modify and thus to characterize surfaces [127]. In this way, a further evidence of the presence of the SiH_2 groups near the a-Si:H surface is given by the behaviors shown in Fig. 27. The SiH_2 contribution is found to strongly decrease after the exposure of a freshly deposited ≈200-Å-thick a-Si:H sample to low-power H_2 and Ar plasmas. In contrast, the bulk SiH signal is insensitive to plasma exposure.

The weak chemical reactivity of the a-Si:H surface, as compared to most bare crystals, can probably be attributed to the presence of this thin hydrogen-rich overlayer. In particular, the reactivity of the a-Si:H with atmosphere can be investigated from the SiO stretching mode ($\sim$1200 cm^{-1}) evolution. The oxide thickness of the as-deposited sample is estimated to ≈4–5 Å after 1 month [132]. The removal of the hydrogen overlayer by the H_2 plasma treatment corresponding to Fig. 27 leads to a strong enhancement of the SiO signal, as shown in Fig. 28. In the latter case the oxide thickness is estimated to 10–12 Å after one month of air exposure. This result evidences the passivation effect of the hydrogen-rich surface layer.

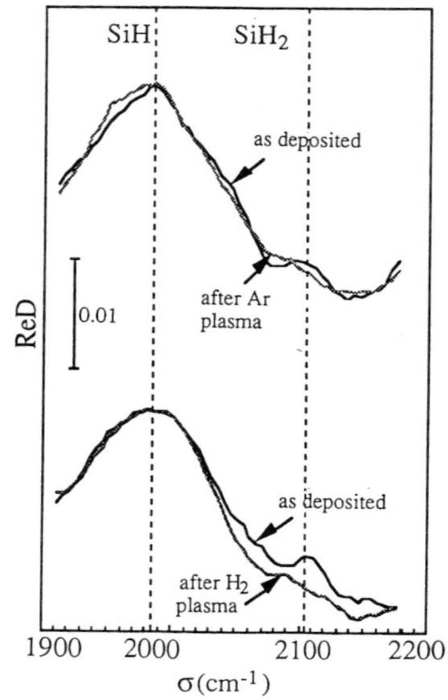

FIGURE 27. Influence of argon and hydrogen plasma treatments on the surface of a-Si:H films. a-Si:H thickness: 240Å (top) and 200Å (bottom); plasma exposure conditions (Ar: 0.1 W cm^{-2}, 20 mtorr, 5 mn) (H_2: 0.05 W cm^{-2}, 250 mtorr, 1 mn).

D. Microstructure Evolution during Growth of a-Si:H on Smooth Substrates

As already pointed out, a detailed knowledge of the microstructure evolution of a thin film such as a-Si:H is of considerable interest not only from a fundamental point of view but also for improving device performances. Real-time SPME in the UV–visible range appears particularly well adapted to perform such an investi-

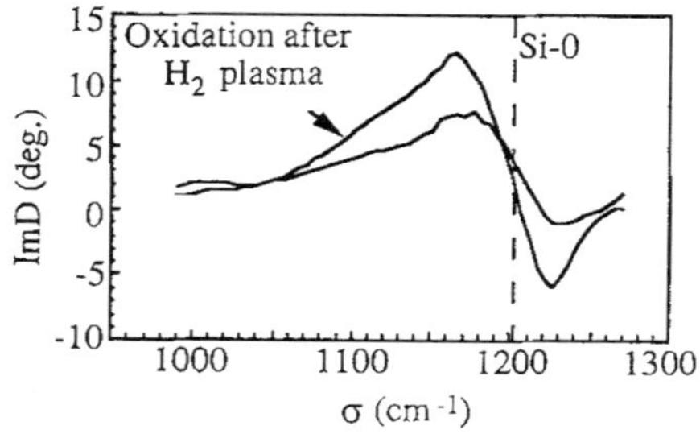

FIGURE 28. Comparison between long-term oxidation of a-Si:H samples exposed to air as deposited and after H_2 surface plasma treatment.

gation. Previous SE experiments revealed that the growth of a-Si:H is influenced by the deposition conditions and the nature of the substrate [98–100]. In particular, transparent conducting oxide and glass substrates were found to be reduced by the silane plasma during the early stage of the a-Si:H deposition [104, 105].

This section describes the results of a detailed study of the nucleation and growth of photoelectronic quality a-Si:H from the deposition of the first monolayers up to a final thickness of more than 1 μm. a-Si:H is deposited by conventional glow discharge (13.56 MHz) of silane on smooth, homogeneous metallic substrates (NiCr films deposited on glass) [133]. Moreover, similar results are obtained when using c-Si substrates [133]. In order to perform a precise analysis of the growth mechanism, real-time ellipsometric trajectories are recorded, using fixed preparation conditions, at various photon energies ranging from 2.2 to 3.6 eV. These photon energies correspond to a light penetration in a-Si:H decreasing from a few thousand angstroms to about 100 Å. For kinetic studies at fixed photon energies, N points ρ_i [as defined by equation (21)] were recorded at regular intervals Δt·Δt ranges from 0.5 s (UV light) up to 1.5 s (2.25 eV), each individual real-time trajectory corresponding typically to N = 3000 points.

The nucleation models are displayed in Fig. 29. The most simple assumption corresponds to the uniform growth of an homogeneous material with a constant refractive index. In the hemispherical nucleation model, a hexagonal network of spherical nuclei of a-Si:H is created with an average distance d between them. The radius of the nuclei increases continuously until the hemispheres come into contact. This results in a density-deficient layer of growing thickness, the void volume fraction of the film decreasing from f = 1 down to 0.39. The index of this layer is continuously deduced from an EMA calculation as described in

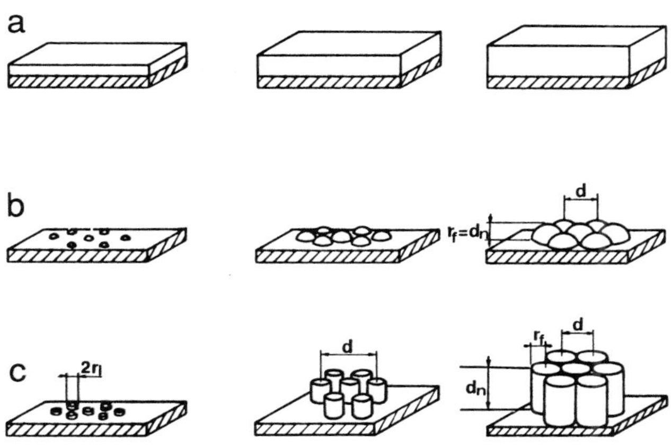

FIGURE 29. Schematic representation of the nucleation models: (a) homogeneous growth; (b) hemispherical nucleation; (c) cylindrical nucleation.

Section II.B [see equation (12)]. In a different way, the third model assumes a columnar microstructural development during the nucleation phase [133, 134]. In the latter case, the a-Si:H nuclei are located on a hexagonal network separated by an average distance $d = 60$ Å. As d is fixed, the free parameters of this last model are the initial and final radii r_i and r_f and the final cylinder height d_n.

The real-time trajectories recorded at 3.1 and 3.54 eV during the deposition of a-Si:H on NiCr substrates are compared in Figs. 30a and 30b to the nucleation models described above. In both cases, the experimental trajectories strongly de-

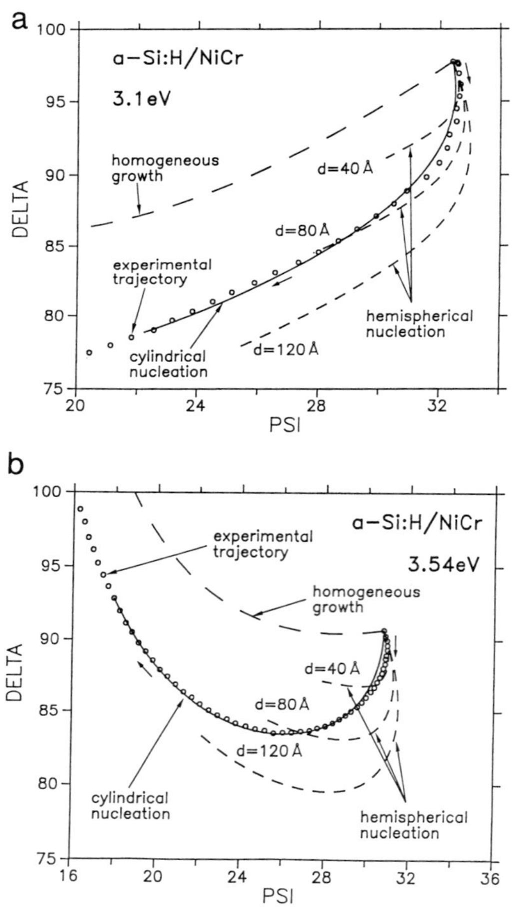

FIGURE 30. Real-time examinations of the beginning of the growth of a-Si:H on NiCr substrates, (a) at 3.1 eV; (b) at 3.54 eV.

part from the uniform growth assumption (long dashes) confirming previous results [104, 129, 135]. In contrast, the (Ψ, Δ) trajectory, recorded at 3.1 eV (Fig. 30a) is in rather good agreement with the hemispherical nucleation assumption corresponding to $d = 80$ Å (short dashes). A similar conclusion was recently deduced from an ellipsometric study (at 3.4 eV) of the early stage of the growth of a-Si:H on Mo and Cr substrates [134]. Nevertheless, the experimental results obtained at 3.54 eV (Fig. 30b) can rule out the hemispherical nucleation model. At 3.54 eV the experimental trajectory is compatible only with the columnar nucleation model; a very good fit with the cylindrical nucleation is also obtained at 3.1 eV (solid lines). Finally, the following average values can be estimated combining the spectroscopic data: $r_i = 20 \pm 4$ Å, $r_f = 29 \pm$ Å, and $d_n = 60 \pm 20$ Å. The high r_f value shows that the cylinders almost come into contact at the end of the nucleation phase.

The subsequent problem is the optical description of the coalescence of the initial columnar microstructure, the presence of a small roughness at the top surface of a-Si:H being well established [98–100]. The schematic coalescence model is shown in Fig. 31. The presence of two layers with different thickness (d_1, d_2) and void volume fractions (f_1, f_2) is considered. The refractive index of these two layers are continuously deduced from EMA calculations. At the end of the nucleation phase, the value of f_1 and f_2 is deduced from the void volume fraction of the columnar layer, thus at this initial point of the coalescence: $d_{1i} + \mathrm{d}_{2i} = d_n$ and $f_{1i} = \mathrm{f}_{2i} = 0.1–0.2$. The overlayer thickness d_1 is fixed during the further growth

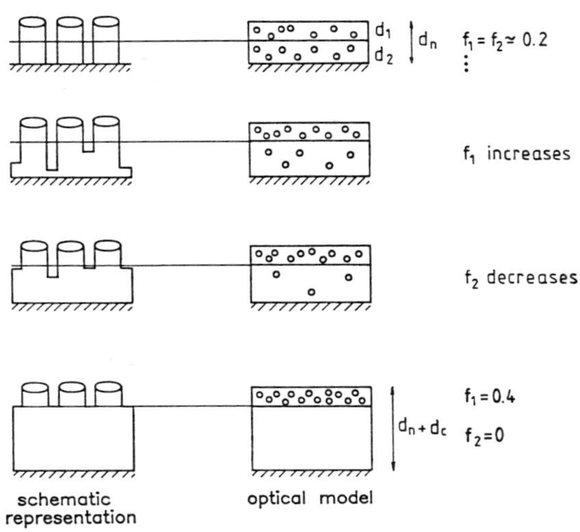

FIGURE 31. Schematic representation of the coalescence model.

while d_2 increases linearly following the deposition rate. The value of d_1 corresponds to the result of the fit to the overall trajectory of the constant surface roughness growth model assuming a void volume fraction $f_1 = 0.39$ (hemispherical surface roughness as displayed in Fig. 24). d_c represents the equivalent thickness of the coalescence phase. At the end of the coalescence phase, $d_1 + d_2 = d_n + d_c$. During this phase, the relative bulk void volume fraction f_2 decreases down to 0 while f_1 increases up to 0.39, which corresponds to the constant surface roughness a-Si:H growth model (Fig. 24). During the coalescence phase, the value of

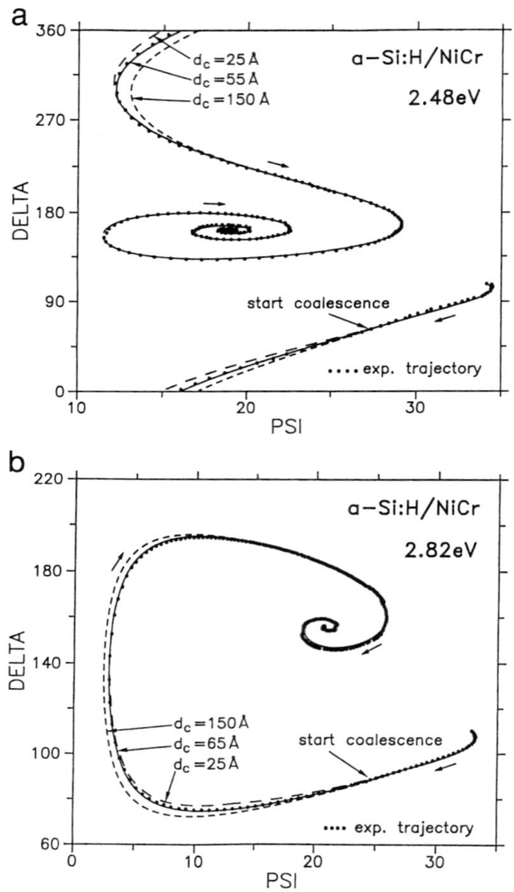

FIGURE 32. Real-time examinations of the long-term growth of a-Si:H deposited on NiCr substrates, (a) at 2.48 eV, and (b) at 2.82 eV. The dashed lines illustrate the influence of the parameter d_c on the coalescence phase.

f_1 as a function of the total film thickness x ($d_n < x < d_n + d_c$) is deduced assuming the following dependence:

$$f_1 = f_{1i} + (f_{1f} - f_{1i})\ ((x - d_n)/d_c).$$

Then the condition of a fixed material flux impinging onto the substrate during growth implies the variation of f_2 as a function of x. After the end of the coalescence phase, the growth models shown in Figs. 24 and 31 are equivalent.

The coalescence model described above is compared to the experimental trajectories recorded, at various photon energies, during the deposition of 800–4500-Å-thick a-Si:H on NiCr substrates. The results are displayed in Fig. 32. An extremely good agreement is obtained with the growth model displayed in Fig. 31, whatever the photon energy. In particular, Fig. 32 illustrates the influence of the parameter d_c on the fit. The beginning of the coalescence phase is indicated on each (Ψ, Δ) trajectory displayed in Fig. 32. Finally, in the case of metallic substrate, the following average value can be estimated to be $d_c = (60 \pm 20)$ Å.

As already pointed out, the pseudodielectric function ε of a-Si:H can be estimated from the spectroellipsometric data by assuming a sharp interface with ambient using equation (22). The long-term evolutions of the imaginary part ε_2 are presented in Fig. 33. A slight decrease of the spectrum together with a slight shift toward lower energies are observed. This behavior is characteristic of an increase of the surface roughness with increasing film thickness [98–100]. Using this assumption, the increase of the surface roughness can be estimated (from EMA calculations) to be 5–10 Å for an increase in thickness of 1.0 μm. Likewise, a

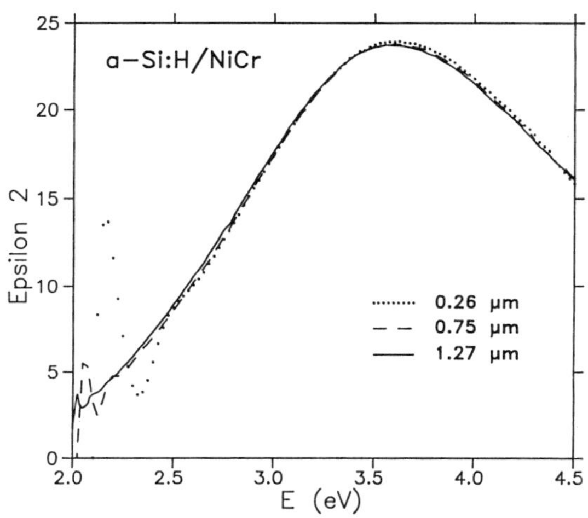

FIGURE 33. Spectroscopic measurements corresponding to a-Si:H films with different thicknesses deposited on NiCr substrates.

3–4-Å increase of the surface roughness was observed in a previous experiment [136]. Finally, the surface roughness thickness can be estimated to (14 ± 2) Å during the growth of the first 3–4000 Å on NiCr substrate.

In summary, the present study reveals that the nucleation mechanism on NiCr substrate is accurately described assuming a columnar microstructural development during the early stages of the growth of a-Si:H. The disk-like geometry corresponding to the deposition of the first monolayers can be probably understood from a thermodynamic standpoint, as already suggested [134]. In the substrate/oxide/a-Si:H configuration considered here, the sum of the a-Si:H surface and the oxide/a-Si:H interfacial free energies must be less than the oxide surface energy. Thus the free energy of the system is minimized by a two-dimensional growth process that maximizes interface area. Then, as a consequence of the incomplete coalescence of the initial nuclei, a surface roughness on a 10–15-Å scale is identified during the further growth of a-Si:H on smooth substrates [133]. Finally, an increase of this surface roughness is evidenced when dealing with the deposition of rather thick films. However, the thickness of this surface roughness is found larger than the hydrogen-rich overlayer evidenced in the previous section.

E. Influence of Preparation Conditions on Growth of a-Si:H

This section is devoted to the influence of the preparation conditions on the film morphology, considering the growth of a-Si:H on smooth substrates. More precisely, the variations of the nucleation parameters, film density, and surface roughness are investigated.

1. Substrate Temperature

The influence of the deposition temperature T_s on the growth of a-Si:H has extensively been investigated [99, 129, 137, 138]. Generally, the growth model described in Fig. 24 provides an acceptable description of a-Si:H deposition. However, the various morphological parameters are found to be strongly affected by T_s. Photoelectronic-quality a-Si:H samples are generally deposited above 150–200°C. However, a further increase above 300°C leads to the hydrogen effusion from a-Si:H.

As discussed in Section IV.D, SE measurements are very sensitive to a-Si:H microstructures [see in particular equation (33)]. As compared to a-Si:H deposited at 250°C by glow discharge, a relative density deficiency of 30% has been reported for a-Si:H films deposited at room temperature (RT) [137]. Likewise, films deposited at 100°C can display a relative density deficiency of 10%. In the same way, the surface roughness thickness, evaluated from kinetic measurements increases by a factor of 2 when T_s decreases from 250°C to RT [129]. This improvement of film morphology with rising T_s is generally attributed to the increase of the surface mobility of the reactive species.

2. *Ion Bombardment*

The growth of a-Si:H can also be affected by ion bombardment [99, 129, 137, 139–141]. In order to enhance the ion flux impinging the growing surface, a-Si:H films were deposited using a low-pressure ($p_t < 5$ mtorr) multipole discharge [139]. In the multipole discharge, the relative ion flux going onto the substrate can reach up to $\Phi_+/\Phi_{tot} = 0.8$, with the ion bombardment energy E_{ion} controlled electrostatically [139]. In this case, moderate ion bombardment ($E_{ion} = 50$–100 eV) induces a very efficient rearrangement at the surface of growing a-Si:H producing high-density films [139, 140]. The latter trend is evidenced by the high value of $\varepsilon_{2,max}$ measured by SE [see equation (33)]. In a 13.56-MHz RF discharge ($p_t > 30$ mtorr) the relative ion flux Φ_+/Φ_{tot} decreases by about one order of magnitude, resulting in a weaker effect of the ion bombardment on the film density [137].

The influence of the ion bombardment is studied by comparing two samples corresponding respectively to $\Phi_+/\Phi_{tot} = 0.2$ (multipole plasma) and 0.03 (RF plasma). The ion bombardment during growth was carefully investigated by means of an electrostatic analyzer as described elsewhere [129, 139]. In the multipole discharge, E_{ion} closely follows the value of the (metallic) substrate bias V_b, where the ion flux is quasimonoenergetic. In contrast, a considerable broadening of the ion energy distribution (due to ion–neutral collisions) is observed in the RF discharge when a bias is applied.

Figure 34 shows the SPME kinetic measurements recorded during the early stage of the growth of a-Si:H on metallic substrate under strong ion bombardment ($\Phi_+/\Phi_{tot} = 0.2$, $V_b = -150$ V). The trajectories displayed in Fig. 30 (weak ion flux, no substrate bias) and Fig. 34 displayed similar behaviors. The shift between the initial points of both trajectories is related to the difference between the substrates (Cr in Fig. 34 instead of NiCr in Fig. 30). Likewise, the experimental trajectory of Fig. 34 departs from the homogeneous growth assumption, even if a density deficiency is taken into account. The simple hemispherical nucleation model (as shown in Figs. 24 and 29) qualitatively reproduces the experimental behavior (Fig. 34). A better agreement is obtained with the cylindrical assumption, as described in Section IV.D (not shown). Thus, it can be inferred that nucleation phase of a-Si:H is weakly affected by the ion bombardment.

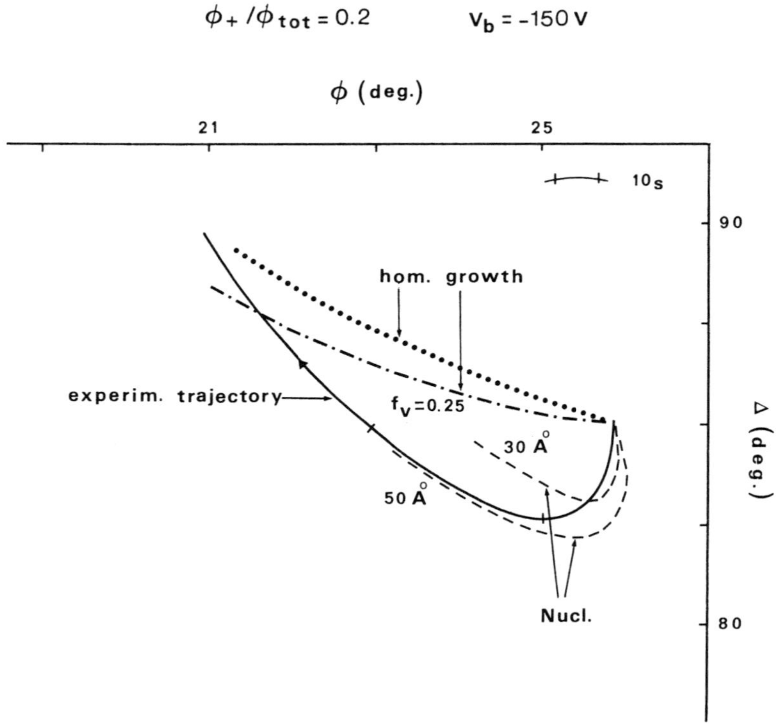

FIGURE 34. Real-time examination (at 3.54 eV) of the beginning of the growth of a-Si:H deposited on Cr substrate using a multipole discharge.

Nevertheless, the latter conclusion cannot be extended to the bulk growth of a-Si:H. In particular, it has been shown in Section IV.D that the presence of a surface roughness is needed to describe the growth of a-Si:H deposited by RF discharge. More precisely, the influence of the density-deficient overlayer in the residual of the fit to the kinetic measurements [as defined by equation (34)] is displayed in Fig. 35, the SPME trajectories being recorded at 2.2 eV. The introduction of a surface roughness (model B) results in a reduction of $\delta^2_{\min}$ by about one order of magnitude, except when the ion bombardment during growth is strong ($\phi_+/\Phi_{\text{tot}} = 0.2$, $V_b = -150$ V). The latter trends are confirmed by the spectroscopic measurements displayed in Fig. 36. A decrease of the ion bombardment induces a shift of the ε_2 curve toward lower energies, which corresponds to an increase of surface roughness (see Section IV.D).

Thus the correlation between various SPME measurements reveals that a strong ion bombardment (multipole plasma) produces high-density a-Si:H films with

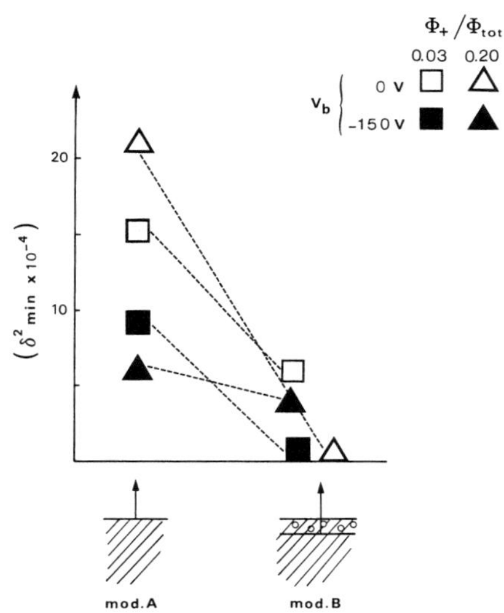

FIGURE 35. Comparison between the least-square-fit residuals obtained assuming an homogeneous growth (A) and a growth with surface roughness (B).

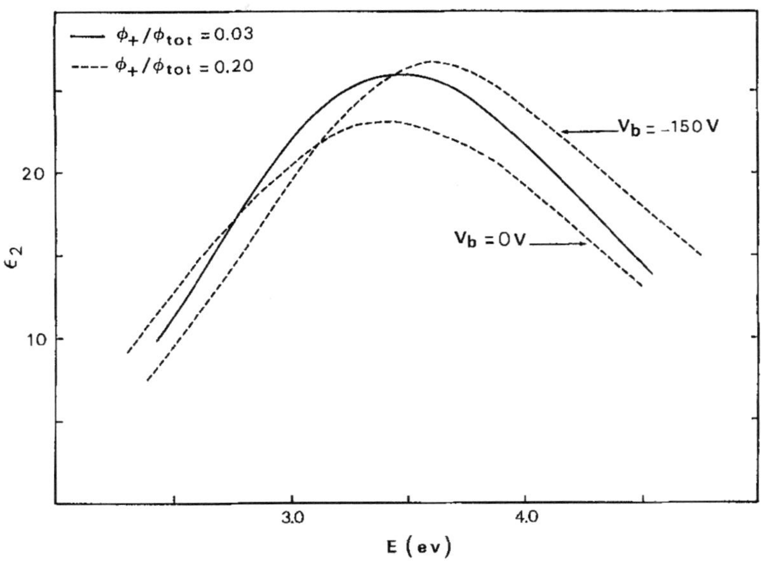

FIGURE 36. Spectroscopic ellipsometry measurements on a-Si:H samples as functions of the relative ion flux (Φ_+/Φ_{tot}) and substrate bias V_b.

smooth interface with ambient. This sharp interface can be considered as a consequence of the coalescence of the initial nuclei. Thus a strong ion bombardment leads to an increase of the surface mobility of the reactive species, then the nuclei fully coalesce in this case. In this frame, the surface roughness (at the 10–20-Å scale) observed in the photoelectronic-quality a-Si : H films (RF discharge) can be interpreted as a consequence of the incomplete coalescence of the nuclei. Besides, Fig. 36 shows that without ion bombardment the a-Si : H surface roughness decreases from the multipole (low pressure) to the RF glow discharge (high pressure). An increase of the pressure leads to a decrease of the relative SiH_2/SiH_3 precursor flux [71, 142, 143]. SiH_2 radical, being more reactive than SiH_3, is expected to have a significantly lower surface diffusion length. Then a-Si : H films produced with dominant SiH_3 precursors (RF plasma) display a smaller surface roughness.

3. Pressure and RF Power

A technological solution for reducing the cost of a-Si : H-based devices is to achieve higher deposition rates without reducing film quality. Several approaches have been proposed to satisfy this requirement. The easiest way to enhance the deposition rate in a PECVD system is an increase of both the silane pressure and the RF power. However, increases of pressure and RF power can lead to gas-phase polymerization reactions that result in powder formation and deterioration of film properties [144, 145]. At fixed RF power, an increase of the pressure induces a sudden increase in deposition rate together with a modification of the RF discharge parameters. These observations reflect a transition between two regimes (α and γ) of the silane RF discharge [146]. The high-pressure γ regime is characterized by a significant deterioration of a-Si : H quality as compared to the low-pressure α regime [146]. The effects of pressure and RF power have systematically been investigated by SPME [147]. The main results are presented here.

Figure 37 shows the variations of the deposition rate of a-Si : H as a function of the pressure at different values of RF power. Around 20 Pa the deposition rate increases suddenly until reaching a maximum and then slightly decreases. This enhancement of deposition rate becomes stronger as the RF power increases and is correlated with a decrease of both RF voltage amplitude and the DC self-bias voltage [148]. The trends shown in Fig. 37 reveal the α–γ discharge transition. The variations of $\varepsilon_{2\max}$ (near 3.5 eV) with pressure and RF power are displayed in Figs. 38 and 39. At low power (5 W), $\varepsilon_{2\max}$ is found independent of the pressure. In contrast at higher power, the increase of the pressure above 20 Pa is correlated with a decrease of $\varepsilon_{2\max}$. At high pressure, a sharp reduction of $\varepsilon_{2\max}$ is observed around 20 W (Fig. 39), $\varepsilon_{2\max}$ when it becomes independent of the RF power. As

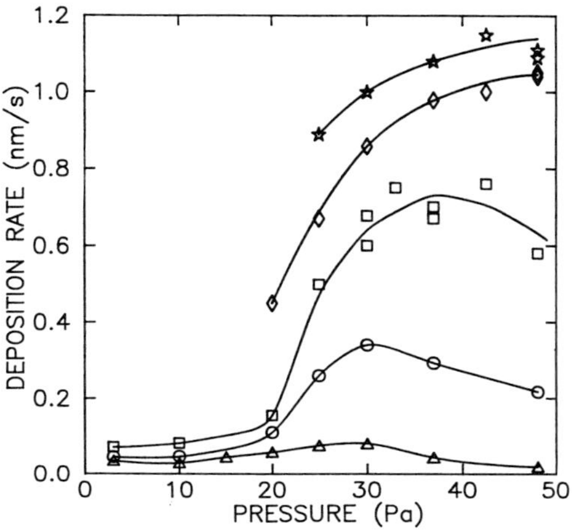

FIGURE 37. Variations of deposition rate of a-Si:H films as functions of pressure at different RF powers: 5 W (triangles), 10 W (circles), 20 W (squares), 50 W (diamonds), and 80 W (stars). (After Andùjar *et al.* [147].)

already discussed, a decrease of $\varepsilon_{2\max}$ can reveal a decrease of film density or an increase of surface roughness. In the present study, the features displayed in Figs. 38 and 39 can be attributed to variations of film density [147]. More pre-

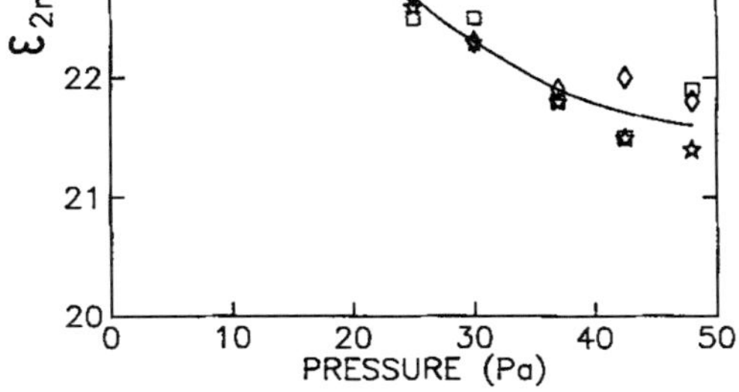

FIGURE 38. Pressure dependence of the maximum value of ε_2 for different RF powers: 5 W (triangles), 10 W (circles), 20 W (squares), 50 W (diamonds), and 80 W (stars). (After Andùjar *et al.* [147].)

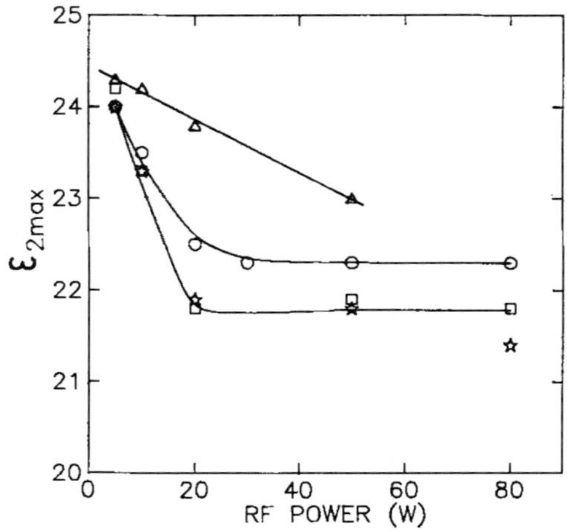

FIGURE 39. Variations of $\varepsilon_{2\mathrm{max}}$ as a function of RF power at different deposition pressures: 20 Pa (triangles), 30 Pa (circles), 37 Pa (squares) and 48 W (stars). (After Andùjar *et al.* [147].)

cisely, the increase of the deposition rate from 0.5 up to 11 Å s^{-1} is related to a 5.5% relative decrease of the film density. Some of the trends shown in Figs. 38 and 39 are interpreted in terms of influence of ion bombardment [147]. However, it has to be noticed that in relation to ion bombardment, the influences of RF power and pressure are opposed [149]. The increase of plasma potential by raising RF power at constant pressure leads to an enhancement of ion bombardment, whereas the increase of pressure at constant RF power leads to the reverse influence because of the increase of collisions within the plasma. The a-Si:H densification due to ion bombardment is revealed by the variations of $\varepsilon_{2\mathrm{max}}$ displayed in Figs. 38 and 39. In particular, in the high-pressure regime (weak ion bombardment), the increase of deposition rate by raising the RF power from 20 to 80 W (Fig. 37) is not correlated with a reduction of $\varepsilon_{2\mathrm{max}}$ (Fig. 39). Moreover, at constant RF power (Fig. 38), the decrease of $\varepsilon_{2\mathrm{max}}$ with pressure can be attributed to the moderation of ion bombardment.

F. Conclusion

Applications of *in situ* SPME, from UV to IR, to study the growth mechanisms of plasma deposited a-Si:H have been reviewed. In the UV–visible range, real-time SPME appears extremely powerful in order to provide information on nucleation and microstructural evolution. Its monolayer capability, together with its sensitivity to film density, allows precise investigations of the preparation conditions on a-Si : H morphology. For photoelectronic-quality a-Si : H (RF discharge) deposited on smooth substrates, a nucleation phase and the presence of a surface roughness at the 10–20 Å scale are evidenced. The coalescence of the initial nuclei under strong ion bombardment leads to the presence of a smooth a-Si : H surface. In the IR, the submonolayer sensitivity of SPME has been illustrated by a precise description of the hydrogen incorporation in a-Si : H. New insights on the constitution and the properties of the hydrogen-rich overlayer are given. In particular, it is shown that the weak chemical reactivity of the a-Si : H surface, as compared to most bare crystals, can be probably attributed to the presence of this hydrogen-rich overlayer. More generally, IRPME appears a promising technique for investigation of plasma–surface interactions.

As a consequence, it can be expected that SPME techniques will be extensively used in the near future in process monitoring and control for preparation of thin-film devices, as well as probing the fundamental growth mechanisms.

Acknowledgments

One of the authors (B. D.) would like to thank A. M. Antoine, N. Blayo, A. Canillas, J. Huc, and P. Roca i Cabarrocas for their technical assistance.

Note added in proof. Recent developments on mass spectrometry detection of ions and radicals in RF silane plamsa can be found in ref (150) and (151).

References

1. J. Perrin and E. Delafosse, *J. Phys. D* **13,** 759 (1980).
2. F. J. Kampas and R. W. Griffith, *J. Appl. Phys.* **52,** 1285 (1981); F. J. Kampas and R. R. Corderman, *J. Non-Cryst. Solids* **59/60,** 683 (1983).
3. A. Matsuda, K. Nakagawa, K. Tanaka, M. Matsumura, S. Yamasaki, H. Okushi, and S. Limiza, *J. Non-Cryst. Solids* **35/36,** 183 (1980); T. Hamasaki, M. Hirose, and Y. Osaka. *J. Phys. (Paris), Colloq.* **42,** C4-807 (1981).
4. S. Ishihara, M. Kitagawa, T. Hirao, and K. Wasa, *J. Appl. Phys.* **62,** 485 (1987).
5. D. Mataras, S. Cavadias, and D. Rapakoulias, *J. Appl. Phys.* **66,** 119 (1989).
6. G. de Rosny, E. R. Mosburg, Jr., J. R. Abelson, G. Devaud, and R. C. Kerns, *J. Appl. Phys.* **54,** 2272 (1983).
7. F. Tochikubo, A. Suzuki, S. Kakuta, Y. Terazono, and T. Makabe, *J. Appl. Phys.* **68,** 5532 (1990).
8. J. P. M. Schmitt, P. Gressier, M. Krishnan, G. de Rosny, and J. Perrin. *Chem. Phys.* **84,** 281 (1984).
9. R. M. Roth, K. G. Spears, and G. Wong, *Appl. Phys. Lett.* **45,** 28 (1984).
10. Y. Takubo, Y. Takasugi, and M. Yamamoto, *J. Appl. Phys.* **64,** 1050 (1988).
11. Y. Matsumi, T. Hayashi, H. Yoshikawa, and S. Komiya, *J. Vac. Sci. Technol. A* **4,** 1786 (1986).

12. G. Inoue and M. Suzuki, *Chem. Phys. Lett.* **105,** 641 (1984).
13. N. Itabashi, K. Kato, N. Nishiwaki, T. Goto, C. Yamada, and E. Hirota, *Jpn. J. Appl. Phys.* **27,** L1565 (1988); N. Itabashi, K. Kato, N. Nishiwaki, T. Goto, C. Yamada, and E. Hirota, *Jpn. J. Appl. Phys.* **28,** L325 (1989); N. Itabashi, N. Nishiwaki, M. Magane, S. Naito, T. Goto, A. Matsuda, C. Yamada, and E. Hirota, *Jpn. J. Appl. Phys.* **29,** L505 (1990).
14. N. Hata and K. Tanaka, *J. Non-Cryst. Solids* **77/78,** 777 (1985); N. Hata, A. Matsuda, and K. Tanaka, *Jpn. J. Appl. Phys.* **25,** 108 (1986).
15. N. Hata, A. Matsuda, and K. Tanaka, *J. Non-Cryst. Solids* **59/60,** 667 (1983).
16. J. M. Jasinski, E. A. Whitaker, G. C. Bjorlund, R. W. Dreyfus, R. D. Estes, and R. E. Walkup, *Appl. Phys. Lett.* **44,** 1155 (1984).
17. J. C. Knights, J. P. M. Schmitt, J. Perrin, and G. Guelachvili, *J. Chem. Phys.* **76,** 3414 (1982).
18. M. Sakakibara, M. Hiramatsu, and T. Goto, *J. Appl. Phys.* **69,** 3467 (1991).
19. P. Chollet, G. Guelachvili, and M. Morillon, *Bull. Soc. Chim. Belges* **92,** 512 (1983); P. B. Davies, N. A. Isaacs, S. A. Johnson, and D. K. Russell, *J. Chem. Phys.* **83,** 2060 (1985).
20. A. Loret and L. Abouaf-Marquin, *Chem. Phys.* **107,** 139 (1986).
21. H. U. Lee and J. P. Deneufville, *Chem. Phys. Lett.* **99,** 394 (1983).
22. H. U. Lee and J. P. Deneufville, *J. Non-Cryst. Solids* **66,** 39 (1984).
23. G. Bruno, P. Capezzuto, and G. Cicala, *J. Appl. Phys.* **69,** 7256 (1991).
24. G. Bruno, P. Capezzuto, G. Cicala, P. Manodoro, and V. Tassielli, *IEEE Trans. Plasma Sci.* **PS-18,** 934 (1990).
25. K. Kajiyama, K. Saito, K. Usuda, S. S. Kano, and S. Maeda, *Appl. Phys. Part B* **B38,** 139 (1985).
26. Y. Asano, D. S. Baer, R. Hernberg, and R. K. Hanson, *Plasma Chem. Plasma Process.* **8,** 1 (1988).
27. J. Perrin, Thèse d'Etat, Univ. de Paris VII, 1983.
28. A. Gallagher, *J. Appl. Phys.* **63,** 2406 (1988).
29. R. M. Roth, *Mater. Res. Soc. Symp. Proc.* **98,** 209 (1987).
30. D. Mataras, S. Cavadias, and D. Rapakoulias, *J. Vac. Sci. Technol. A* **11,** 664 (1993).
31. P. Kounavis, E. Mytilineou, D. Mataras, S. Cavadias, and D. Rapakoulias, *Proc. Int. Conf. Phys. Semicond. 20th,* (E. M. Anastasakis and J. D. Joannopoulos, eds.), p. 2115 (1990).
32. D. Mataras, S. Cavadias, and D. Rapakoulias, *Mater. Res. Soc. Symp. Proc.* **165,** 55 (1990).
33. V. I. Kuznetsov, R. C. van Oort, and J. W. Metselaar, *J. Appl. Phys.* **165,** 55 (1990).
34. Y. Asano, D. S. Baer, and R. K. Hanson, *J. Non-Cryst. Solids* **94,** 5 (1987).
35. C. Bohm and J. Perrin, *J. Phys. D.* **24,** 865 (1991).
36. V. A. Godyak and A. S. Kanneh, *IEEE Trans. Plasma Sci.* **PS-14,** 112 (1986).
37. J. P. Boeuf and P. Belenguer, *J. Appl. Phys.* **71,** 4751 (1992).
38. T. Kokubo, F. Tochikubo, and T. Makabe, *Appl. Phys. Lett.* **56,** 818 (1990); F. Tochikubo, T. Kokubo, S. Kakuta, A. Suzuki, and T. Makabe, *J. Phys. D* **23,** 1184 (1990).
39. G. Turban, B. Grolleau, P. Launay, and P. Briaud, *Rev. Phys. Appl.* **20,** 609 (1985).
40. G. Turban and M. Rapeaux, *J. Electrochem. Soc.* **130,** 2231 (1983).
41. R. R. Corderman and P. E. Vanier, *J. Appl. Phys.* **54,** 3987 (1987).
42. R. Robertson, D. Hils, H. Chatham, and A. Gallagher, *Appl. Phys. Lett.* **43,** 544 (1983).
43. F. Okuyama, H. Hayashi, and H. Iwai, *Surf. Sci.* **246,** 408 (1991).
44. P. Potzinger and F. W. Lampe, *J. Phys. Chem.* **73,** 3912 (1969).
45. J. P. Morrison and J. C. Traeger, *Int. J. Mass Spectrom. Ion Phys.* **11,** 289 (1973).
46. H. Chatham, D. Hils, R. Robertson, and A. Gallagher, *J. Chem. Phys.* **81,** 1770 (1984).
47. G. Turban, Thèse d'Etat, Univ. de Nantes, 1981.
48. G. Turban, Y. Catherine, and B. Grolleau, *Plasma Chem. Plasma Process* **2,** 61 (1982).
49. G. Turban, Y. Catherine, and B. Grolleau, *Thin Solid Films* **67,** 309 (1980).
50. I. Haller, *Appl. Phys. Lett.* **37,** 282 (1980).

51. H. A. Weakliem, *Plasma Process. 3rd Symp.* (J. Dielman, R. G. Frieser, and G. S. Mathad, eds.), p. 14. Electrochem. Soc., Pennington, New Jersey, 1982.
52. A. Matsuda and K. Tanaka, *Thin Solid Films* **92,** 171 (1982).
53. I. Haller, *J. Vac. Sci. Technol., A* **1,** 1376 (1983).
54. H. A. Weakliem, *in* "Semiconductors and Semimetals, Vol. 21: Hydrogenated Amorphous Silicon," Part A (J. Pankove, ed.), Chap. 10. Academic Press, New York, 1984.
55. M. L. Mandich, W. D. Reents, Jr., and K. D. Kolenbrander, *J. Chem. Phys.* **92,** 437 (1990).
56. B. Drevilon, J. Huc, A. Lloret, G. de Rosny, and J. P. M. Schmitt, *Appl. Phys. Lett.* **37,** 646 (1980).
57. J. Perrin, A. Lloret, G. de Rosny, and J. P. M. Schmitt, *Int. J. Mass Spectrom. Ion Phys.* **57,** 249 (1984).
58. H. Chatham and A. Gallagher, *J. Appl. Phys.* **58,** 159 (1985).
59. H. A. Weakliem, R. D. Estes, and P. A. Longeway, *J. Vac. Sci. Technol, A* **5,** 29 (1987).
60. D. J. Koch and W. N. G. Hitchon, *J. Appl. Phys.* **70,** 4139 (1991).
61. J. C. Knights, R. A. Lujan, M. P. Rosenblum, R. A. Street, D. K. Bieglesen, and J. A. Reimer, *Appl. Phys. Lett.* **38,** 331 (1981).
62. H. Chatham, D. Hils, R. Robertson, and A. Gallagher, *J. Chem. Phys.* **79,** 1301 (1983).
63. E. R. Fisher and P. B. Armentrout, *J. Chem. Phys.* **93,** 4858 (1990).
64. G. Turban, Y. Catherine, and B. Grolleau, *Thin Solid Films* **77,** 287 (1981).
65. Y. Catherine, G. Turban, and B. Grolleau, *Thin Solid Films* **76,** 23 (1981).
66. Y. Catherine, G. Turban, and B. Grolleau, *Plasma Chem. Plasma Process.* **2,** 81 (1982).
67. T. M. H. Cheng, T. Y. Yu, and F. W. Lampe, *J. Phys. Chem.* **77,** 2587 (1973).
68. M. P. Delplancke, J. M. Powers, G. J. Vandentop, M. Salmeron, and G. A. Somerjai, *J. Vac. Sci. Technol., A* **9,** 450 (1991).
69. A. Lloret and G. de Rosny, *Int. J. Mass Spectrom. Ion Phys.* **62,** 89 (1984).
70. G. Turban, Y. Catherine, and B. Grolleau, *Thin Solid Films* **60,** 147 (1979).
71. R. Robertson and A. Gallagher, *J. Appl. Phys.* **59,** 3402 (1986).
72. G. Nolet, *J. Electrochem. Soc.* **122,** 1030 (1975).
73. J. J. Wagner and S. Veprek, *Plasma Chem. Plasma Process.* **2,** 95 (1982).
74. P. A. Longeway, H. A. Weakliem, and R. D. Estes, *J. Appl. Phys.* **57,** 5499 (1985).
75. S. Veprek and M. Heintze, *Plasma Chem. Plasma Process* **10,** 3 (1990).
76. M. Heintze and S. Veprek, *Appl. Phys. Lett.* **54,** 1320 (1989).
77. S. Veprek and N. G. J. Veprek-Heijman, *Appl. Phys. Lett.* **56,** 1766 (1990).
78. J. P. Conde, K. K. Chan, J. M. Blum, M. Arienzo, and J. J. Cuomo, *J. Appl. Phys.* **71,** 3981 (1992).
79. K. Ensslen and S. Veprek, *Plasma Chem. Plasma Process.* **7,** 139 (1987).
80. J. R. Doyle, D. A. Doughty, and A. Gallagher, *J. Appl. Phys.* **68,** 4375 (1990).
81. G. Turban, *Pure Appl. Chem.* **56,** 215 (1984).
82. J. Perrin, J. P. M. Schmitt, G. de Rosny, B. Drévillon, J. Huc, and A. Lloret, *Chem. Phys.* **73,** 383 (1982).
83. P. A. Longeway, R. D. Estes, and H. A. Weakliem, *J. Phys. Chem.* **88,** 73 (1984).
84. P. A. Longeway, H. A. Weakliem, and R. D. Estes, *J. Phys. Chem.* **88,** 3282 (1984).
85. J. M. Jasinski, *Mater. Res. Soc. Symp. Proc.* **165,** 45 (1990).
86. D. A. Doughty, J. R. Doyle, G. H. Lin, and A. Gallagher, *J. Appl. Phys.* **67,** 6220 (1990).
87. D. A. Doughty and A. Gallagher, *Phys. Rev.* **42,** 6166 (1990).
88. J. R. Doyle, D. A. Doughty, and A. Gallagher, *J. Appl. Phys.* **71,** 4771 (1992).
89. J. R. Doyle, D. A. Doughty, and A. Gallagher, *J. Appl. Phys.* 69, 4169 (1991).
90. J. R. Doyle, D. A. Doughty, and A. Gallagher, *J. Appl. Phys.* **71,** 4727 (1992).
91. G. H. Lin, J. R. Doyle, M. He, and A. Gallagher, *J. Appl. Phys.* **64,** 188 (1988).
92. S. Veprek, F. A. Sarrott, S. Rambert, and E. Taglauer, *J. Vac. Sci. Technol. A* **7,** 2614 (1989).
93. J. Andreu, G. Sardin, A. Lloret, J. Estève, and J. L. Morenza, *J. Appl. Phys.* **63,** 1230 (1988).

94. E. R. Mosburg, R. C. Kerns, and J. R. Abelson, *J. Appl. Phys.* **54,** 4916 (1983).
95. G. Bruno, P. Capezzuto, G. Cicala, P. Manodoro, and V. Tassielli, *IEEE Trans. Plasma Sci.* **PS-18,** 934 (1990).
96. G. Cicala, G. Bruno, P. Capezzuto, P. Manodoro, and V. Pische, *Proc. Int. Symp. Plasma Chem. 9th (ISPC–9) (IUPAC), Pugnochiuso, Italy* (R. d'Agostino, ed.), Vol. 3, p. 1405 (1989).
97. J. C. Knights, *in* "The Physics of Hydrogenated Amorphous Silicon" (J. D. Joannopoulos and G. Lucovsky, eds.), p. 6. Springer-Verlag, Berlin, 1984.
98. B. Drévillon, *Thin Solid Films* **163,** 157 (1988).
99. R. W. Collins, *in* "Advances in Disordered Semiconductors" (H. Fritzsche, ed.), Vol. 1, p. 1003. World Sci. Publ. Co., Teaneck, New Jersey, 1989.
100. B. Drévillon, *J. Non-Cryst. Solids* **114,** 139 (1989).
101. R. A. Street and M. J. Thompson, *Appl. Phys. Lett.* **45,** 769 (1984).
102. K. Hiranaka, T. Toshimura, and T. Yamaguchi, *Jpn. J. Appl. Phys.* **28,** 2197 (1989).
103. M. Stchakovsky, B. Drévillon, and P. Roca i Cabarrocas, *J. Appl. Phys.* **70,** 2132 (1991).
104. A. M. Antoine and B. Drévillon, *J. Appl. Phys.* **63,** 360 (1988).
105. S. Kumar and B. Drévillon, *J. Appl. Phys.* **65,** 3023 (1989).
106. B. Drévillon, S. Kumar, P. Roca i Cabarrocas, and J. M. Siéfert, *Appl. Phys. Lett.* **54,** 2088 (1988).
107. P. Roca i Cabarrocas, M. Stchakovsky, B. Drévillon, F. Fortuna, and H. Bernas, *J. Non-Cryst. Solids* **137/138,** 1055 (1991).
108. R. M. A. Azzam and N. M. Bashara, "Ellipsometry and Polarized Light." North-Holland Publ., Amsterdam, 1977.
109. B. Drévillon, J. Perrin, R. Marbot, A. Violet, and J. L. Dalby, *Rev. Sci. Instrum.* **53,** 969 (1982).
110. B. Drévillon, J. Y. Parey, M. Stchakovsky, R. Benferhat, Y. Josserand, and B. Schlayen, *SPIE Symp. Proc.* **1188,** 174 (1990).
111. N. Blayo and B. Drévillon, *Appl. Phys. Lett.* **59,** 950 (1991).
112. N. Blayo, B. Drévillon, and R. Ossikovski, *SPIE Symp. Proc.* **1681,** 116 (1992).
113. F. Abéles, *in* "Physics of Thin Films" (G. Hass and R. E. Thun, eds.), Vol. 6, p. 151. Academic Press, New York, 1971.
114. R. W. Stobie, B. Rao, and M. J. Dignam, *Surf. Sci.* **56,** 334 (1976).
115. B. Drévillon and R. Benferhat, *J. Appl. Phys.* **63,** 5088 (1988).
116. M. Cardona, *Phys. Status Solidi B* **118,** 463 (1983).
117. R. W. Collins, *Rev. Sci. Instrum.* **61,** 2029 (1990).
118. B. E. Hayden, W. Wyrobisch, W. Oppermann, S. Hachicha, P. Hofman, and A. M. Bradshaw, *Surf. Sci.* **109,** 207 (1981).
119. A. Röseler, *Infrared Phys.* **21,** 349 (1981).
120. S. N. Jasperson and S. E. Schnatterly, *Rev. Sci. Instrum.* **40,** 761 (1969).
121. O. Acher, E. Bigan, and B. Drévillon, *Rev. Sci. Instrum.* **60,** 65 (1989).
122. A. T. M. Wilbers, G. M. W. Kroesen, C. J. Timmermans, and D. C. Schram, *Meas. Sci. Technol.* **1,** 1326 (1990).
123. D. E. Aspnes, *Thin Solid Films* **89,** 249 (1982).
124. J. C. M. Garnett, *Philos. Trans. R. Soc. London, Ser. A* **205,** 237 (1906).
125. D. A. G. Bruggeman, *Ann. Phys. (Leipzig)* **24,** 636 (1935).
126. L. Ley, J. Reichardt, and R. L. Johnson, *Proc. Int. Conf. Phys. Semicond, 17th* (J. D. Chadi and W. A. Harrison, eds.), p. 811. Springer-Verlag, New York, 1985.
127. G. H. Lin, J. R. Doyle, M. He, and A. Gallagher, *J. Appl. Phys.* **64,** 188 (1988).
128. N. Blayo and B. Drévillon, *J. Non-Cryst. Solids* **137/138,** 771 (1991).
129. A. M. Antoine, B. Drévillon, and P. Roca i Cabarrocas, *J. Appl. Phys.* **61,** 2501 (1987).
130. N. Maley, I. Szafranek, L. Mandrell, M. Katiyar, J. R. Abelson, and J. A. Thornton, *J. Non-Cryst. Solids* **114,** 163 (1989).

131. Y. Toyoshima, K. Arai, A. Matsuda, and K. Tanaka, *Appl. Phys. Lett.* **57,** 1028 (1990).
132. N. Blayo and B. Drévillon, *Surf. Sci.* **260,** 37 (1992).
133. A. Canillas, E. Bertran, J. L. Andùjar, and B. Drévillon, *J. Appl. Phys.* **68,** 2752 (1990).
134. R. W. Collins and B. Y. Yang, *J. Vac. Sci. Technol., B* **7,** 1155 (1989).
135. R. W. Collins and J. M. Cavese, *J. Non-Cryst. Solids* **97/98,** 269 (1987).
136. R. W. Collins and J. M. Cavese, *J. Appl. Phys.* **61,** 1662 (1987).
137. A. M. Antoine, B. Drévillon, and P. Roca i Cabarrocas, *J. Non-Cryst. Solids* **77/78,** 769 (1985).
138. E. Bertran, J. L. Andùjar, A. Canillas, C. Roch, J. Serra, and G. Sardin, *Thin Solid Films* **205,** 140 (1991).
139. B. Drévillon, J. Perrin, J. M. Siéfert, J. Huc, A. Lloret, and J. P. M. Schmitt, *Appl. Phys. Lett.* **42,** 801 (1983).
140. B. Drévillon, *Thin Solid Films* **130,** 165 (1985).
141. A. M. Antoine and B. Drévillon, *Mater. Res. Soc. Symp. Proc.* **75,** 333 (1987).
142. B. Drévillon, J. Huc, A. Lloret, J. Perrin, G. de Rosny, and J. P. M. Schmitt, *Appl. Phys. Lett.* **37,** 646 (1980).
143. J. Perrin, *J. Non-Cryst. Solids* **137/138,** 639 (1991).
144. M. H. Brodsky, *Thin Solid Films* **10,** L23 (1987).
145. R. C. Ross and J. Jaklik, *J. Appl. Phys.* **55,** 3785 (1984).
146. J. Perrin, P. Roca i Cabarrocas, B. Allain, and J. M. Freidt, *Jpn. J. Appl. Phys.* **27,** 2041 (1988).
147. J. L. Andùjar, E. Bertran, A. Canillas, C. Roch, and J. L. Morenza, *J. Vac. Sci. Technol., A* **9,** 2216 (1991).
148. J. L. Andùjar, A. Canillas, E. Bertran, and J. L. Morenza, *Proc. Int. Symp. Plasma Chem. 9th (ISPC–9) (IUPAC), Pugnochiuso, Italy* (R. d'Agostino, ed.), **3,** 1323 (1989).
149. P. Roca i Cabarrocas, *Mater. Res. Soc. Symp. Proc.* **149,** 33 (1989).
150. A. A. Howling, L. Sansonnens, J. L. Dorier, Ch. Hollenstein, *J. Appl. Phys.* **75,** 1340 (1994).
151. P. Kae-Nune, J. Perrin, J. Guillon, J. Jolly, *Jpn. J. Appl. Phys.* **33,** 4303 (1994).

3 Deposition Conditions and the Optoelectronic Properties of a-Si:H Alloys

C. M. Fortmann
Electrical Engineering Department
Pennsylvania State University
University Park, Pennsylvania

I. Introduction

The industry based on amorphous silicon devices has grown into a multi-billion-dollar-per-year business. As in any other quickly changing technology, the improvements in device performance are only partially understood. The descriptions of the underlying physics of amorphous materials are not yet complete. Adding to the difficulty in understanding these materials and devices is the fact that many critical parameters change as function of time when voltage or illumination stresses are applied [1, 2]. This is relevant because most commercial viable amorphous based devices employ one of these stresses. In order to fully exploit amorphous-silicon-based materials for device applications, it is necessary to gain a greater understanding of the properties of amorphous materials and how these properties relate to the deposition process used to create them.

In this chapter the relationship between deposition conditions and the measured material parameters is discussed. Also, where possible, the stress-induced changes in material parameters will be related back to the initial deposition process. In those areas where a consistent picture is beginning to emerge we will describe the underlying principals that govern these correlations. For example, at this juncture we can begin to make connections between deposition conditions (and alloy species) with the carrier mobilities and lifetimes. We can also partially connect these parameters to the time-dependent and saturated characteristics as a result of light and voltage stresses.

The spectrum of material parameters includes those that are intrinsic to the amorphous materials as well as those that are controlled; for example, by deposition parameters (such as deposition technique, deposition temperature, and gas mixtures). The categorization of parameters and the range of parameters possible is at present far from complete. Although the composition can be relatively well determined, there can be difficulties with reproducibility in some cases because of the critical manner in which the growth species interact to form these alloy materials. Moreover, all these materials must be considered an alloy of hydrogen as well (i.e., amorphous silicon is an alloy of silicon and *hydrogen;* amorphous silicon–germanium is an alloy of silicon, germanium, and *hydrogen*).

In this chapter a review of the techniques used to extract an electrical or optical property from a measurement will be provided where necessary to provide the sought-after link between deposition conditions and material parameters. The industrial interest is always focused on those defects that ultimately limit the performance of the device. For this reason device analysis is a powerful tool for quantifying the relevant defects. However, the analysis of devices can often be complicated by nonbulk (device-design-related) defects, for example, interfaces and surfaces. Therefore, device analysis cannot completely eliminate the need for other material characterization techniques. In addition, a number of nondevice measurements require less complex depositions and/or measurements that are easier to perform.

It is important to recognize that although the development presented here is based on the familiar crystalline concepts such as optical band gaps, electron diffusion coefficient, and mobility, these concepts are not fully justified in amorphous materials. These concepts will be modified when it is apparent that an alternate description is necessary.

II. General Comments on Amorphous Alloy Growth

In this section we will consider how the deposition parameters determine the macroscopic material properties of a-Si-based films and solar cells. Deposition parameters include the growth temperatures, feed gases, reactor power, and total reactor pressure. Macroscopic material properties include the optical band gap, and the microstructure (microvoid fraction and hydrogen-bonding-related IR absorption), film, and solar cell stability.

A. Growth and Hydrogen-Related Microstructure in a-Si:H

Details of reactor design and deposition are considered in Chapter 4. In this chapter the relation between identifiable structure resulting from a given fabrication condition and the optical and electrical properties of the materials is sought. It is necessary to focus on several key aspects of the relationship between deposition processes and the structure and defects that control the optoelectronic properties of the amorphous materials. For this purpose it is interesting to consider the underpinnings of the kinetic theory of crystal growth. Classical kinetic theory is based on the concept that the growth rate and defect generation rates can be determined by considerations related to small and definable departures from thermodynamic equilibrium during the growth process. The thermodynamic equilibrium not only generates the structure but also defines other important parameters, including the densities of dislocations and point defects. These point defects in turn are related to the optoelectronic properties of the crystal.

It is clear from the work presented in Chapter 4 that the growth of a-Si-based materials cannot be cast in terms of thermodynamic equilibrium. Rather, a kinetic-limited growth description is applicable. Kinetic-limited growth takes into account the fact that a number of possible reactions are occurring simultaneously. The rate at which the various reactions occurs ultimately determines the composition and structure of the materials. Even though amorphous materials themselves are not equilibrium materials in some circumstances, it is useful to describe the defects in amorphous silicon in terms of the relative thermodynamic free energies of the defected and nondefected states.

Descriptive engineering models have been developed that describe the simpler photo-CVD process [3] of amorphous silicon–alloy deposition. For the purpose of building a useful intuitive concept of how amorphous alloy materials obtain their composition and structure it is worth reviewing some of these findings. In the photo-CVD process silicon and hydrogen radicals are produced by the transfer of energy from an excited mercury atom (having absorbed a UV photon) to a silane molecule through a vapor-phase collision. The energy transfer causes the silane molecule to split into a SiH_3 and a H radical. If the pressure in the reactor is low enough, it is unlikely that the SiH_3 radical will undergo any further reactions until it encounters the growth surface. At the growth surface the SiH_3 radicals undergo further deposition reactions whereby much of the hydrogen is evolved.

At high reactor pressures (>1 torr) the probability of gas-phase reactions increases substantially. In cases where no hydrogen dilution is used, gas-phase reactions include collisions between two or more silane radicals to form more complex silicon polymeric molecules. The formation of polymeric species is expected to scale with the density of SiH_3 radicals, which increase with increasing reactor pressure as shown in Fig. 1. At the higher deposition pressures without dilution gases the amorphous materials typically have higher hydrogen content which is bonded in polymeric (SiH_2 and $[SiH_2]_n$) configurations as seen in Fig. 2. It is interesting to note that these films were grown at the same rate and temperature.

In standard PECVD higher growth rates can be used as an indication of increased SiH_3 radical densities. Figure 3 shows the increase in higher-order silanes present in the effluent as a function of growth rate [4]. Analysis of the IR absorption of these films shows that substantial amounts of the extra hydrogen is incorporated in polymeric bonding configurations, $(SiH_2)_n$. Both the PECVD and the photo-CVD results suggest that once formed, the complex vapor molecules do not

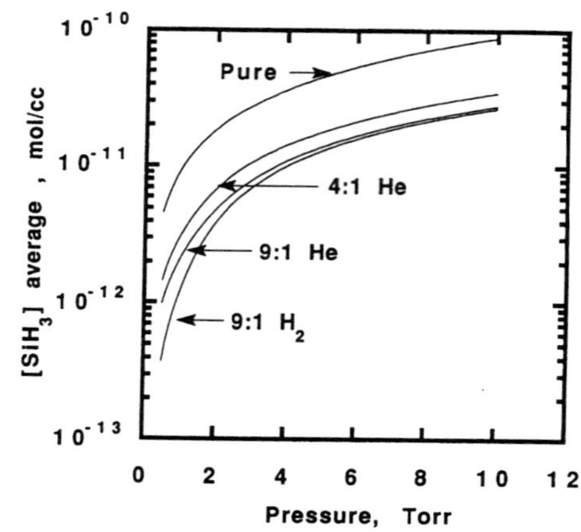

FIGURE 1. SiH_3 radical density as a function of the photo-CVD total reactor pressure.

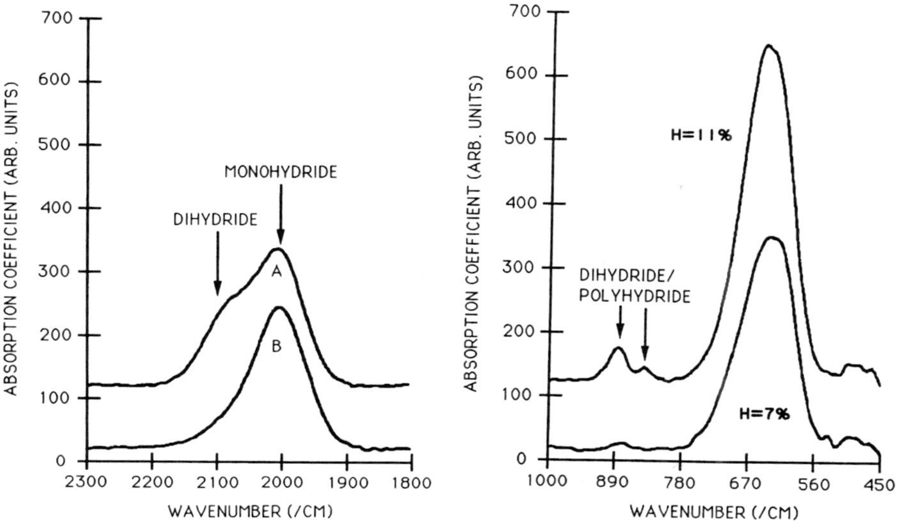

FIGURE 2. The IR absorption spectrum of two a-Si:H samples grown at two different photo-CVD total reactor pressures.

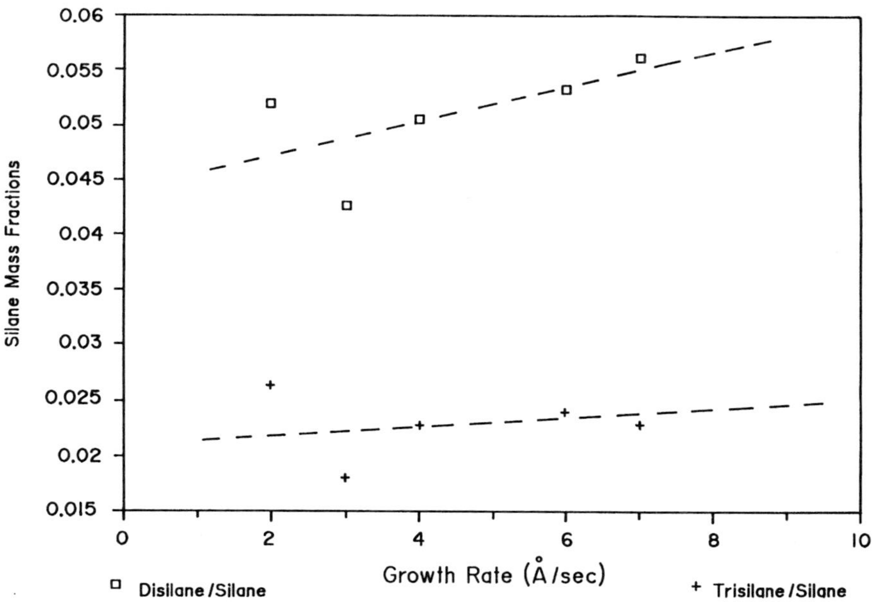

FIGURE 3. Disilane and trisilane concentrations in the PECVD reactor affluent as a function of growth rate (or RF power level).

reconfigure before being incorporated during growth. In extreme cases the gas-phase reactions produce macroscopic powders as described elsewhere.

When dilution gases are used, a complex set of side reactions can occur. For example, in a high-pressure reactor where gas collisions are likely, hydrogen dilution can accelerate the dissociation of silane molecules as the hydrogen radical (formed by a collision between a hydrogen atom and an excited mercury atom) can strip a hydrogen atom from the silane molecule, thereby producing a new hydrogen molecule and a silane radical. Although the hydrogen radical produces larger densities of silane radicals that promote gas-phase polymerization, hydrogen dilution suppresses these reactions as the hydrogen radical tends to rehydrogenate the polymeric species producing silane and silane radicals.

At low pressures where gas-phase collisions are less probable, it is more likely that the hydrogen radical will not undergo any reactions until it reaches the growth surface. At the growth surface there are several possible reactions involving the hydrogen radical. If the film is growing quickly, the radical could bond to a surface atom, and be covered by the arriving silicon containing species and, thereby, incorporated into the film. This extra hydrogen in the film can be observed as an increased hydrogen content in the film. When the arrival rate of silicon containing species is low compared to that of the hydrogen radicals, the surface silicon dangling bonds will become hydrogenated.

If the arrival rate of silicon containing species is low compared to that of the hydrogen radical, the surface silicon atoms can be completely saturated by hydrogen; the hydrogen then converts these silicon atoms back into silane molecules (vapor species). When hydrogen atoms remove surface silicon atoms, the film growth is slowed or reversed (etching). Alternatively, the hydrogen atoms could occupy a large number of surface sites, thereby altering the surface chemistry and enhancing the diffusion of silicon radicals on the growth surface [5]. (The surface diffusion process might involve the creation of silane molecules that evolve from the surface, but before diffusing away from the proximity of the growth surface, these molecules undergo a collision with an incoming hydrogen radical that again forms a new silane radical that has a new opportunity to deposit on the surface; these processes are repeated until an etch resistant, crystal-like, bonding position is obtained by the silicon atom.) The more rapid etching of poorly bonded silicon surface atoms (relative to atoms bonded in a crystal configuration) explains the formation of microcrystalline films by photo-CVD [6] deposition, and it appears necessary to include some form H-enhanced surface diffusion of silicon species to explain the PECVD deposition of microcrystalline films [5].

The effect of these hydrogen surface reactions hydrogen content of amorphous on the photo-CVD growth is summarized in Fig. 4. Under hydrogen dilution as the pressure is reduced, the increased hydrogen radical flux at the growth surface increases as shown in Fig. 5. At a pressure of approximately 2 torr the hydrogen

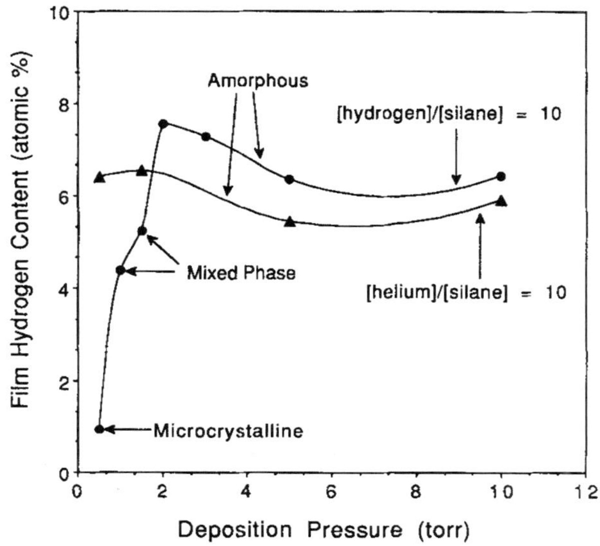

FIGURE 4. Hydrogen content of a-Si:H materials as a function of total photo-DVD reactor pressure and gas dilutions.

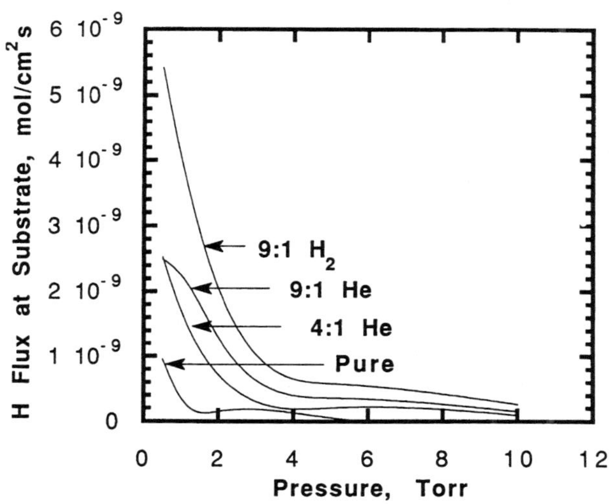

FIGURE 5. The hydrogen radical flux incident on the growth surface in photo-CVD as a function of total reactor pressure.

radical flux is large enough to contribute "extra hydrogen" to the amorphous silicon film. At even lower pressures the hydrogen radicals arrive fast enough to selectively etch away poorly bonded silicon atoms from the film through the formation of SiH_4 molecules, slowing the growth rate and causing the transition to microcrystalline structure.

B. Relationship between Growth Conditions and Alloy Microstructure

During alloying the role of hydrogen etching reactions become more complex. The influence of hydrogen selective etching on the composition of a-Si germanium alloys was studied by Albright [7, 8]. These etching reactions rates depend on the bonding configurations, atomic species, and growth temperature. The hydrogen etching rates of carbon, germanium, and silicon surface atoms are all different and seem to scale with the bond hydrogen energies. That is, carbon etches more easily than silicon, which measures more easily than germanium. These considerations explain the observed shift in alloy composition as function of dilution (when the ratios of silicon and alloy gas sources are held constant). Hydrogen radicals selectively etch silicon atoms (relative to germanium atoms), particularly at low temperatures, resulting in a larger C_{Gc} for a given flow rate of GeH_4 and SiH_4 as pressure and temperature are decreased.

The silicon–carbon alloys behave in a manner analogous to that of silicon germanium alloys. As the hydrogen dilution is increased at low pressures (where the probability of the hydrogen radical reaching the growth surface is large), the carbon concentration in the film is reduced as a result of the selective etching of the carbon atoms relative to the silicon [9, 10]. It must be stressed that because of the selective hydrogen radical etching the composition of materials in the alloy system is not determined solely by the CH_4 to SiH_4 feed-gas ratios but on the dilution species and partial pressure as well.

The hydrogen content, the hydrogen bonding configuration, and the related microvoid content of materials are strong functions of growth conditions. Figure 6 shows the increase in microvoid (determined by *in situ* ellipsometery [11]) content with decreasing deposition temperature [12] for PECVD-grown films. Also shown is the increase in microvoid fraction as the RF power (or growth) rate is increased. Dopant species are also seen to increase the microvoid fraction.

The effect of growth rate and temperature on the silicon–germanium–carbon alloys has a similar effect on the microvoid fraction and/or the degree of polymeric hydrogen bonding. Figure 7 shows the increase in IR absorption related to these structures as the growth rate is increased from 1 to 4 Å/s as well as the

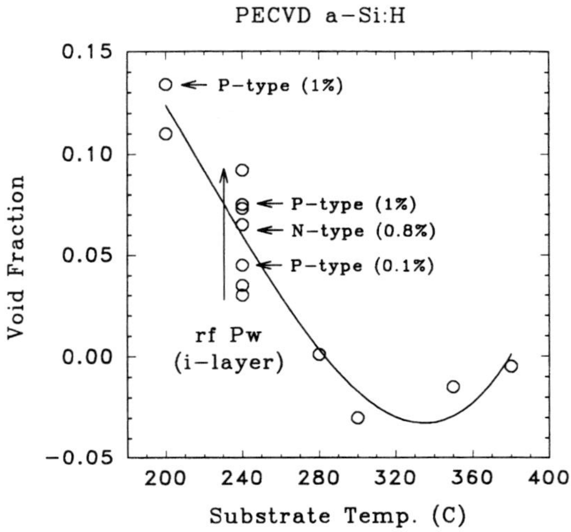

FIGURE 6. Microvoid fraction as a function of growth temperature, RF power, and doping (Y. M. Li and R. W. Collins, unpublished data).

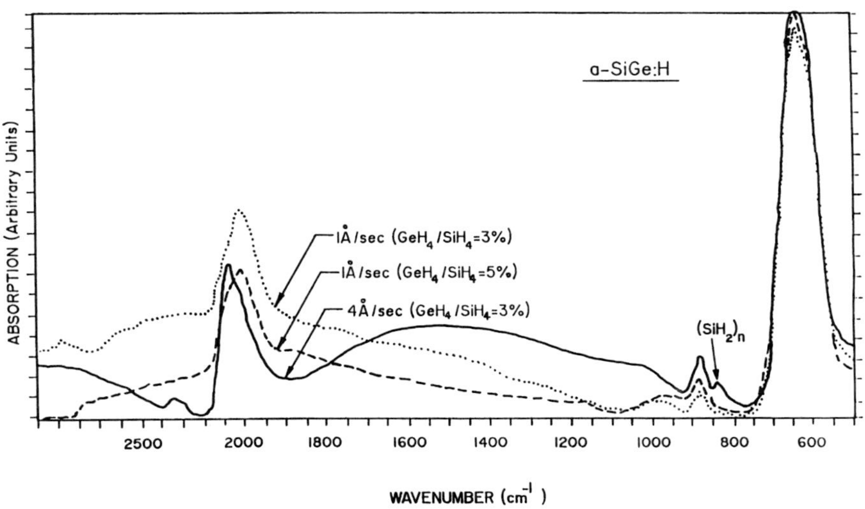

FIGURE 7. The IR absorption spectrum of a-SiGe:H alloys as a function of growth rate and feed-gas mixtures.

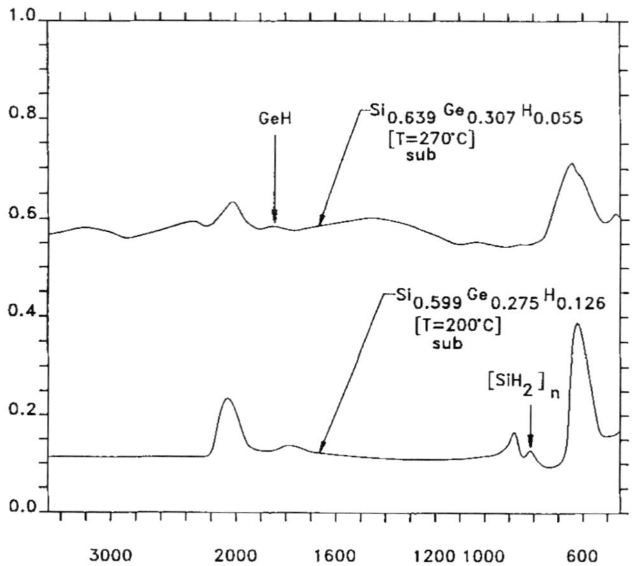

FIGURE 8. The IR absorption of a-SiGe:H samples grown at substrate temperatures of 200 and 270°C.

increase in GeH_4/SiH_4 ratio of the feed gases. Decreasing the growth temperature from 270 to 200°C can also be seen in Fig. 8 to increase the absorption of microvoid–related hydrogen-bonding configurations in the film.

Figure 9 shows the effect of growth rate on the microvoid-related hydrogen-bonding IR absorption in amorphous silicon–carbon alloys. Again increasing growth rate increases the microvoid fraction; however, it should be noted that the microvoid content appears to be much greater in the silicon carbon alloys. The observation that increasing the RF power (or growth rate increases the hydrogen content, as noted by other researchers as well as in ref. [13]) and microvoid density.

C. Relationship between Microstructure and a-Si:H Film Properties

Since T_s affects the structure and electronic properties of hydrogenated amorphous silicon (a-Si:H) whereas T_s increases, dark conductivity (σ_d) increases [14], the optical band gap shrinks [15], the hydrogen content decreases [16], and the void fraction decreases [17]. However, the quantitative correlation between

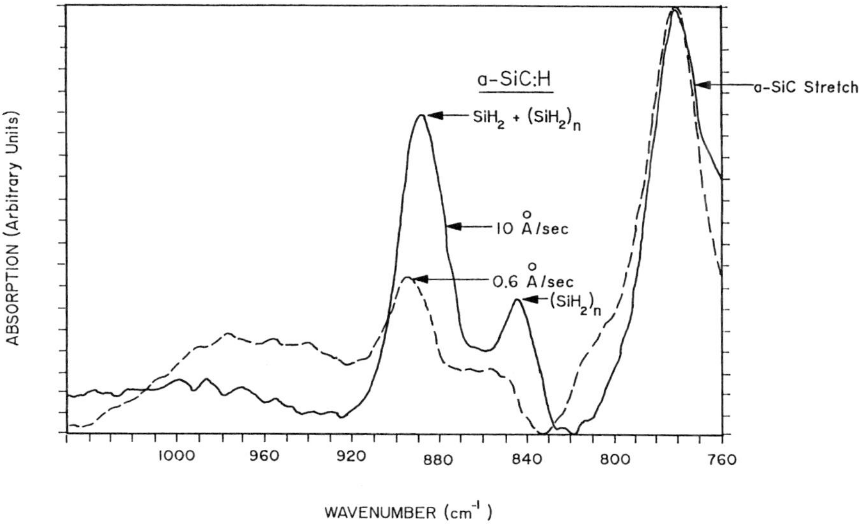

FIGURE 9. The IR absorption spectrum of a-SiC:H samples grown by PECVD at deposition rates of 0.6 and 10 Å per second.

the microstructural changes characterized by hydrogen content and void fraction with electronic transport mechanisms [18] is not complete. In a recent study [19] deposition temperature (T_s) was used to vary the hydrogen content and microstructure of a-Si films where the observed effects of microstructure on transport mechanisms in undoped a-Si:H deposited at T_s between 200° to 350°C were studied.

As T_s increased from 200 to 350°C, the dihydride content decreased from 9 to 2 atom. % the monohydride content decreased from 8 to 4 atomic %, and the void fraction decreased from 11 to -3% relative to a standard material [20]. All the films used in this study had low subgap (sub-band-gap) absorption corresponding to low densities of midgap states ($< 10^{16}$ cm^{-3}).

The dark (σ_d) and photo- (σ_{ph}) conductivities in the annealed state were measured as a function of temperature; σ_{ph} was measured as a function of generation rate (G) using volume absorbed monochromatic light whose intensity was varied with neutral-density filters. To maintain the same value of G with temperature, the corresponding changes in optical absorption with temperature [21] were taken into account; $G = 1 \times 10^{19}$ cm^{-3} s^{-1} was chosen in order to maintain a large signal to noise ratio.

The results of σ_d versus temperature for the different materials are shown in Fig. 10a, where—when measured at 40°C—the conductivities of the films vary

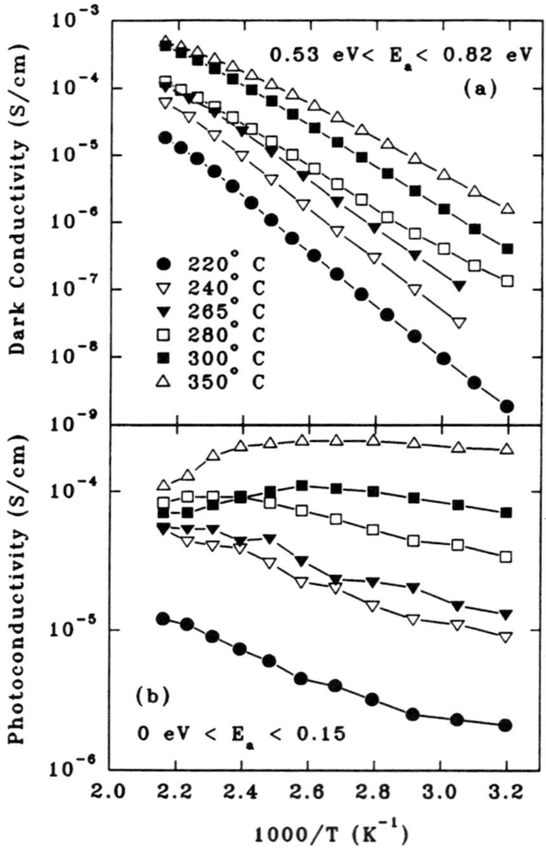

FIGURE 10. The dark conductivity (a) and photoconductivity (b) as a function of measurement temperature for a series of PECVD a-Si:H films grown at different substrate temperatures.

by three orders of magnitude and by 1.5 orders of magnitude at 190°C. In agreement with previously reported results, the activation energies extracted from the linear regions at measurement temperatures (T_m) < 115°C decrease from 0.82 to 0.53 eV as T_s increases from 220 to 350°C and the σ_d prefactors (σ_{0d}) decrease from 2×10^4 to 5×10^2 S/cm in accordance with the Meyer–Neldel rule [22]. This is true even though at higher T_m, the Arrhenius plots have a distinct curvature that becomes more pronounced as T_s increases.

The values of σ_{ph} for the materials with different T_s are shown in Fig. 10b. At 40°C, they vary by two orders of magnitude, but converge at higher temperatures so that distinctly different activation energies E_{ph} are obtained. The values of E_{ph}

change systematically with T_s where for the material with $T_s = 220°C$, with $E_{ph} \cong 0.15$ eV, and for $T_s = 350°C$, $E_{ph} \cong 0.0$ eV. It can also be seen that the Arrhenius plots of σ_{ph} display distinct curvature, similar to σ_d, where the curvature increases with T_s and T_m. It is important to note here that the materials with the highest σ_{ph} at 40°C have the lowest SiH_2 content and void fraction, but the highest subgap (sub-band-gap) absorption signal, which, at 1.2 eV, increases systematically with T_s from 0.3 to 1 cm^{-1}. If σ_{ph} were determined predominantly by lifetime τ, such increases in the midgap densities of states would lead to a lower τ and hence, lower values of σ_{ph}. Since the opposite is observed, the results suggest that σ_{ph} is not entirely controlled by midgap defects.

The values of σ_d/σ_{ph} versus $1000/T$ are shown in Fig. 11, where the curvature present in the Arrhenius plots of σ_d and σ_{ph} is now completely eliminated. The activation energies derived from the slopes of the σ_d/σ_{ph} ratios of Fig. 11 monotonically increase from 0.53 eV for $T_s = 350°C$ to 0.65 eV for $T_s = 220°C$. It is interesting to note that, unlike the prefactors of σ_d and σ_{ph}, the extrapolated prefactors of the σ_d/σ_{ph} ratio (σ_{0dph}) vary by only 40% and have no systematic dependence on T_s. This much smaller difference in the σ_{0dph} indicates that the factors that cause the large variations in σ_{0d} (2×10^4–5×10^2 S/cm) are not present in σ_{0dph}.

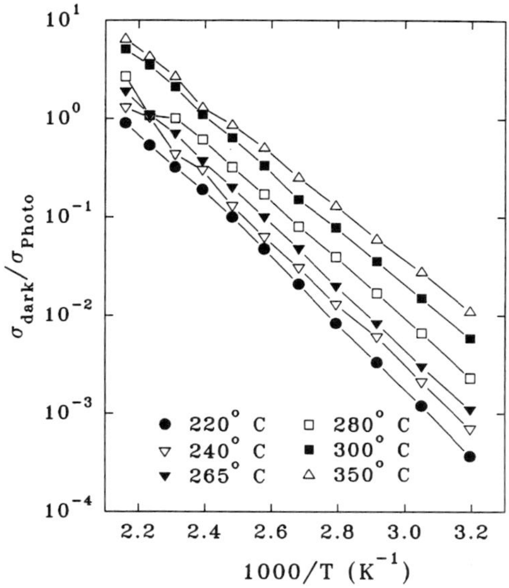

FIGURE 11. The dark/photoconductivity ratio as a function of measurement temperature for a series of films grown at different substrate temperatures.

These results can be used to better understand the carrier transport in these a-Si:H materials. Since intrinsic a-Si:H is slightly n-type because the electron mobility is greater than that for holes [23], these results can be discussed in terms of electrons only. The dark conductivity as a function of temperature is given by

$$\sigma_{\mathrm{d}}(T) = q\mu_n(T)N_{\mathrm{c}}(T)\exp\left(-\frac{E_{\mathrm{c}}(T) - E_{\mathrm{f}}(T)}{kT}\right) \tag{1}$$

where q is the electronic charge, $\mu_n(T)$ is the mobility, $N_{\mathrm{c}}(T)$ is the effective density of states at the conduction band edge, kT is the thermal voltage, and $E_{\mathrm{f}}(T)$ and $E_{\mathrm{c}}(T)$ are the Fermi level and the conduction band edge, respectively. The photoconductivity σ_{ph} can be expressed as

$$\sigma_{\mathrm{ph}}(T) = q\mu_n(T)\tau_n(T)G(T), \tag{2}$$

where $\tau_n(T)$ is the carrier lifetime and $G(T)$ is the carrier generation rate. Dividing σ_{d} by σ_{ph} yields

$$\frac{\sigma_{\mathrm{d}}(T)}{\sigma_{\mathrm{ph}}(T)} = \frac{N_{\mathrm{c}}(T)\exp\left(-\dfrac{E_{\mathrm{c}}(T) - E_{\mathrm{f}}(T)}{kT}\right)}{\tau_n(T)G(T)}. \tag{3}$$

In the following discussions the temperature dependence of N_{c} is neglected since a $T^{3/2}$ dependence results in only about a 50% change in σ_{d} and the $\sigma_{\mathrm{d}}/\sigma_{\mathrm{ph}}$ for the T_{m} range studied. As discussed earlier, $G(T)$ was kept constant in the experiments discussed here.

Since both $\sigma_{\mathrm{d}}(T)$ and the $\sigma_{\mathrm{d}}(T)/\sigma_{\mathrm{ph}}(T)$ ratio exhibit exponential behavior over the T_{m} range studied, it is convenient to express σ_{ph}, σ_{d}, $\tau(T)$ and $\mu(T)$ in an exponential form and use equations (1), (2), and (3), respectively to obtain the relationships between their activation energies. The measured activation energies for the dark conductivity (E_{dark}), photoconductivity (E_{photo}), and the dark/photoconductivity ratio ($E_{\mathrm{dark/photo}}$) are listed in Table 1 for the $T_{\mathrm{s}} = 220°\mathrm{C}$, 280°C and 350°C films. Self-consistency between the various activation energies of the different materials can be obtained with nonzero values of both the activation energies of μ, (E_{μ}); and τ, (E_{τ}). However, semiconductors with defect distributions such as those found in a-Si:H are not expected to have temperature-activated τ [24] and preliminary results of numerical calculations of σ_{ph}, based on a previously derived midgap density of states distribution (expected in annealed amorphous silicon films) and capture cross sections [25], indicate that τ is virtually temperature-independent over the T_{m} range studied; note that this may not be the case in degraded a-Si:H materials. Furthermore, the experimentally observed curvature in both σ_{d} and σ_{ph} cannot be explained in terms of τ since there is no curvature in the $\sigma_{\mathrm{d}}/\sigma_{\mathrm{ph}}$ ratios. This curvature can be explained only in terms of a temperature-dependent μ, so it is reasonable to assume that $E_\tau \cong 0$ and that μ is temperature-dependent with an activation energy E_μ as suggested previously [26, 27].

With this assumption, the following relationships are valid: $E_{photo} = E_\mu$, $E_{dark/photo} = E_c - E_f$, and $E_{dark} = (E_c - E_f) + E_\mu$. Consequently, μ can be determined using $E_c - E_f$ extracted from the Arrhénius plots of $\sigma_d(T)/\sigma_{ph}(T)$, the measured $\sigma_d(T)$, and equation (1). Note that the values obtained for $E_c - E_f$ (0.53–0.65 eV) are consistent with the n-type nature of the materials as previously assumed. The mobilities obtained using these values of $E_c - E_f$, the measured $\sigma_d(T)$ and $N_c = 1.9 \times 10^{20}$ cm^{-3} [25] are shown as a function of temperature in Fig. 12 and are in general agreement with earlier results [28]. At 40°C, the extended-state mobilities systematically increase with T_s from $\simeq$0.8 cm^2 V^{-1} s^{-1} (T_s = 220°C) to $\simeq$30 cm^2 V^{-1} s^{-1} (T_s = 350°C) as the SiH_2 content and void fractions decrease. For the low-microvoid material (T_s = 350°C), μ is in good agreement with the estimates from drift mobility measurements, 13 cm^2 V^{-1} s^{-1} [29]. The large drop in $\mu(T)$ found for the T_s = 220°C material, on the other hand, can be associated with the sharp increase in the SiH_2 content from about 5 to 10 atom. %. A further confirmation of the self-consistency of the μ calculations is indicated by the excellent agreement between the measured values of $E_{photo}(T)$

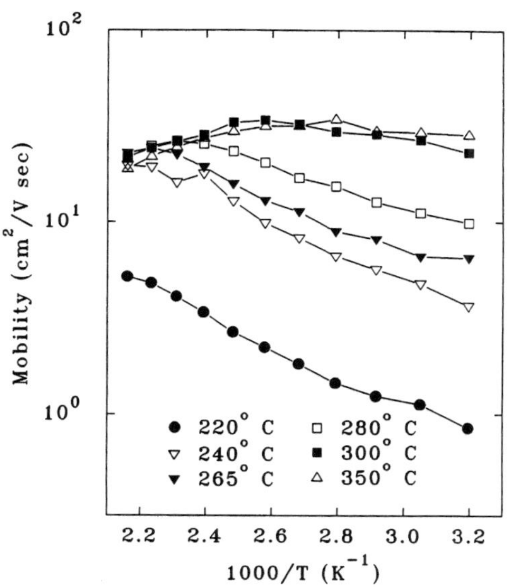

FIGURE 12. Mobility as a function of measurement temperature for a series of films grown at different substrate temperatures.

and $E_{\mu}(T)$, where for all the materials, and at each T_{m} interval, $E_{\mathrm{photo}}(T) \cong E_{\mu}(T)$ (Figs. 10b and 12). There is a corresponding decrease in the temperature dependence of these mobilities may be related to the changes in the microstructure and potential fluctuations [26]. Furthermore, the effects of microstructure on the mobility of a-Si:H reported here have important consequences on the carrier transport in the more disordered a-Si:H-based alloy materials. The thermally activated mobilities obtained in this work can explain the Meyer–Neldel rule in a-Si:H materials with high activation energies ($E_{\mathrm{a}} > 0.4$ eV), in which this rule cannot be explained by the statistical shift of the Fermi level [30].

D. Relationship between Growth and Alloy Film Properties

The possibly range of materials in the a-Si alloy system is broader as not only does the hydrogen content affect the material parameters but the alloy content also contributes. For example, it has been shown that a range of optoelectronic properties could be obtained for a-SiGe:H materials even when the band gap is held fixed [31]. However, several groups have developed a means to deposit relatively good a-SiGe:H using a variety of feedstocks and conditions [32]. The properties of the a-SiC:H and a-SiN:H alloys are much less well known. Both grading of the solar cell i-layer [33, 34] and the overall i-layer thickness [31] are considerations that enter into relatively efficient solar cell designed for use with a-SiGe:H. Grading has also been employed in solar cells using an a-SiC:H i-layer [35].

The optical band gap of amorphous silicon depends on C_{H} on the hydrogen content and therefore on the deposition conditions (including temperature and power). A detailed discussion of the relation between optical band gap and hydrogen content can be found in reference [36] where an empirical relationship between C_{H} and E_{g} is described by, $E_{\mathrm{g}} \sim 1.5 + 0.015$, where C_{H} is in atomic percent yielding E_{g} in electronvolts. Experimentally significant deviations (± 0.1 eV) from this relation are noted. Better correlations have been found by relating the E_{g} to the SiH_2 content of the amorphous materials [37]. Figure 13 shows the complex relationship between deposition temperature, the hydrogen content, and the band gap as a function of germanium content in a-SiGe alloys. The hydrogen content increases markedly at lower substrate temperatures for the alloys of intermediate germanium content. The band gap decreases with increasing germanium content (see Fig. 13), but the increase is germanium content dependent.

The change in hydrogen and germanium content, and in the related optical gap, of a-SiGe alloys grown by PECVD affects the electronic transport of the alloys as seen in Fig. 14, where the electron mobility–lifetime product ($\mu_{\mathrm{c}}\tau_{\mathrm{c}}$) derived from photoconductivity ($\Delta\sigma_{\mathrm{p}} = \sigma_{\mathrm{p}} - \sigma_{\mathrm{d}} = \mathrm{e}\mu_{\mathrm{c}}G\tau_{\mathrm{c}}$) measurements is shown. The

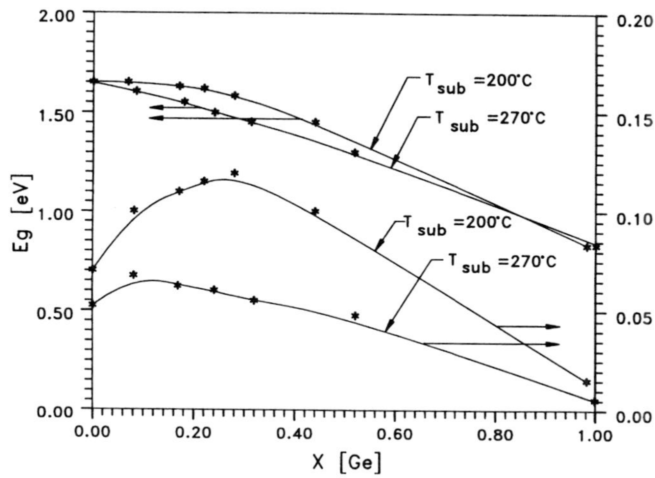

FIGURE 13. Effect of microstructure composition on the optoelectronic properties of amorphous silicon–germanium alloys.

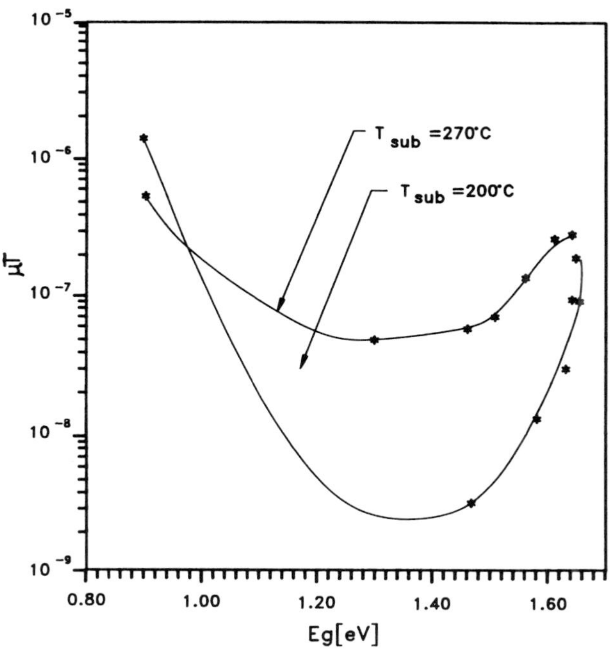

FIGURE 14. The $\mu_c \tau_c$ of a-SiGe:H alloys as a function of energy gap and substrate temperature.

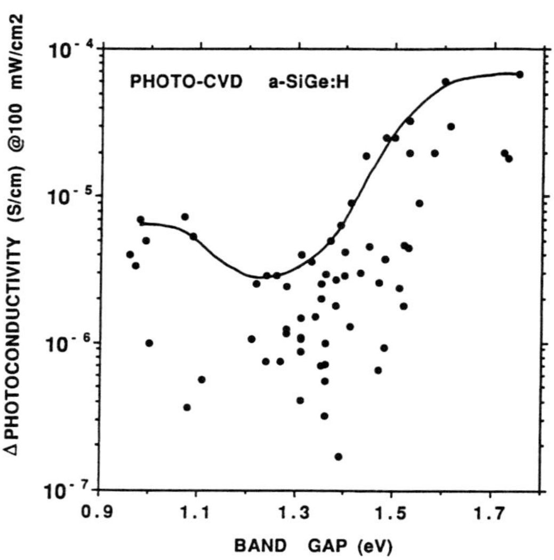

FIGURE 15. The $\Delta\sigma_p$ of a-SiGe:H films as a function of band gap.

lowest $\mu_c\tau_c$ is seen to occur in the intermediate alloys, with $\mu_c\tau_c$ becoming more markedly reduced with increasing hydrogen content (or lower substrate temperature). The effect of band gap on the photoconductivity of a large number of films grown by photo-CVD [38] is shown in Fig. 15.

All films in this photo-CVD study had activation energies for the dark conductivity within a few kT of $E_g/2$ at room temperature. In general the largest $\Delta\sigma_p$ for a given band gap decreases with decreasing band gap reaching a minimum at band gaps of ~1.3 eV as was seen also in Fig. 14. The $\Delta\sigma_p/\sigma_d$ ratios R for the same database, (Fig. 16) shows far less scatter. This is interesting because R, the inverse of equation (3), is independent of the electron mobility.

$$R = \Delta\sigma_p/\sigma_d = eG\tau\mu_c/(e\mu_c N_c \exp[-\{E_f + E_\mu\}/kT]) = G\tau_c/(N_c \exp[-\{E_f + E_\mu\}/kT]), \quad (4)$$

where G is the generation rate, μ_c is the effective electron mobility, τ_c is the effective electron lifetime, N_c is the density of states in the conduction band, and E_μ allows for the possibility of a thermally activated mobility ($\mu_c\tau_c = \mu_0\exp\{-E_\mu/kT\}$) as would be the case for conduction through localized states [39].

If $E_\mu = 0$, then E_f is equal to the E_a, and multiplying R by $\exp[-E_a/kT]$ yields $G\tau_c/N_c$ (Fig. 17); thus the band gap dependence that enters through E_a is removed. It is interesting to note that both a-Si:H and a-Ge:H has the same value of G_{tc}/N_c. The relatively poor $\Delta\sigma_p$ of a-Ge:H (less than 10% that of a-Si:H) together with a value of $G\tau/N_c$ that is equal to that of a-Si:H is consistent with

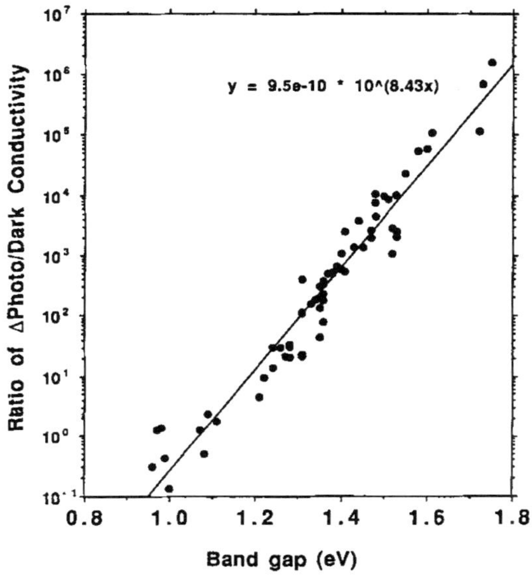

FIGURE 16. The $\Delta\sigma_p/\sigma_d$ ratio for the films of Fig. 1.

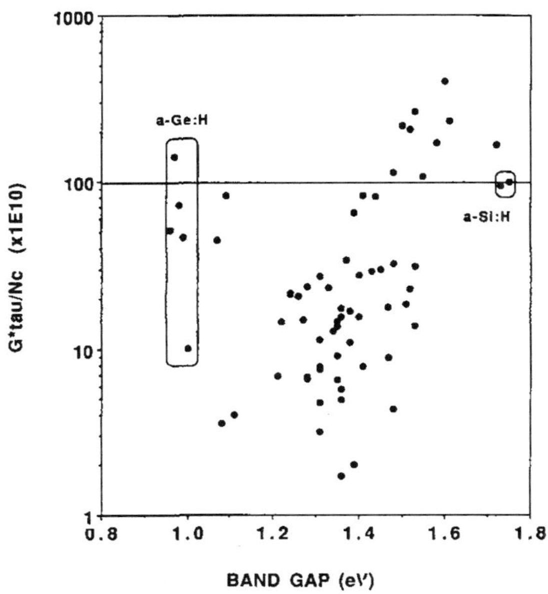

FIGURE 17. $G\tau_c/N_c$ vs. E_g for the films of Fig. 1.

μ_c of a-Ge:H being only 10% of that of a-Si:H. Over this range of alloys the minimum $G\tau_c/N_c$ occurs at the intermediate compositions ($1.1 < Eg < 1.4$ eV) but even here the change can be relatively small (~30% of the values for either a-Si:H or a-Ge:H).

The decrease in $G\tau_c/N_c$ for the intermediate-band-gap alloys can be due to either a reduction in τ_c for these alloys or a nonzero value for E_μ (i.e., the mobility of the intermediate alloys has a larger thermal activation energy than either a-Si:H or a-Ge:H). The temperature dependence of $\Delta\sigma_p$ is in fact greater for the intermediate alloys than that of either a-Si:H or a-Ge:H (Fig. 18), resulting in a larger activation energy. Although a greater activation energy of $\Delta\sigma_p$ could in principle be due to either a thermally activated lifetime or mobility, we have yet to successfully model it in terms of lifetime changes resulting from recombination or shallow trapping [equation (10)]. The conclusion that μ_c in a-SiGe:H alloys was reduced (and thermally activated) compared to that of a-Si:H as recognized by Karg *et al.* [40], which—as was the case described above—in a-Si:H materials is consistent with the Meyer–Neldel rule. Others, including Nebel *et al.* [41], also recognized that the electron mobility in the a-SiGe:H alloys was suppressed compared to that in a-Si:H materials.

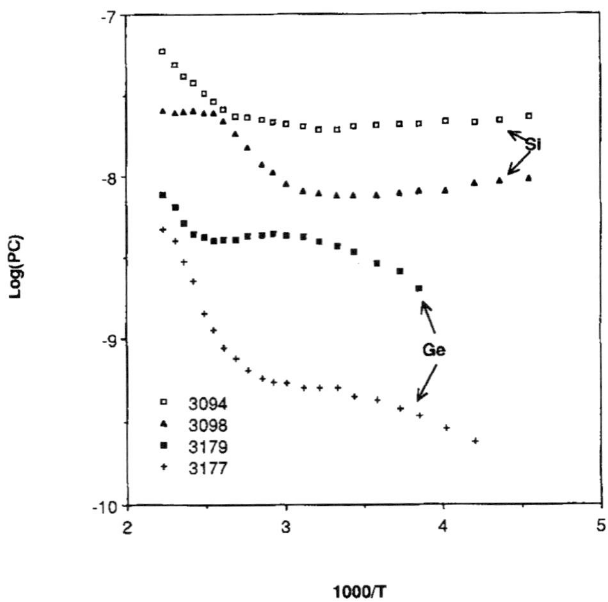

FIGURE 18. The excess of photoconductivity of a-Si:H, a-Ge:H, and a-SiGe:H films as a function of temperature.

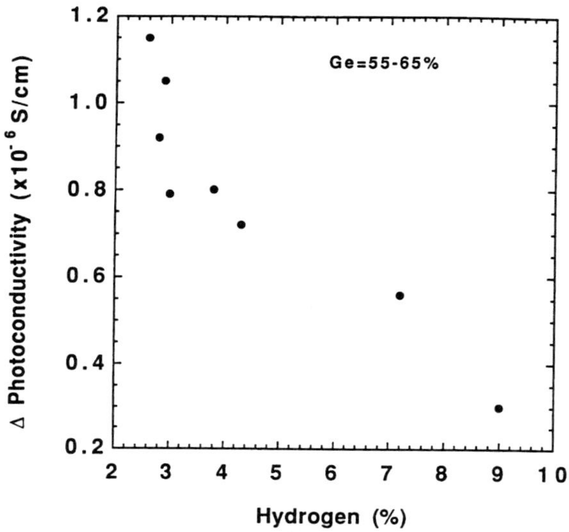

FIGURE 19. $\Delta\sigma_p$ vs. C_H for films containing ~60% Ge.

Figure 19 shows the photo conductivity of amorphous films containing ~60% germanium as a function of hydrogen content. It is important to note that for these set of films $\Delta\sigma_p$ decreases with increasing band gap (band gap increases with C_H) for a fixed Ge content (C_{Ge}) as seen in Fig. 20. For the films of Figs. 19 and 20 the optical band gap increases with hydrogen content as shown in Fig. 21. It is important to note that hydrogen content does not appear to reduce the electron lifetime in these same a-SiGe:H materials (Fig. 22). In the amorphous alloy system the hydrogen content has a detrimental effect on electronic mobility and, along with C_{Ge} content, provides a range of transport properties for a given band gap. Films grown under H radical etching (at levels not sufficient to result in detectable microcrystalline structure), films grown from polymeric species as well as those grown at lower temperatures are expected to have relatively large C_H and therefore poorer electronic transport.

The effect of hydrogen content and alloying on hole mobility is less well known because hole transport is not directly probed by conductivity measurements. However, it is unlikely that the hole mobility is reduced to the extent that the electron is. Bifacial illumination studies of solar cells concluded that the hole mobility could not be as reduced as that of the electron [38, 40]. Capacitance studies [42, 43] have revealed that the electron and hole mobilities have converged to the degree that the electron mobility is only a factor of 3 larger than that of the hole (compared to more than a factor of 10 for electrons and hole in a-Si:H materials).

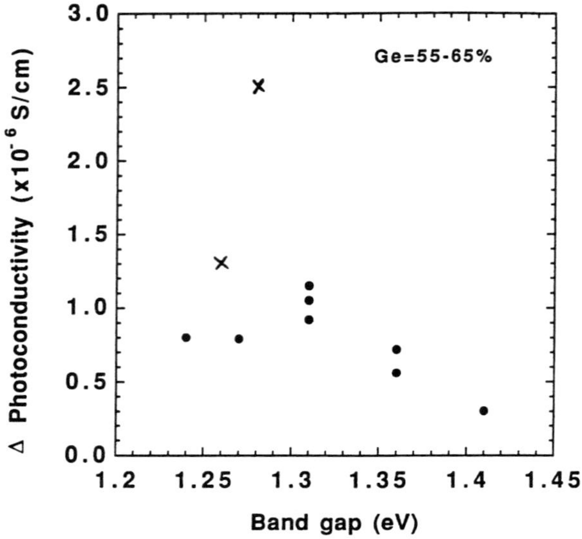

FIGURE 20. $\Delta\sigma_p$ vs. E_g for the same films as in Fig. 19.

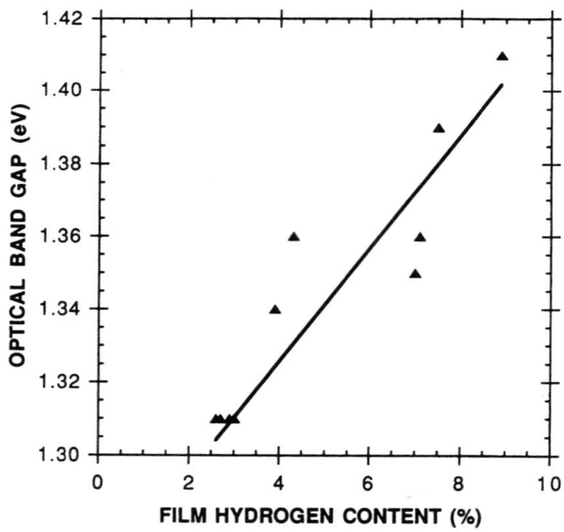

FIGURE 21. The optical band gap as a function of hydrogen content for a-SiGe:H films with ~60% germanium content.

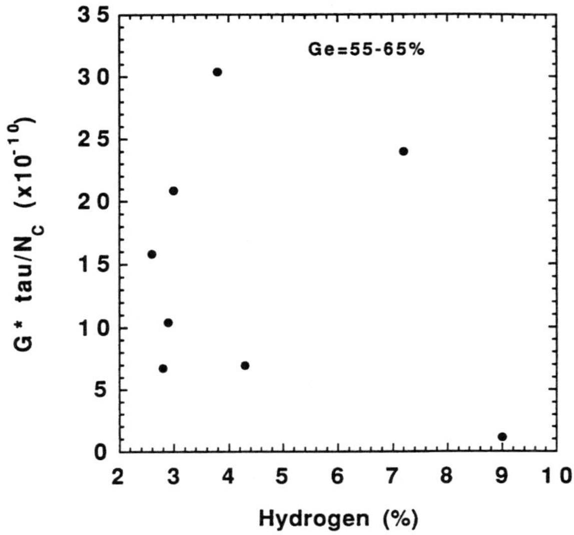

FIGURE 22. The value of $G\tau_c/N_c$ determined from photoconductivity and dark conductivity as a function of film hydrogen content.

Diffusion length determined by optical grating technique in a-SiGe:H alloys with compositions ranging from 0 to 20% germanium have hole diffusion lengths that are relatively unaffected by alloying [44].

These results are consistent with an alloy transport model [45] that proposes that conduction band disorder contributes to both the reduction in electron mobility and its thermal activation ($E_\mu > 0$). In terms of this model, conduction band barriers result from the clustering of Si atoms, while wells result from the clustering of Ge atoms [46, 47] as illustrated in Fig. 23. Both Ge and Si clusters are expected from a statistical distribution of Si–Ge, Si–Si, and Ge–Ge bonds in a randomly mixed alloy. Since H bonds preferentially to the silicon atoms [48] and H increases the average band gap in a-Si:H, it is assumed that the conduction band barriers will be heightened by H. The result is that alloys with intermediate band gaps are expected to have the smallest $\Delta\sigma_p$ and that $\Delta\sigma_p$ decreases further with increasing C_H as found experimentally. The small reduction in $G\tau_c/N_c$ seen in these alloys would be explained by a violation of the assumption that $E_\mu \sim 0$, rather than even a small decrease in τ_c.

The effect of potential fluctuations in the conduction band tail [49] has been probed by the intentional deposition of multilayer structures where a thin a-Ge:H (<5-Å) layer was deposited every 1000 Å in an a-Si:H matrix resulting in a two-

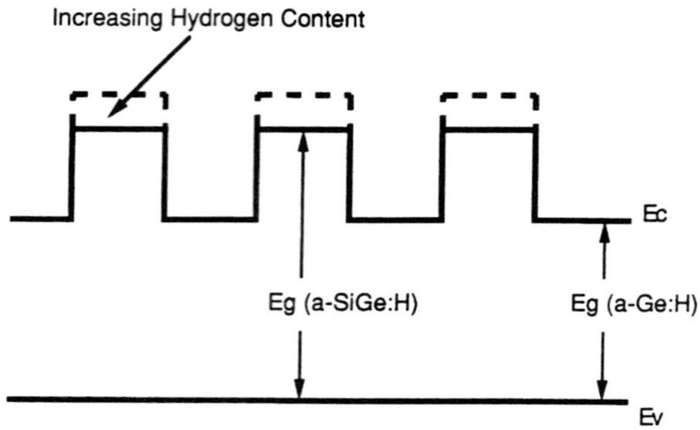

FIGURE 23. Idealized diagram of band-gap fluctuation due to clustering and hydrogen on Si atoms in a-SiGe:H.

dimensional band-gap fluctuation [48]. These multilayer structures were found to have very poor electron transport. Three-dimensional band-gap fluctuations having a more complex structure composed of several distinct phases, including microcrystalline Si, microcrystalline Ge, and amorphous SiGe were created (Fig. 24). The $\Delta\sigma_p$ and $G\tau_c/N_c$ of SiGe mixed-structure alloys are both less than

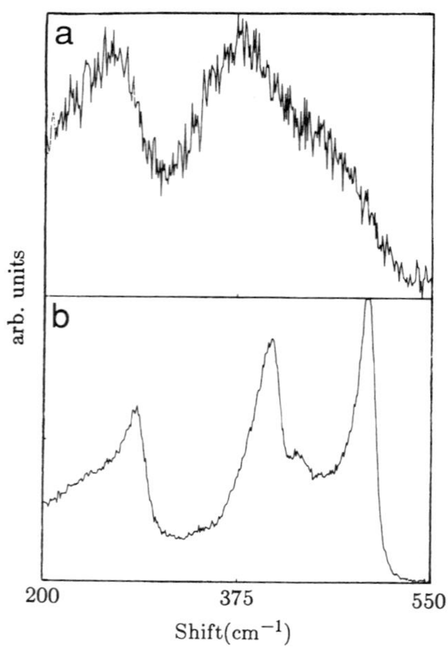

FIGURE 24. Raman spectrum of mixed crystalline phase SiGe (a) and amorphous SiGe (b).

that of the analogous amorphous alloys. The reduction in electronic transport is predicted by the alloy transport model as band-gap fluctuations arise as a result of the Si and Ge microcrystalline clusters.

E. Relationship between Alloying and Hot-Carrier Lifetime in a-Si:H, a-SiGe:H, and a-SiC:H

The picosecond carrier dynamics of the alloys prepared by photo-CVD have been investigated. In these experiments a high-intensity laser is used to produce a large population of energetic carriers [50]. The relaxation of this energetic carrier density is probed by a second beam that is attenuated by free-carrier absorption associated with the pump-beam-induced carrier population.

Interestingly, it was found that the electron thermolysis in a-Si:H, a-SiC:H, and a-SiGe:H materials occurs with approximately the same lifetime [51] as shown in Fig. 25. Since the electron mobility is reduced in a-SiGe:H materials, this could indicate that the predominate recombination center density has increased such that the time it takes for a carrier to travel from the position where it was photogenerated to the recombination centers remains roughly constant.

The same considerations might be applied to a-SiC:H where time-of-flight measurements [52] show that the electron mobility is reduced in these carbon alloys, although the behavior of the electron mobility appears to be more complex than that in the a-SiGe:H materials.

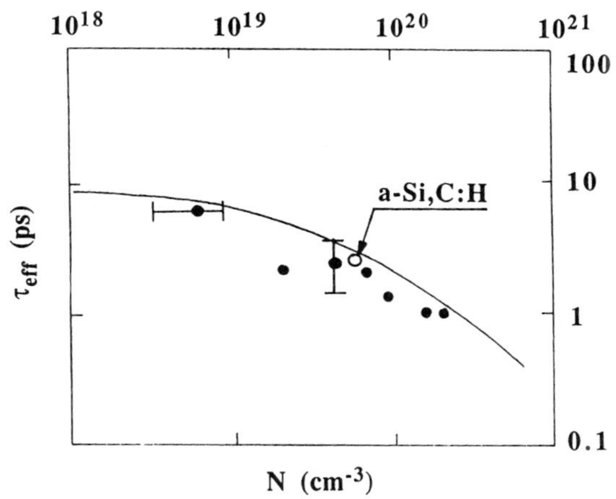

FIGURE 25. Effective lifetime in a-SiGe:H (closed circles) and in a-SiC:H (open circle) as a function of photoinduced carrier density. The full line is a fit described in Fauchet *et al.* [50].

F. Optical Properties of Alloy Materials

With germanium alloying the band gap (and the optical absorption vs. photon energy shifts to lower energy as shown in Fig. 26, the so-called Urbach energy does not appear to be function of germanium alloying [43].

The photoemission spectrum shifts to lower energy with germanium alloying, and can be scaled to the band gap. The nonradiative recombination is apparently not controlled by the density of germanium dangling bonds. The defect band scales with the band gap, suggesting a constant energy difference the conduction band edge and the negatively charged silicon-like dangling bond [53].

The situation in a-SiC:H materials is more complex. The band gap does not vary systematically with carbon content but is more closely correlated with the hydrogen content of the alloy [54]. However, the photoemission spectrum of a-SiC:H [55] moves to higher energy with increasing carbon content (and increasing band gap); however, as the band gap is increased in a-Si:H through increased hydrogen content (by growth at lower substrate temperatures), the photoemission spectrum moves to lower energies [56].

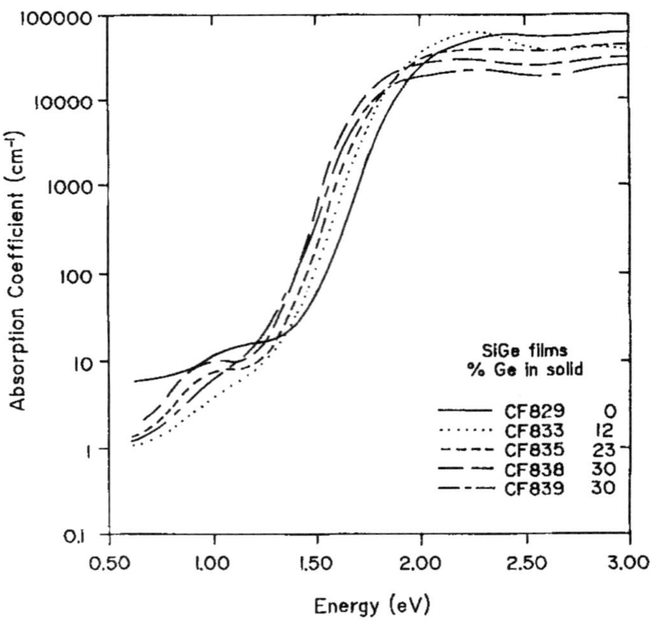

FIGURE 26. Optical absorption of a series of a-SiGe:H films as a function of germanium content.

III. Relationship between Mobility and Device Performance

The electron mobility appears to be one of the factors that is strongly dependent on growth conditions. Although many parameters enter into the device performance and stability, it is important to demonstrate how this one parameter, mobility, influences device performance and the stability. If progress is to be maintained in the development of new and better a-Si-based devices, a comprehensive understanding of how various parameters are linked to each other and to the deposition conditions will have to be established.

A. Relative Stability as a Function of Hydrogen Content and Alloying

The hydrogen content has been related to the stability of solar cells and other devices [57]. The degradation rate of solar cells increases as the hydrogen content is increased in amorphous silicon and its alloys [58]. Also, the dihydride content has been related to the degradation rate with increasing dihydride content being correlated to increase [59]. The saturated defect density in solar cells can be quantitatively determined using the short-wavelength quantum-efficiency method [60, 61].

Despite some interesting reports of improved stability in films grown by novel techniques [62, 63], the saturated defect density does not appear to be a function of hydrogen content [64]. The saturated degraded solar cell defect density (determined using the short-wavelength analysis) does appear to be function of the i-layer hydrogen content [65]. Figure 27 shows the saturated defect density as a function of the current density used to degrade the film and the hydrogen content of the i-layer (the IR absorption spectrum of films grown under the same conditions is shown in Fig. 2).

B. Relationship between Mobility, Recombination Kinetics, and Device Performance

Early in the study of electronic transport of amorphous materials it was recognized that the poor electron (and hole) mobilities implied that the motion of the charge carriers was a diffusive (or Brownian) motion process. Calculations [66, 67] based on the extended state mobilities lead to a hop distance as small as 2–4 Å. More recently, diffusive transport has been used [68, 69] to interpret experimental observations. Two new facets of diffusive transport in a-Si-based materials are im-

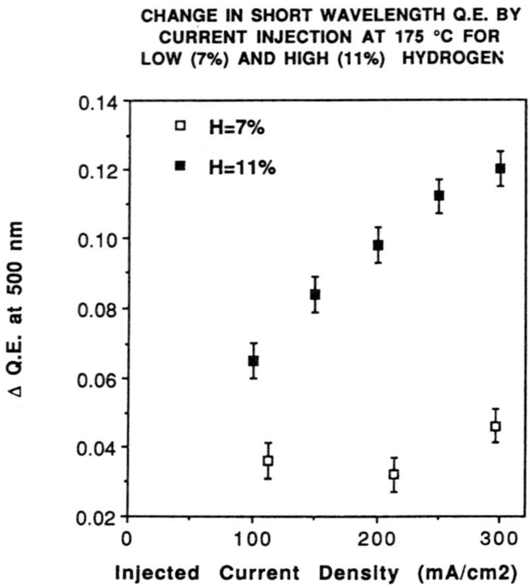

FIGURE 27. Equilibrium-degraded change in short-wavelength quantum efficiency as a function of current injection density and i-layer hydrogen content.

portant: (1) the effect of diffusive velocity ν_d on recombination kinetics; and (2) ν_d, which must be considered to be a function of material deposition conditions and alloying.

Recently numerical models of solar cell performance have become increasingly useful for understanding the myriad factors that influence the performance of a-Si solar cell performance. Here a previously described and tested numerical model [70], its material parameter inputs, and experimental results are used to probe the underlying assumptions concerning carrier transport in low-mobility a-Si-based materials.

It has been reported that the mobility of a-Si i-layer films is a function of deposition conditions, including the deposition temperature, and that p–i–n solar cells prepared experimentally with these same i-layers show that the solar cell performance does not decrease with decreasing electron mobility [71].

The lifetime of photogenerated carriers in amorphous intrinsic materials is a function of many parameters, including the sample configuration used for the measurement. Furthermore, in p–i–n solar cells the carrier lifetime is a function of position within the i-layer as well. The numerical model enables the direct input of the appropriate velocity, and then calculates the recombination rate as a function of position within the i-layer.

The capture rate R of electrons by a recombination center is given by

$$R = n\beta_n N_r \nu, \tag{5}$$

where n is the local electron concentration, β_n is the electron capture cross section, N_r is the density of recombination centers, and ν is the appropriate velocity as discussed below. The recombination lifetime is roughly proportional to $1/R$ in those portions of the i-layer where the electron is the minority carrier (the numerical model does not assume a lifetime, but rather calculates it self-consistently). The recombination rate at interfaces or free surfaces is typically expressed as an interface recombination velocity S, which can be cast in terms of the carrier velocity as

$$S = \beta_i \nu, \tag{6}$$

Where β_i can be thought of as the effective transmission rate of the interface (or free surface) with values ranging from zero to unity, i.e., no carrier loss or complete reflection back into the i-layer ($\beta_i = 0$), total carrier loss at the interface, or transmission beyond the interface to be lost to p-layer recombination ($\beta_i = 1$). The effective interface recombination velocity can also be a function of applied bias [72]. It is not physically possible for electrons to be lost at an interface at rates exceeding their velocity, nor could they be lost at bulk recombination centers at rates exceeding these velocities.

Now we must determine the appropriate velocity at which carriers move toward interfaces and recombination centers. During the actual hop process the speed with which electrons transit the hop distance might approach the speed of light as it involves tunneling between adjacent atomic sites. However, it is clear that the hoping speed is not that which is measured when distances exceed a hop distance and times are longer than the hop time. To obtain the large-scale (distances greater than several hop distances) velocity of a carrier, the macroscopic carrier mobility and diffusivity must be considered as all these quantities are directly related. First, the Einstein relationship relates the diffusivity to the mobility

$$D = kT\mu/q, \tag{7}$$

and from diffusion theory

$$D = kT\mu/q = (1/6)d^2\Gamma, \tag{8}$$

where D is the diffusivity, μ is the mobility, Γ is the hop frequency, and d is the hop distance. The large-scale, or diffusive, velocity can now quantified if either the hop distance or the frequency are known. In the limiting case of a group of materials where the hop distance is variable and the hop frequency is constant, it is convenient to express the diffusive velocity in terms of the mobility using equation (8) to remove the explicit hop distance dependence:

$$v_d = d\Gamma = (D\Gamma)^{0.5} = (6kT\mu\Gamma/q)^{0.5}, \tag{9}$$

where ν_d is the diffusive velocity. Alternatively, if the hop frequency is variable and the distance is constant among the materials under consideration, then equation (9) could be rewritten as

$$\nu_d = d\Gamma = D/d = 6kT\mu/qd. \tag{10}$$

Since we wish to determine the diffusive velocity in order to compare the recombination kinetics of a group of materials with a range of macroscopic mobilities, we must work "backward" to determine ν_d. The range of mobilities corresponds to a range of hop distances and/or frequencies in accordance with equations (8)–(10). The corresponding range of ν_d is bounded by linear and square root relations depending on whether it is the hop distance causing the mobility change [equation (9)] or the hop frequency [equation (10)]. The values of ν_d range from $\sim 10^5$ to 10^7 cm/s for mobilities ranging from 1 to 10 $cm^2\ V^{-1}\ s^{-1}$ with hop distances ranging from 2 to 50 Å. Since the carrier velocity enters directly into the recombination kinetics [equations (5) and (6)], it is apparent that a reduction in velocity will correspondingly produce a reduction in the recombination rate for a given defect capture cross section and density.

In order to illustrate the impact the diffusive velocity has on device performance, a numerical solar cell model was used [70]. To enhance the effect of electron recombination kinetics, only the properties of the front half of the i-layer were varied. Figure 28 shows the quantum efficiency of a p–i–n solar cell in which the front half of the i-layer is assumed to have variable material properties. Figure 28a is the reference case, and Fig. 28b shows the decreased short-wavelength quantum efficiency (QE) resulting from an increase in recombination rate when the electron mobility is reduced by an order of magnitude. The decrease in QE corresponds to an increase in electron concentration in the front half of the i-layer as seen in Fig. 29. Now reducing ν_d by an order of magnitude [the same amount as the mobility reduction, i.e., assuming equation (6) to hold] does not reduce the electron concentration but does effectively reduce the interface and bulk recombination losses, resulting in a QE that is nearly equivalent to the reference case as seen in Fig. 28c. If the relationship of equation (5) had been assumed, the QE would be intermediate between the reference and poor mobility case.

The concept of diffusive velocity when applied to holes offers an explanation for the order of magnitude lower hole effective interface recombination velocity at the n/i interface compared to that of the electron at the p/i interface [72]. This result is readily explained in terms of the smaller hole mobility (at least an order of magnitude smaller than that of the electron) and therefore smaller hole diffusive velocity rather than a lower interface defect concentration or transmission factor.

The direct link among the steady-state mobility, recombination kinetics, and the diffusive velocity is an important consideration for a-Si-based devices. The velocity that controls carrier motion and therefore recombination rates is related to

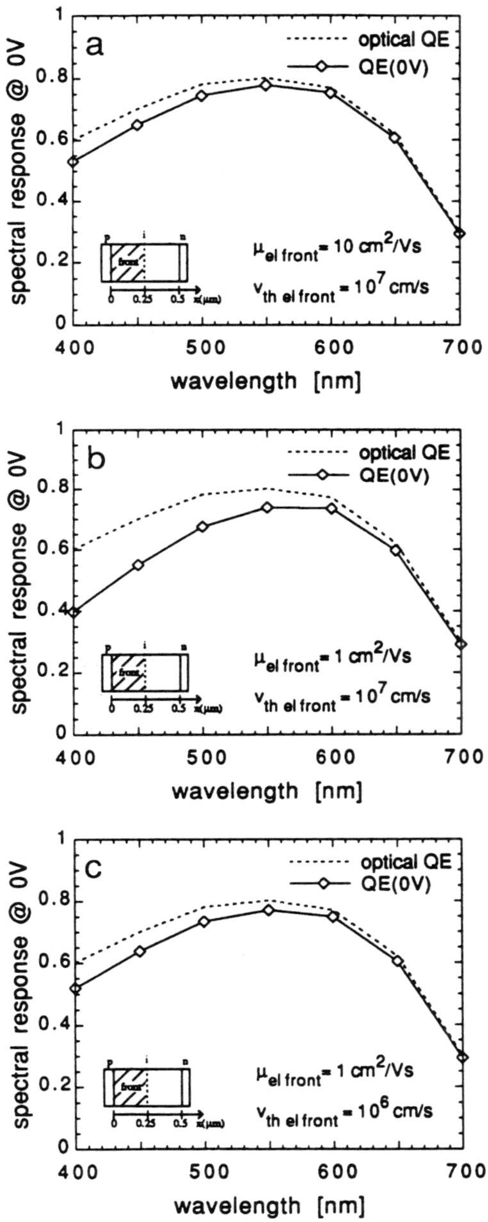

FIGURE 28. Numerical model calculated quantum efficiency versus wavelength for amorphous silicon p–i–n reference solar cell (a), where the electron mobility is reduced by an order of magnitude in the front half of the i-layer (b); and where the electron mobility and the diffusive velocity both have been reduced by an order of magnitude in the front half of the i-layer (c).

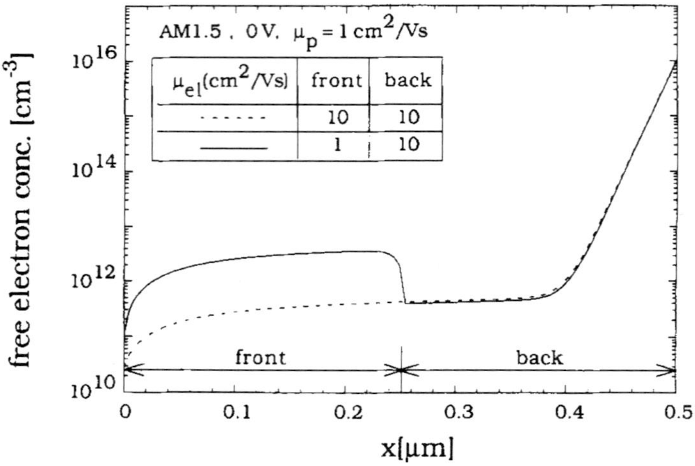

FIGURE 29. Electron concentration as a function of the mobility in the front half of the i-layer for the cases considered in Fig. 1.

the steady-state carrier mobility. This relationship can be expressed as $\nu_d = A\mu^B$, where A is a constant and B is a number between 0.5 and 1. The analysis will not hold if the mobility becomes so low that an equilibrium carrier density is not maintained a hop distance away from the recombination centers [73].

It is noteworthy that diffusive velocity considerations resolve an apparent discrepancy [74]. Increasing hydrogen content in a-SiGe:H reduces the mobility [71], whereas it is reported elsewhere that the increased hydrogen content does not reduce the initial performance of solar cells but does reduce the stability of the solar cell [75] and that increasing hydrogen dilution enhances the stability of a-SiGe:H alloys [76]. Part of the discrepancy can be explained that the hydrogen dilution used in ref. [76] actually reduced the hydrogen content of the film by suppressing gas-phase polymerization. The corresponding reduction in diffusive velocities in the materials with relatively poor electron mobilities masks the poor mobility in the annealed state. The coupling between mobility and stability will be considered below.

C. Relationship between Mobility, Carrier Density, and Stability

There have been numerous studies of the rate of change or kinetics [1, 77–81] of light-induced degradation, yet many of the details have eluded description. In previous work [82, 83] it was shown that the high-temperature steady-state thermal defects and light-induced defect densities could be concisely described in terms of classical thermodynamics.

It is the incorporation of the net energy and entropy changes including lattice components that sets this thermodynamic model apart from the currently popular defect-pool models [84]. Defect-pool models assume the significant energetics to be solely electronic in nature and other factors such as strain energy and entropy to be negligible. Figures 32 and 33 contain the data used to develop the considerations discussed below.

It is an interesting feature of the Staebler–Wronski effect that illumination or any increase in charge carriers due to other stresses can induce a change in dangling-bond densities. The equilibrium proposed here involves charge carriers. Studies of the saturated defect densities resulting from intense illumination indicate that the greatest possible dangling-bond (DB) density ($\sim 10^{18}$ cm^{-3}) is far less than density of lattice sites ($\sim 10^{22}$ cm^{-3}). Therefore, the proposed equilibrium incorporates a unique lattice site, a weak bond (WB), that is susceptible to bond rupture. Note, that the entropy change determination made below is not highly dependent on [WB] (brackets indicate concentration). There is some debate over whether one, two, or many dangling bonds are produced by the rupture of a weak bond. At the onset of the development of the thermodynamic description it is impossible to know the number of dangling bonds in equilibrium with weak bonds or for that matter whether the description itself is appropriate. Since the quantity of dangling bonds determine the order of the mass action relations, the treatment of the data will ultimately determine the order of the relationships.

Thermally and light-induced defect formation can be described by an equilibrium between two DBs as a susceptible site or a WB and a DB that have three possible charge states: neutral (DB^0), negative (DB$^-$), and positive (DB$^+$). It must be stressed, however, that the ability of any set of equilibrium relations to predict the equilibrium concentration of the species does not guarantee that the chosen set of relations are a unique set. Rather, it would be more correct to view these relations as the basis set. The basis equilibrium expressions and the associated mass action relations used by this work are

$$\mathrm{WB} + \mathrm{e} \rightleftarrows \mathrm{DB}^- + \mathrm{DB}^0,\ \Delta G_1:\ [\mathrm{DB}^0][\mathrm{DB}^-]/([\mathrm{WB}][\mathrm{e}]) = K_1, \quad (11)$$

$$\mathrm{WB} + \mathrm{h} \rightleftarrows \mathrm{DB}^+ + \mathrm{DB}^0,\ \Delta G_2:\ [\mathrm{DB}^+][\mathrm{DB}^0]/([\mathrm{WB}][\mathrm{h}]) = K_2, \quad (12)$$

$$\mathrm{DB}^0 + \mathrm{e} \rightleftarrows \mathrm{DB}^-, \quad (13a)$$

$$\mathrm{DB}^0 + \mathrm{h} \rightleftarrows \mathrm{DB}^+, \quad (13b)$$

$$\mathrm{DB}^- + \mathrm{DB}^+ \rightleftarrows 2\mathrm{DB}^0,\ \Delta G_3;\ [\mathrm{DB}^0]^2/([\mathrm{DB}^-][\mathrm{DB}^+]) = K_3, \quad (13c)$$

$$\mathrm{e + h + 2WB \rightleftarrows 2DB^0 + DB^+ + DB^-},\ \Delta G_4\text{: } [\mathrm{DB^0}]^2[\mathrm{DB^+}][\mathrm{DB^-}]/([\mathrm{e}][\mathrm{h}][\mathrm{WB}]^2) = K_4, \quad (14)$$

$$[\mathrm{e}]*[\mathrm{h}] = N_v N_c \exp(-(E_g - \gamma T)/kT). \quad (15)$$

Equations (11) and (12) represent the equilibrium between DB pairs, charge carriers, and WB. Equations (13) allow for interaction among charged DB; equation (13c) is an expression for the correlation energy, which is known to be a small quantity (absolute value <0.2 eV) [85]. Equation (14) is the summation of equations (11) and (12). Each K is an equilibrium constant that is related to the corresponding free-energy change ΔG by

$$K = \exp(-\Delta G/kT) = \exp(-E/kT)\exp(\Delta S/k), \quad (16)$$

where each E and ΔS is the enthalpy and the entropy change (respectively) for the creation of the corresponding DB pair [e.g., in equation (11) K_1 is dependent on the enthalpy and entropy of formation of a pair of DBs in which one is neutral and the other is negatively charged]. Equation (15) is the mass action relation for holes (h) and electrons (e), where E_g is the band gap and γ is the temperature dependence of the band gap (band gap decreases with increasing temperature). Under illumination the hole and electron concentrations, [h] and [e], are taken as functions of $\mathrm{DB^0}$ [86]:

$$[\mathrm{e}]\tilde{}F\tau_c \text{ and } [\mathrm{h}]\tilde{}F\tau_h, \quad (17)$$

$$\tau_c = 1/(A_1*[\mathrm{DB^0}]) \quad \text{and} \quad \mathrm{t_h} = 1/(A_2[\mathrm{DB^0}]), \quad (18)$$

where τ_c and τ_h are the electron and hole lifetimes, respectively; A_1 and A_2 are the product of the carrier thermal velocity and capture cross section respectively, by a neutral DB; and F is the photoinduced generation rate of electrons and holes. *Note:* It is assumed that at this point the electron mobility is assumed to be not temperature dependent. In the interest of clarity the resulting simple relation between photoconductivity, σ_p, and $\mathrm{DB^0}$ will be used. Note that previously [82, 83] σ_p was related to $\mathrm{DB^+}$, the simpler $\mathrm{DB^0}$ approach taken here does not alter the major conclusions.

$$\sigma_p = \mathrm{q}\mu_c F\tau_c = q\mu_c F/(A_1*[\mathrm{DB^0}]), \quad (19)$$

where μ_c is the electron mobility. Since the electron mobility is far greater than that of the hole, photoconductivity is principally a measurement of electron concentration and lifetime.

Substituting equation (18) into equation (17) and equation (17) into equation (14), we see that equation (14) becomes

$$[\mathrm{DB^+}][\mathrm{DB^-}][\mathrm{DB^0}]^4 = [\mathrm{WB}]^2(1/A_1A_2)F^2K_4. \quad (20)$$

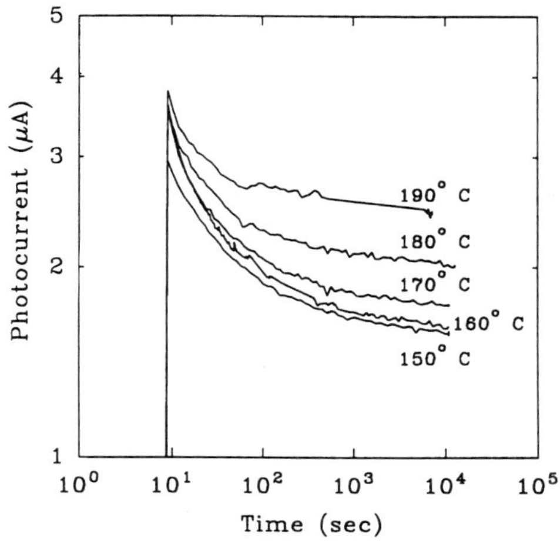

FIGURE 30. Photoconductivity of a standard amorphous silicon film versus time for various temperatures.

The equilibrium assumption requires that both the left-hand and right-hand sides of equation (20) have the same temperature dependence. Therefore, the measured temperature dependence of any one of the species of equation (20) is related to the others by

$$4E^0 = E_4 - E^- - E^+, \qquad (21)$$

where E^+, E^- and E^0 are the energies determined by the slope of the natural logarithm of the density of the particular charged DB (positive or negative) or neutral DB versus $1/T$. Since the photoconductivity does not probe the charged dangling-bond densities, it is not possible to determine E^- and E^+ separately.

The photoconductivity was measured [83] as a function of temperature over the range 150–225°C. Figure 30 shows that the steady-state (achieved in ~30 min.) defected condition shows σ_p increasing with increasing temperature. The steady-state photoconductivity versus $1/T$ is shown in Fig. 31 and has an activation energy of +0.2 eV. The DB^0 density affects σ_p through [e] as given by equation (19); therefore, $[DB^0]$ is decreasing with increasing T. Alternatively, ESR measurement of light-induced DB^0 defects [87] versus $1/T$ determined E^0 to be −0.13 eV, since from the photoconductivity measurements

$$4E^0 = 4(-0.20\text{ eV}) = E_4 - E^- - E^+. \qquad (22)$$

The two measurements of E^0 can be resolved if we assume that the charged dangling bonds combine, during the ESR probe of the neutral dangling-bond den-

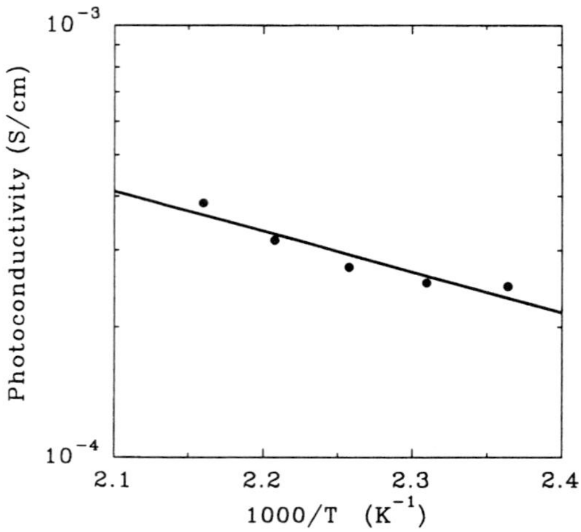

FIGURE 31. Steady-state photoconductivity versus $1/T$ for the sample of Fig. 1.

sity, which is done in the dark, in accordance with equation (13) and that the correlation energy E_3 is close to zero:

$$6E^0 = 6(-0.13 \text{ eV}) = E_4 - E_3. \tag{23}$$

ESR studies [81, 88] measured the DB as a function of temperature without any illumination. The $[DB^0]$ increased with increasing temperature with an activation energy (E_a) ranging from 0.35 to 0.42 eV; this work assumes that the 0.4-eV value is material-preparation-dependent as shown in Fig. 32. Substituting equation (15) into equation (14) and rearranging yields

$$[DB^0]^2[DB^+][DB^-] = \omega \exp(-(E_g - E_4/kT), \tag{24}$$

where ω is $\{WB^2 N_c N_v \exp([\Delta S_4 + \gamma]/k)\}$. Since temperature dependencies of both the left- and right-hand sides must be equivalent, the apparent activation energy (E_a) for the thermal creation of DB^0 is

$$2E_a = 2(+0.4 \text{ eV}) = E_g + E_4 - E^- - E^+ \tag{25}$$

With E_g equal to 1.7 eV, the values that satisfy equations (22) and (23) also satisfy (25), that is, $E_4 = -0.80$, $E^0 = -0.20$, and $E^- + E^+ < 0.20$ eV. Whereas the slopes of Figs. 3 and 4 are used to determine the energetics of defect formation, the intercepts of Fig. 4 are used to provide insight into the entropy. The intercept of the neutral dangling-bond density of Fig. 4 has values ranging from $\sim 10^{18}$ to 10^{19} cm^{-3}. The intercept A can be written

$$A = [DB^0]_{1/T=0} = [WB]^*(N_c N_v)^{0.5} {}^* exp([\Delta S_4 + \gamma]/2k)^*([DB^+]^*[DB^-])^{-0.5} \tag{26}$$

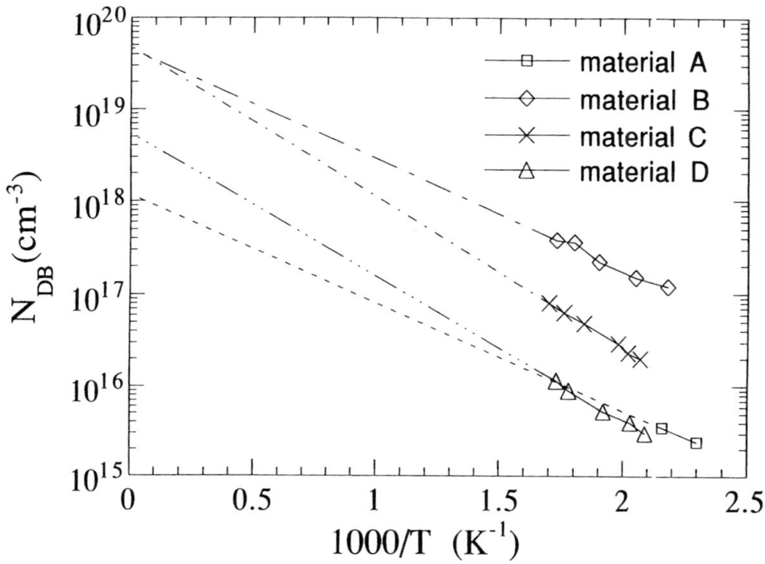

FIGURE 32. N_{DB} versus $1/T$ for a number of a-Si:H samples deposited at different substrate temperatures. (From Zarfar and Schiff [81].)

and

$$[WB]*([DB^+]*[DB^-])^{-0.5} \geq 1. \quad (27)$$

Therefore

$$A*(N_c*N_v)^{-0.5}*\exp(-\gamma/2k) \geq \exp(\Delta S_4/2k). \quad (28)$$

The values of $N_v \sim N_c$ range from 10^{19} to 10^{21} cm^{-3} and γ is about 10^{-4} eV/°C. The quantity on the left-hand side of equation (28) is less than unity for any reasonable value of N_c and N_v, and γ is positive (band gap decreases with increasing temperature); thus, ΔS_4 must be negative. Values ranging between $-3.8 \cdot 10^{-4}$ to $-3 \cdot 10^{-3}$ eV K^{-1} are consistent with the values given above. This magnitude of entropy change cannot be accounted for by electronic transitions. Rather, a structural rearrangement is suggested. Perhaps the entropy reduction in defect formation is associated with an increase in crystal-like character of the amorphous lattice (fewer bond distortions). The total free-energy change (using an intermediate ΔS) at $T = 450$ K is ~ -0.8 eV. Note that if the equilibrium relation was chosen to be e + p + 2WB <--> 2DB^0 [rather than as in equation (24)] as was done in Winer [84], the considerations of the intercepts would yield an entropy change (for two neutral DBs) that is too large to explain.

It is important to note that the entropy is also one of the parameters that estab-

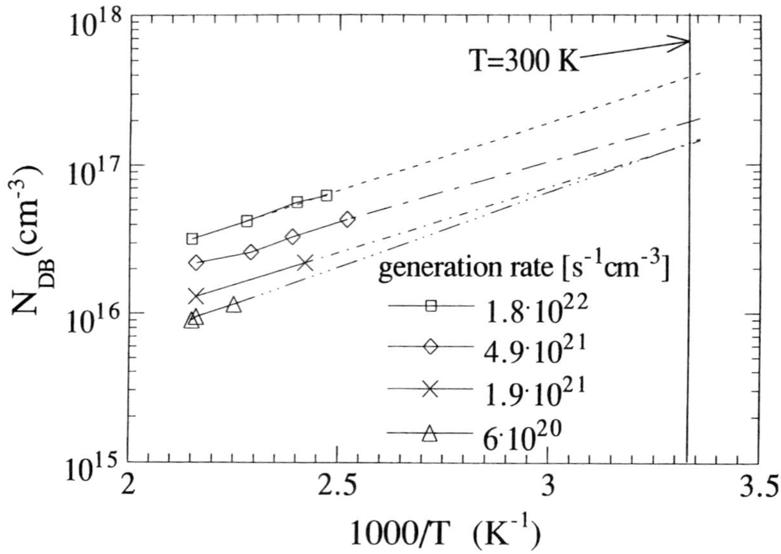

FIGURE 33. Dangling-bond density as a function of illumination intensity and temperature. (From Benatar *et al.* [90].)

lishes the intercept (DB concentration at $1/T = 0$) of the saturated illuminated dangling-bond density seen in Fig. 33. From equation (20) it can be seen that unlike the dark (conductivity) case, the exponential prefactor is additionally dependent on the square of the generation rate (and therefore illumination intensity) and the excess carrier lifetimes. If these parameters are known, the generation rate is determined from the optical absorption, and the illumination intensity the lifetimes of the electrons and holes must have values in the range of 10^{-9} to 10^{-10} seconds to account for the observed magnitude of the photo- and dark conductivities. Therefore, the entropy values derived from the thermal data of Fig. 32 are consistent with the intercepts of the illuminated data of Fig. 33.

D. Relationship between Saturated Defect Density, Deposition Conditions, and Mobility

Starting with equation (18) in order to examine the coupling between mobilities and defect densities, it is necessary to express A_1 and A_2 in terms of the diffusive velocity:

$$A_1 = 1/(S_c \nu_c) \quad \text{and} \quad A_2 = 1/(S_h \nu_h), \tag{29}$$

where S_c and S_h respectively are the electron and hole capture cross sections by the oppositely charged light induced defects. Now equation (20) can be recast as

$$[DB^+]^2[DB^-]^2[DB^\circ]^2 = [WB]^2(S_c\nu_c S_h\nu_h)^{-1}F^2K_4. \tag{30}$$

From this expression it can be seen that for two materials with similar weak-bond densities, but different mobilities (and diffusive velocities) the material with the smaller diffusive velocity will have a larger density of defects when the equilibrium or saturated defected state is obtained. The reason for this is that the materials with the lower mobility has a smaller diffusive velocities, which for a given illumination condition, leads to a larger standing carrier concentration. These larger carrier densities, in turn, through the equilibrium expressed in equation (30), induce larger dangling-bond populations.

The photoconductivity in the equilibrium degraded state can be expressed in terms of equations (19), (29), and (30):

$$\sigma_p = \mu_c F\tau_e = \mu_c F(1/\nu_d S_c[DB^+])) \\ = \mu_c(S_h\nu_h)^{1/2}/[(\nu_c S_c)^{1/2}[WB]K_4^{1/2}]. \tag{31}$$

There are several interesting features of this expression; the first is that under light soaking the photoconductivity will tend toward an equilibrium value that does not depend on the light intensity. On the other hand, the equilibrium degraded $\mu\tau$ products, which can be derived from the photoconductivity by dividing σ_p by the generation rate F, will be proportional to the inverse of the intensity. Also, since μ_c is related to ν_c, $\nu_c = A\mu^B$, both σ_p and $\mu_c\tau_c$ for a collection of films with different μ_c values will end up in their equilibrium-degraded states with values—of σ and $\mu\tau$—that are much closer together than might be expected, as the difference will only be proportional to $\sim\mu^{1/2}$.

An understanding of the coupling between mobility and the stability of materials is necessary in order to design better devices. The hydrogen content of the films reported in Wronski [64] ranged from a few percent to almost 20%. The saturated defect concentration is not a strong function of hydrogen content. This result can be explained in two alternative ways: (1) the observed saturation in films represents a complete conversion of weak bonds into dangling bonds, and the weak-bond density increases slightly with C_H; and (2) the apparent saturation is due to the establishment of a degraded equilibrium state in accordance with equation (3), with ν_c being proportional to μ_c (as derived in this work) and μ_c decreases with increasing C_H; therefore, the equilibrium defect concentration increases because of the resulting increase in electron concentration.

These reports must be reconciled with the finding that the stability of solar cells *does* appear to be a stronger function of the hydrogen content (or microstructure).

It was noted above that the equilibrium degraded state of solar cells under current injection was a function of i-layer hydrogen content. Solar cells, unlike films, are affected by field redistributions resulting from charged dangling bonds [89]. The change in short-wavelength quantum-efficiency determination of dangling-bond concentrations takes the effect of charged dangling bonds into account [63]. The change in short-wavelength quantum efficiency in solar cells resulting from a given light-soaking dose is a function of the i-layer deposition temperature (Fig. 34) and light soaking. This could be understood to result directly from the poor mobility in the low-temperature i-layer being smaller than that in the high-temperature i-layer material. The electron concentration as a function of electron mobility is shown in Fig. 35. The speed at which the materials degrade toward their equilibrium condition may be as in other processes roughly proportional to the departure from equilibrium; therefore, the low-mobility materials would not only degrade to a more defected state but may initially degrade more quickly if the starting state is further from their eventual equilibrium state. Films degrade to a saturated degraded state that is different than that of solar cells as detailed in [90].

IV. Concepts of Electronic Transport in Amorphous Semiconductors

It must be stressed that the use of concepts such as free-carrier mobility and optical band gap are useful, the fundamental understanding of these concepts in

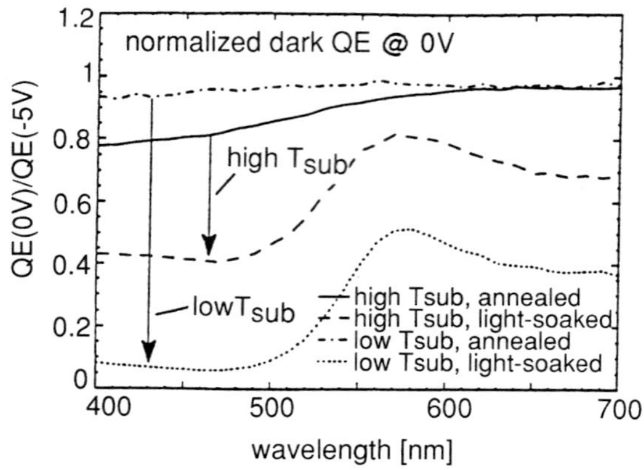

FIGURE 34. Quantum efficiency of standard and low T_s VFH–CVD solar cells as a function of wavelength and light-soaking time.

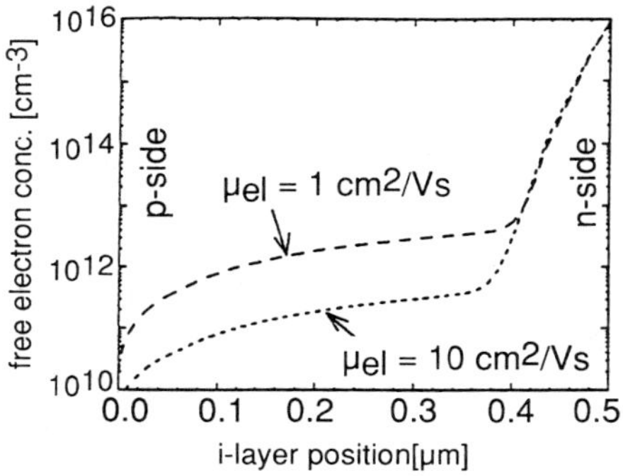

FIGURE 35. Numerical modeled free election concentration as a function of i-layer position for different free-electron mobilities. The generation rate is $2 \cdot 10^{21}$ $cm^{-3}S^{-1}$, at short-circuit current.

amorphous materials is not fully applicable and could legitimately be challenged. The concepts developed and discussed above side-stepped these problems by avoiding consideration of carrier hop frequencies and hop distances as a function of hydrogen content and other material parameters. The range of mobilities encountered in this work and their relationship to amorphous silicon structures (which also come in a wide range) will ultimately have to be addressed in order to understand how to prepare materials with improved properties.

The concept of extended-state mobility does not answer as many questions as it raises. For example, the temperature dependence of the mobility [91] does not fit a simple description of extended-state transport, and the electron mobility may not be linear in voltage [92]. Table 1 shows a number of optical and electrical phenomena that lack a complete description.

Table 1 Unusual phenomena in amorphous materials

Temperature dependence and activation energy of photoconductivity and drift mobility electron mobility [93, 94]
Difference between thermal electric power and activation energy of dark conductivity [95]
Magnetoresistance (and its temperature dependence) in amorphous silicon materials [96–98]
Time dependence of IR quenching of the photoluminescence [99]
Sign of Hall voltage [100–102]
Thermal quenching of photoluminescence [103]
Pressure dependence of the resistivity [104]
Electrical quenching of photoluminescence [105]
Optical absorption as a function of energy and temperature [106]

V. Summary and Conclusions

Electronic transport properties have been investigated in undoped hydrogenated amorphous silicon (a-Si:H) materials whose microstructure and void fraction are changed by deposition temperature (T_s). The hydrogen content in these materials decreases from 15 to 5 atomic % and the void fraction to 14% as T_s is raised from 200 to 350°C. The photo- and dark conductivities are measured from 40 to 190°C and extended-state electron mobilities are derived from a self-consistent analysis. The room-temperature mobilities are found to increase from 0.8 to 30 $cm^2\ V^{-1}\ s^{-1}$ and become less temperature-dependent as T_s increases. These temperature-activated mobilities explain the Meyer–Neldel rule in a-Si:H materials whose dark-conductivity activation energies are greater than 0.4 eV where it cannot be explained by the statistical shift of the Fermi level.

The transport properties of intrinsic amorphous SiGe films are to have a comparatively poor electron mobility. The relationship between the electronic transport and composition (Si, Ge, and H content) is established. The electron mobility decreases with increasing hydrogen content for a given Ge content. The optical band gap shifts smoothly to lower energies, with germanium alloying without an increase in the Urbach energies as has also been noted by others [107, 108]. (In particular, Urbach *et al.* [108] noted nonvarying Urbach energies, with low DOS as well as some evidence of inhomogeneity.) Growth conditions that reduce the extent of gas-phase polymerization as well as the flux of hydrogen radical to the growth surface yield the best electronic transport.

A link between the observed steady state mobility and the diffusive velocity, the speed at which carriers move to distances greater than a single hop distance (100 Å), is found to be an important transport link between deposition conditions and stability. However, the carrier mobilities are not expected to impact solar cell performance in the annealed state. The electron diffusive velocity is a function of material fabrication technique. The diffusive velocity affects device performance through the recombination kinetics, as it is this velocity at which carriers move toward interfaces and recombination centers.

By using a thermodynamic descriptions for the saturated density of dangling bonds in equilibrium with a given light stress it is possible to link deposition parameters to the weak-bond density and the carrier mobilities. The higher the weak-bond density, the greater the ultimate defect density for a given light stress. Also, as the carrier mobilities are reduced, the saturated defect density increases for a given light stress, the carrier densities increase in proportion to the mobility decrease, and therefore the equilibrium defect concentrations increase.

References

1. D. L. Staebler and C. R. Wronski, *Appl. Phys. Lett.* **31,** 292 (1977).
2. M. J. Powell, S. C. Deane, I. D. French, J. R. Hughes, and W. I. Milne, *Philos. Mag., Part B* **63,** 325 (1991).
3. D. E. Albright, N. Saxena, C. M. Fortmann, R. E. Rocheleau, and T. W. F. Russell, *AIChE J.* **36,** 1555 (1990).
4. C. M. Fortmann, S. Lange, M. Farley, and J. O'Dowd, *Proc. IEEE Photovoltaic Spec. Conf., 19th* (1987).
5. A. Asano, *Appl. Phys. Lett.* **56,** 533 (1990).
6. N. Saxena, D. E. Albright, C. M. Fortmann, T. W. F. Russell, P. M. Fauchet, and I. M. Cambell, *J. Non-Cryst. Solids* **144,** 801 (1989).
7. D. E. Albright, C. M. Fortmann, and T. W. F. Russell, *Mater. Res. Soc. Symp. Proc.* **149,** 521 (1989).
8. D. E. Albright, Ph.D. Thesis, Univ. of Delaware, Newark, 1990.
9. N. Saxena, Ph.D. Thesis, Univ. of Delaware, Newark, 1990.
10. N. Saxena, C. M. Fortmann, and T. W. F. Russell, *Mater. Res. Soc. Symp. Proc.* **149,** 99 (1989).
11. R. W. Collins, *in* "Amorphous Silicon and Related Materials" (H. Fritzsche, ed.), p. 939. World Sci. Publ. Co., Teaneck, New Jersey, 1988.
12. Y. M. Li and R. W. Collins, unpublished data.
13. G. Lucovsky and T. M. Hayes, *in* "Amorphous Semiconductors" (M. H. Brodsky, ed.), Vol. 36, p. 236. Springer-Verlag, New York, 1979.
14. P. G. LeComber, A. Madan, and W. E. Spear, *J. Non-Cryst. Solids* **11,** 219 (1972).
15. R. H. Klazes, M. H. L. M. Van Den Broek, J. Bezemer, and S. Radelaar, *Philos. Mag.* **45,** 377 (1982).
16. M. H. Brodsky, M. A. Frisch, J. F. Ziegler, and W. A. Landford, *Appl. Phys. Lett.* **30,** 561 (1977).
17. R. W. Collins, *in* "Amorphous Silicon and Related Materials" (H. Fritzsche, ed.), p. 939. World Sci. Publ. Co., Teaneck, New Jersey, 1988.
18. M. Silver and V. Cannella, *in* "Tetrahedrally-Bonded Amorphous Semiconductors" (D. Adler and H. Fritzsche, eds.), p. 389. Plenum, New York, 1985.
19. R. M. Dawson, C. M. Fortmann, M. Gunes, Y. M. Li, S. S. Nag, R. W. Collins, and C. R. Wronski, *Appl. Phys. Lett.* **63,** 955 (1993).
20. R. M. Dawson, Youming Li, M. Gunes, S. Nag, R. W. Collins, C. R. Wronski, M. Bennett, and Y.-M. Li, *Proc. E. C. Photovoltaic Sol. Energy Conf., 11th* (L. Guimaraes, W. Palz, C. De Reyff, H. Kiess, and P. Helm, eds.), p. 680. Harwood Academic Publ., Montreux, Switzerland, 1992.
21. G. Weiser and H. Mell, *J. Non-Cryst. Solids* **114,** 298 (1989).
22. D. E. Carlson and C. R. Wronski, *in* "Amorphous Semiconductors" (M. H. Brodsky, ed.), pp. 287–329. Springer-Verlag, Berlin, 1985.
23. T. Tiedje, *in* "Physics of Disordered Materials" (D. Adler, H. Fritzsche, and S. R. Ovshinsky, eds.), pp. 461–467. Plenum, New York, 1985.
24. R. H. Bube, "Photoconductivity of Solids," p. 72. Wiley, New York, 1970.
25. M. Gunes and C. R. Wronski, *Appl. Phys. Lett.* **61,** 678 (1992).
26. D. Hauschildt, W. Fuhs, and H. Mell, *Phys. Status Solidi B* **111,** 171 (1982).
27. W. Beyer and H. Overhof, *Solid State Commun.* **31,** 1 (1979).
28. Y. Tang, R. Braunstein, and B. von Roedern, *Mater. Res. Soc. Symp. Proc.* **258,** 735 (1992).
29. T. Tiedje, *in* "Semiconductors and Semimetals, Vol. 21: Hydrogenated Amorphous Silicon," Part C (J. Pankove, ed.), p. 207. Academic Press, New York, 1984.
30. D. I. Jones, P. G. LeComber, and W. E. Spear, *Philos. Mag.* **36,** 541 (1977).

31. C. M. Fortmann and J. Tu, *Proc. IEEE Photovoltaic Spec. Conf., 20th, Las Vegas, Nev., 1988* p. 139. IEEE, New York, 1989.
32. D. E. Carlson, *IEEE Trans. Electron Devices* **ED-36,** No. 12 (1989).
33. S. Guha, J. Yang, A. Pawlikiewicz, T. Glatfelter, R. Ross, and S. R. Ovshinsky, *Proc. IEEE Photovoltaic Spec. Conf. 20th* p. 79 (1989).
34. C. M. Fortmann, S. S. Hegedus, and W. A. Buchanan, *J. Non-Cryst. Solids* **115,** 21 (1989).
35. Y. Nakata, H. Sannomiya, S. Moriuchi, A. Yokota, Y. Inoue, M. Itoh, and H. Itoh, *Mater. Res. Soc. Symp. Proc.* **192,** 15 (1990).
36. L. Ley, *in* "Hydrogenated Amorphous Silicon II" (J. D. Joannopoulos and G. Lucovsky, eds.), Vol. 56. Springer-Verlag, New York, 1984.
37. B. von Roedern and W. Paul, unpublished data.
38. C. M. Fortmann, *Mater. Res. Soc. Symp. Proc.* **192,** 27 (1990).
39. M. H. Brodsky, *Top. Appl. Phys.* **36,** 124 (1985).
40. F. Karg, W. Kruhler, M. Moller, and K. V. Klitzing, *J. Appl. Phys.* **60,** 2016 (1986).
41. C. E. Nebel, H. C. Weller, and G. H. Bauer, *Mater. Res. Soc. Symp. Proc.* **118,** 507 (1988).
42. T. Unold, J. D. Cohen, and C. M. Fortmann, *J. Non-Cryst. Solids* **137/138,** 809 (1991).
43. T. Unold, J. D. Cohen, and C. M. Fortmann, *Mater. Res. Soc. Symp. Proc.* **258,** 499 (1992).
44. G. H. Bauer, C. E. Nebel, and H.-D Mohring, *Mater. Res. Soc. Symp. Proc.* **118,** 679 (1988).
45. C. M. Fortmann, D. E. Albright, I. H. Campbell, and P. M. Fauchet, *Mater. Res. Soc. Symp. Proc.* **164,** 315 (1990).
46. B. von Roedern and A. Madan, *Philos. Mag., Part B* **63,** 293 (1991).
47. W. Paul, *in* "Amorphous Silicon and Related Materials" (H. Fritzsche, ed.), pp. 63–79. World Sci. Publ. Co., Teaneck, New Jersey, 1988.
48. C. M. Fortmann, *Mater. Res. Soc. Symp. Proc.* **118,** 691 (1988).
49. W. Paul, D. K. Paul, B. von Roedern, J. Blake, and S. Oguz, *Phys. Rev. Lett.* **46,** 1016 (1981).
50. P. M. Fauchet, D. A. Young, W. L. Nighan, Jr., and C. M. Fortmann, *IEEE J. Quantum Electron.* **27,** 2714 (1991).
51. D. A. Young, P. M. Fauchet, Y. M. Liu, W. L. Nighan, Jr., and C. M. Fortmann, *Mater. Res. Soc. Symp. Proc.* **258,** 807 (1992).
52. Q. Wang, E. A. Schiff, and Y.-M. Li, *Mater. Res. Soc. Symp. Proc.* **297,** 419 (1993).
53. R. Carius, *in* "Amorphous Silicon and Related Materials" (H. Fritzsche, ed.), p. 939. World Sci. Publ. Co., Teaneck, New Jersey, 1988.
54. W. Beyer, *J. Non-Cryst. Solids* **97/98,** 1027 (1987).
55. R. Fischer, *in* "Amorphous Semiconductors" (M. H. Brodsky, ed.), Vol. 36, p. 176. Springer-Verlag, New York, 1985.
56. R. Fischer, *in* "Amorphous Semiconductors" (M. H. Brodsky, ed.), Vol. 36, p. 163. Springer-Verlag, New York, 1985.
57. P. G. LeComber, *J. Non-Cryst. Solids* **115,** 1 (1989).
58. C. M. Fortmann, J. O'Dowd, J. Newton, and J. Fischer, in *AIP Conf. Proc.* No. 157, p. 103 (1987).
59. N. Nakamuru, T. Takahama, M. Isomura, M. Nishikuni, K. Yoshida, S. Tsuda, S. Nakano, M. Ohnishi, and Y. Kuwano, *Jpn. J. Appl. Phys.* **28,** 1762 (1989).
60. C. M. Fortmann, S. Lange, M. Hicks, and C. R. Wronski, *J. Appl. Phys.* **64,** 4219 (1988).
61. C. M. Fortmann and D. Fischer, *Proc. IEEE Photovoltaic Spec. Conf., 23rd, Louisville, Ky., 1993.*
62. M. Vanecek, A. H. Mahan, B. P. Nelson, and R. S. Crandall, *Proc. E. C. Photovoltaic Sol. Energy Conf., 11th* (L. Guimaraes, W. Palz, C. De Reyff, H. Kiess, and P. Helm, eds.), p. 96. Harwood Academic Publ., Montreux, Switzerland, 1992.
63. K. S. Stevens and N. M. Johnson, *Mater. Res. Soc. Spring Meet., San Francisco, 1993.*
64. C. R. Wronski, N. Maley, T. M. Peterson, J. R. Abelson, P. K. Bhat, R. Biswas, V. L. Dalal, A.

E. Delahoy, S. J. Fonash, C. M. Fortmann, S. Guha, W. Luft, T. McMahon, D. Redfield, P. C. Taylor and S. Wagner *Proc. E. C. Photovoltaic Sol. Energy Conf.*, Montreux, *11th* (L. Guimaraes, W. Palz, C. De Reyff, H. Kiess, and P. Helm, eds.), p. 72. Harwood Academic Publ., Montreux, Switzerland, 1992.

65. C. M. Fortmann, S. S. Hegedus, T. X. Zhou, and B. N. Baron, *Sol. Cells* **30,** 255 (1991).
66. N. F. Mott and E. A. Davis, "Electronic Processes in Non-Crystalline Materials," 2nd Ed., p. 219. Oxford Univ. Press, London, 1979.
67. P. Nagels, *Top. Appl. Phys.* **36,** 122 (1985).
68. M. Silver and V. Cannella, *in* "Tetrahedrally Bonded Amorphous Semiconductors" (D. Adler and H. Fritzsche, eds.), pp. 389–396. Plenum, New York, 1985.
69. E. Schiff, personal communication (1992).
70. D. Fischer, N. Pelleton, H. Keppner, A. Shah, and C. M. Fortmann, *Mater. Res. Soc. Symp. Proc.* **258,** 887 (1992).
71. C. M. Fortmann, *Proc. IEEE Photovoltaic Spec. Conf., 21st, Orlando, Fla., 1989* p. 1493. IEEE, New York, 1990.
72. F. A. Rubinelli, S. J. Fonash, and J. K. Arch, *Tech. Proc. Int. Photovoltaic Sci. Eng. Conf., 6th.* p. 851. Vedams Books Int., New Delhi, 1992.
73. C. M. Fortmann and D. Fischer, *Appl. Phys. Lett.* **62,** 3147 (1993).
74. P. Menna, private communication, 1989.
75. C. M. Fortmann, T. Zhou, C. Malone, M. Gunes, and C. R. Wronski, *Proc. IEEE Photovoltaic Spec. Conf., 21st, Orlando, Fla., 1989* p. 1648. IEEE, New York, 1990.
76. M. Bennett, *Proc. IEEE Photovoltaic Spec. Conf., 21st, Orlando, Fla., 1989* p. 1653. IEEE, New York, 1990.
77. M. Stutzmann, W. B. Jackson, and C. C. Tsai, *Phys. Rev. B* **32,** 23 (1985).
78. D. Redfield and R. H. Bube, *Mater. Res. Soc. Symp. Proc.* **192,** 273 (1990).
79. D. Adler, *in* "Semiconductors and Semimetals, Vol. 21: Hydrogenated Amorphous Silicon," Part A (J. Pankove, ed.), p. 291. Academic Press, New York, 1984.
80. H. Branz and M. Silver, *Mater. Res. Soc. Symp. Proc.* **192,** 261 (1990).
81. S. Zarfar and E. A. Schiff, *Phys. Rev. Lett.* **66,** 1493 (1991).
82. C. M. Fortmann, R. M. Dawson, and C. R. Wronski, *J. Non-Cryst. Solids* **137/138,** 207 (1991).
83. C. M. Fortmann, R. M. Dawson, and C. R. Wronski, *Mater. Res. Soc. Symp. Proc.* **219,** 63 (1991). And more recently; C. M. Fortmann, R. M. Dawson, H. Y. Liu, C. R. Wronski, *J. Appl. Phys.* **76,** 768 (1994).
84. K. Winer, *Phys. Rev. B* **41,** 12150 (1991).
85. A. V. Gelatos, J. D. Cohen, and J. P. Harbison, *Appl. Phys. Lett.* **49,** 722 (1986).
86. H. M. Branz and M. Silver, *Phys. Rev. B* **42,** 7420 (1990).
87. M. Stutzmann, W. B. Jackson, and C. C. Tsai, *Phys. Rev. B* **32,** 23 (1985).
88. T. J. McMahon, *Sol. Cells* **30,** 235 (1991).
89. D. Fischer, N. Wysch, C. M. Fortmann, and A. Shah, *Proc. IEEE Photovoltaic Spec. Conf., 23rd, Louisville, Ky., 1993.*
90. C. M. Fortmann, R. M. A. Dawson, M. Gunes, C. R. Wronski, *J. Non-Crystalline Solids* **164–166,** 509.
91. H. Fritsche, *J. Non-Cryst. Solids* **114,** 1 (1989).
92. E. A. Schiff and M. Silver, *in* "Amorphous Silicon and Related Materials" (H. Fritzsche, ed.), p. 825. World Sci. Publ. Co., Teaneck, New Jersey, 1988.
93. P. G. Le Comber and W. E. Spear, *Phys. Rev. Lett.* **25,** 509 (1970).
94. W. E. Spear, R. J. Loveland, and A. Al-Sharbaty, *J. Non-Cryst. Solids* **15,** 410 (1974).
95. W. Beyer, H. Mell, and H. Overhof, *Proc. Int. Conf. Amorphous Liq. Semiconduct., 7th, Univ. Edinburgh* (W. E. Spear, ed.), p. 328 (1977).
96. K. Wang, Y. Sawan, and M. Silver, *J. Non-Cryst. Solids* **97/98,** 631 (1987).

97. N. F. Mott and E. A. Davis, "Electronic Processes in Non-Crystalline Materials," 2nd Ed., p. 242. Oxford Univ. Press, London, 1979.
98. P. E. Vanier, *in* "Semiconductors and Semimetals, Vol. 21, Hydrogenated Amorphous Silicon," Part B (J. J. Pankove, ed.), pp. 329–357. Academic Press, New York, 1984.
99. 12 W. E. Spear, *Adv. Phys.* **26,** 811 (1977).
100. P. G. Le Comber, D. I. Jones, and W. E. Spear, *Philos. Mag.* **35,** 1173 (1977).
101. D. Emin, *Philos. Mag.* **35,** 1189 (1977).
102. R. A. Street, "Hydrogenated in Amorphous Silicon," p. 302. Cambridge Univ. Press, Cambridge, England, 1991.
103. S. Minomura, *in* "Semiconductors and Semimetals, Vol. 21: Hydrogenated Amorphous Silicon," Part A (J. I. Pankove, ed.), pp. 284–285. Academic Press, New York, 1984.
104. R. A. Street, *in* "Semiconductors and Semimetals, Vol. 21, Hydrogenated Amorphous Silicon," Part B (J. I. Pankove, ed.), p. 237. Academic Press, New York, 1984.
105. G. D. Cody, *in* "Semiconductors and Semimetals, Vol. 21: Hydrogenated Amorphous Silicon," Part B (J. I. Pankove, ed.), p. 42. Academic Press, New York, 1984.
106. N. F. Mott and E. A Davis, "Electronic Processes in Non-Crystalline Materials," 2nd Ed., p. 113. Oxford Univ. Press, London, 1979.
107. N. Bernhard and G. H. Bauer, *Proc. E. C. Photovoltaic Sol. Energy Conf.* (L. Guimaraes, W. Palz, C. De Reyff, H. Kiess, and P. Helm, eds.), p. 92. Harwood Academic Publ., Montreux, Switzerland, 1992.
108. T. Unold, J. D. Cohen, and C. M. Fortmann, *Mater. Res. Soc. Symp. Proc.* **258,** 499 (1992).

4 Reactor Design for a-Si:H Deposition

Jérôme Perrin
Laboratoire de Physique des Interfaces et des Couches Minces
Ecole Polytechnique
Palaiseau, Cedex, France

I. Introduction

The concept of plasma-enhanced chemical vapor deposition (PECVD) is derived from the original term chemical vapor deposition (CVD) corresponding to the deposition of solid films from the thermal decomposition (or pyrolysis) of a reactive gas. In PECVD, the plasma "assistance" is first discussed in terms of temperature and yield, as the deposition can be obtained at a lower wall and gas temperature than conventional CVD, and second in terms of materials properties since the low-temperature material (PECVD) has properties different from those of the high-temperature material (CVD). In some cases, the main argument for choosing PECVD is the reduction of the process temperature to avoid a deterioration of the substrate. However, in the case of a-Si:H PECVD prepared from pure SiH_4 gas (or SiH_4 diluted in H_2 or rare gases), the second argument prevails since a-Si:H, obtained between 200 and 300°C, has unique semiconducting optoelectronic properties that have led to the development of the a-Si:H device technology. Unfortunately, from a technological point of view, these properties are quite sensitive to the deposition conditions and require a careful design and operation of the PECVD reactor.

Historically, the early versions of a-Si:H PECVD reactors were derived from the simple planar diode structures of DC or RF glow discharges used for thin-film deposition by reactive cathodic sputtering or for plasma chemical etching in the microelectronics industry. However, many other reactor concepts have been empirically developed and optimized, involving triode or remote plasma configurations, magnetic or electrostatic confinement, very high frequency (VHF) or microwave (MW) excitation, electron cyclotron resonance (ECR), and even plasma jets. Moreover, it has appeared that the plasma "assistance" could be replaced by external photon sources (lamps or lasers) operating either in the UV or in the IR spectral range. It should be noted that nonisothermal CVD techniques such as homogeneous CVD (homo-CVD) or hot-filament CVD can also be used to deposit a-Si:H material. In this wide typology of reactors, the reasons for the claimed advantage of one type versus the others are often unclear and the choice might be troublesome, if one remains at the level of the reactor technology and external control parameters and considers the plasma as a "blackbox."

In order to understand or compare the different types of reactors and their operating conditions, one should relate the external control parameters (electrode geometry, gas flow management, pressure, electrical excitation frequency, mode of coupling, power, etc.) to the energy and material balance governing the a-Si:H yield and to the gas-phase and surface physicochemistry governing the a-Si:H film growth process and its microstructural and optoelectronic properties. In that respect, the accumulated knowledge from the combination of (1) diagnostics of the plasma components (electrons, molecules, positive and negative ions, neutral atoms and radicals, clusters, or powders), (2) basic data on elementary collisions and reactions, (3) discharge modeling, and (4) surface reactions kinetics, now allows a global view of the effects of the control parameters on the a-Si:H deposition performances. A synopsis of this approach is shown in Fig. 1.

This chapter is divided in three parts. Section II is devoted to the electrical

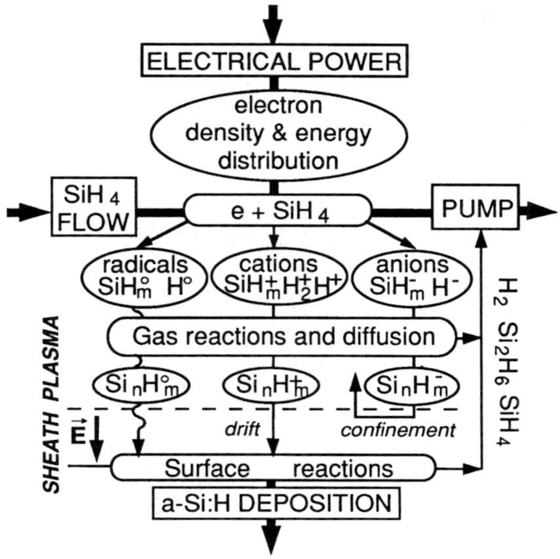

FIGURE 1. Synopsis of the energetic and material balance and the physicochemistry in a-Si:H PECVD from SiH_4.

power dissipation in the discharge and emphasizes the different mechanisms by which energy is coupled to the electrons and released into dissociation, ionization, attachment, and vibrational excitation of the molecules and eventually ends up on the walls via exothermic chemical reactions, gas heating, and thermal diffusion. Section III considers the material balance in terms of gas flow and SiH_4 dissociation efficiency and the basic effects of gas phase and surface physicochemistry on the a-Si:H film quality in relation to the external reactor control parameters. Section IV is a review of the various types of reactors and of their respective advantages and limitations with respect to the effects discussed in Section III. In addition, alternative a-Si:H deposition methods such as UV or IR photo-CVD, homo-CVD, hot-filament CVD, and reactive sputtering are discussed in comparison with PECVD.

II. Power Dissipation Mechanisms in SiH_4 Discharges

A. Structure of a Glow Discharge

Let us consider a simple diode planar structure where a DC voltage V or an RF voltage $V(t) = V_{RF}\cos(\omega t)$ is applied between the electrodes. Once the discharge is ignited, the discharge volume can be decomposed into three regions:

- A central region, the *plasma,* electrically quasineutral, i.e., where the negative charge density [electrons, negative ions, and negatively charged particulates (see Section III.D)] equals the positive ion density:

$$n_e + n_- = n_+. \tag{1}$$

  In the plasma volume, the average electron energy is often characterized by a temperature T_e. Glow discharge plasmas are far from the thermodynamic equilibrium, and T_e is much larger than the gas and ion temperature (see Section II.C.1).
- Two *sheaths* between the plasma and the electrodes, where a space charge develops. Due to the difference in mobility between the electrons and ions, the space charge is mostly positive, and the sheath electric field tends to accelerate the positive ions toward the walls and confine electrons in the plasma to maintain the plasma electroneutrality (consequently, negative ions or particulates are trapped in the plasma). Within the sheath, a small fraction of fast electrons gains energy, which allows ionization of the gas molecules, as discussed in Section II.B.

In a static DC discharge most of the voltage variation occurs across the positive space charge of the cathodic sheath, whereas the space charge of the anodic sheath is negligible (slightly negative or positive depending on discharge conditions). The edge of the cathodic sheath where fast electrons dissipate their energy is called the *negative glow.* Depending on the pressure and interelectrode distance, a small potential gradient may also exist between the negative glow and the anodic sheath, throughout a part of the plasma zone usually called the *positive column.*

In capacitively coupled RF discharges, the sheath dynamics depends on the frequency ω. At low frequency ($\leq$1 MHz) the discharge behaves as a DC discharge with alternative cathodic and anodic sheath on each electrode since both electrons and ions respond to the instantaneous electric field. However, at high frequency (e.g., the usual frequency of 13.56 MHz) only the electrons respond to the instantaneous field, whereas ions "see" the time-averaged field [1]. Moreover, when the RF power is coupled to the discharge via a blocking capacitor, a DC self-bias V_B may appear on the RF-powered electrode depending on the "effective" area ratio S_1/S_2 of the RF-powered surface and the grounded surface facing the plasma ("effective" means that one has to take into account not only the grounded plate facing the RF electrode but also lateral grounded surfaces of the reactor walls toward which the plasma tends to expand). According to simple analytical models and assuming that the sheaths can be represented as a capacitance in the equivalent RF circuit of the discharge, and that both electrodes collect identical ion currents, the ratio of the time-averaged potential drops $V_{s,1}$ and $V_{s,2}$ through the sheaths follows an inverse power law of S_1/S_2 verified experimentally [2, 3].

$$\frac{V_{s,1}}{V_{s,2}} \approx \left(\frac{S_2}{S_1}\right)^n \qquad with \quad 1 \leq n \leq 4. \tag{2}$$

Practically, $n \leq 2$ for most PECVD conditions. Moreover, by ensuring that the instantaneous plasma potential is always positive with respect to each electrode, one obtains the following relations between $V_{s,1}$ and $V_{s,2}$ and V_{RF} and V_B.

$$V_{s,1} \approx \frac{1}{2}(V_{RF} - V_B), V_{s,2} \approx \frac{1}{2}(V_{RF} + V_B). \tag{3}$$

The time-averaged plasma potential V_P with respect to the ground is equal to $V_{s,2}$. From equations (2) and (3), one derives that V_B is negative if $S_1 < S_2$, which is generally the case in conventional RF diode planar discharges. Figure 2a illus-

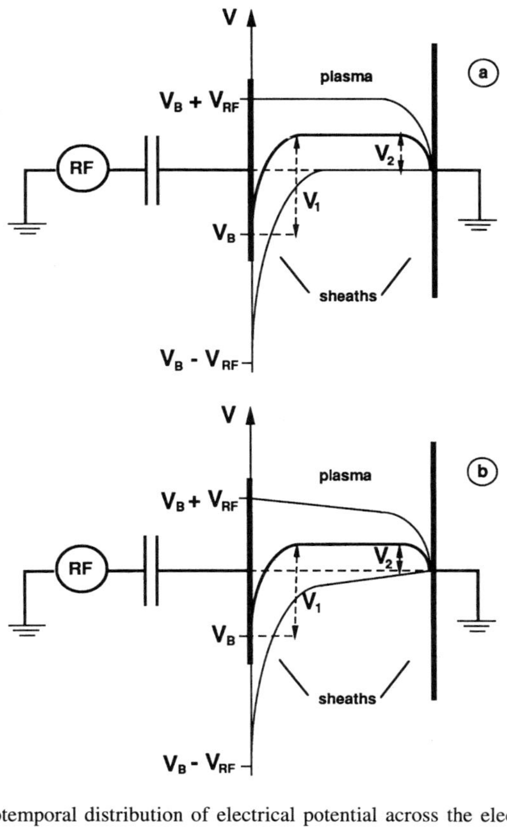

FIGURE 2. Spatiotemporal distribution of electrical potential across the electrodes in a RF discharges: (a) electropositive gas at low pressure where the RF electric field in the plasma bulk is negligible; (b) electronegative gas (SiH_4) at high pressure where electron attachment in the plasma bulk induces the buildup of a RF electric field between the sheaths.

trates the spatiotemporal distribution of the potential in an asymmetrical RF discharge when the potential gradient throughout the plasma zone is negligible. Nevertheless, when the central plasma zone becomes very resistive (see Section II.B.3), the potential gradient through the plasma is no more negligible and the potential distribution resembles the pattern shown in Fig. 2b.

When the discharge involves an inductive coupling of the RF power by an external coil, or a confinement by permanent magnets or electromagnets, the sheath and plasma behaviors are, of course, affected by the magnetic field distribution superimposed on the electric field.

Eventually, as the frequency increases toward the MW range (e.g., the usual frequency of 2.45 GHz), the wavelength appraoches the characteristics dimensions of the discharge ($\lambda = 12$ cm at 2.45 GHz), which affects the spatiotemporal voltage and field distribution. Moreover, the parasitic inductances along the plates or the electrical feedthroughs in the reactor play an increasing role in the equivalent circuit of the discharge.

B. Fast-Electron Generation

1. *Secondary-Electron Emission: γ Regime*

The main mechanism of fast-electron generation in a DC discharge involves secondary-electron emission under ion bombardment of the cathode and acceleration of these secondary electrons through the cathodic sheath as illustrated in Fig. 3a. The accelerated electrons release their energy by ionization, dissociation, and excitation in the negative glow, which is the brighter part of the discharge. In discharge operating conditions, the effective secondary-electron emission coefficient γ (or second Townsend coefficient) involves not only ion-induced emission but also the contribution of metastable atoms or molecules and of UV photons, and depends on the nature of the cathode surface, on the composition of the ion, on metastable and photon flux, and on the cathodic sheath electric field and gas pressure that defines the ion kinetic energy distribution on the electrode. Secondary-electron emission also takes place in RF discharges on each electrode, especially on the negatively biased one. For PECVD in SiH_4 gas, the relevant value of γ corresponds to the growing a-Si:H film surface. The only measured value of γ in these conditions was found to be ≈ 0.03, which is approximately constant with surface temperature between 25 and 250°C [4]. A value of $\gamma = 10^{-3}$ was used in a modeling study of pure SiH_4 DC discharges [5].

2. *Sheath Heating: α Regime*

In RF discharges, the successive contraction and expansion of the sheath is accompanied by the movement of plasma electrons incoming from the plasma to the

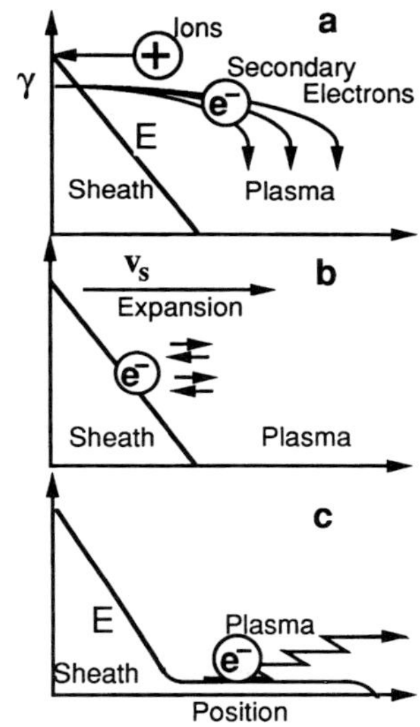

FIGURE 3. Power dissipation mechanisms and discharge regimes: (a) ion-induced secondary-electron emission on the electrode, acceleration through the cathodic sheath, and ionization in the negative glow—γ regime; (b) wave riding of electrons on the expanding sheath edge in RF discharges—α regime; (c) Joule heating due to the buildup of a DC or RF field in the plasma bulk—γ' regime.

wall and receding to the plasma. During the sheath expansion, electrons at the sheath edge gain energy from the sheath electric field. At low pressure this corresponds to a collision of electrons with the sheath moving wall, and has been referred to as "stochastic electron heating" [6]. At high pressure when the electron–molecule collision frequency is much larger than the RF frequency, the sheath heating mechanism is analogous to a surfer on a wave, and has been referred to as collisional "wave riding" [7]. This so-called α regime [8] is illustrated in Fig. 3b. For a given value of the RF voltage V_{RF}, the power dissipated in the α regime is proportional to ω^2. In other words, for a given power fed into the discharge, V_{RF} decreases as a function of ω, and consequently the sheath voltage V_s [equation (3)] also decreases.

3. *Plasma-Bulk Joule Heating: γ' Regime*

At high pressure or large interlectrode distance, and when the electrons suffer losses during their transport between both sheath regions by ambipolar diffusion or by electron attachment in electronegative gases, such as SiH_4, an electric field builds up in the plasma zone to heat the electrons in order to compensate electron losses by additional ionization. This corresponds to the positive column of a DC discharge. The positive column region may also exist in RF discharges. This mechanism illustrated in Fig. 3c has been referred to as "Joule heating" since the positive column is equivalent to an electrical resistance. We also call it γ' regime for personal historical reasons and to distinguish it from the γ regime. In SiH_4 discharges, electron attachment occurs not only onto SiH_4 molecules but also onto radicals and above all on powders trapped in the plasma which tend to increase the contribution of the γ' regime to the power dissipation (see Sections II.B.5 and III.D).

4. *Electron Cyclotron Resonance*

Another mechanism is exploited by combining an external magnetic field B and a microwave (MW) source. The condition of electron cyclotron resonance (ECR) is achieved when the MW angular frequency ω matches the electron cyclotron frequency, $\omega_{ce} = e \cdot B/m_e$, (where e and m respectively are the electron charge and mass values and B the resonant magnetic field value). Therefore, for a given frequency, ω, the required magnetic field is

$$B = \omega \cdot m_e/e, \tag{4}$$

which corresponds to 880 G for $\omega/2\pi = 2.45$ GHz. However, the ECR is efficient only when the electron collision mean free path $l_{eN} = (\sigma_{eN} \cdot N)^{-1}$, where N is the molecule density and σ_{eN}, the collision cross section for momentum transfer, is much larger than the electron Larmor gyroradius, $r_{Le} = v_e/\omega_{ce}$, where v_e is the electron velocity. Practically, for ECR at 2.45 GHz, the maximum pressure for efficient energy coupling is around 10^{-2} torr.

5. *Transition between Discharge Regimes*

The relative contributions of the different power dissipation mechanisms evolve with control parameters of the discharge such as applied voltage V_{DC} or V_{RF}, frequency ω, pressure P and temperature defining the gas density $N = P/k_B T$, nature of the gas, and interelectrode distance d. In a planar diode DC discharge, the cathodic sheath and negative glow (γ regime) tend to shrink as the product Pd increases, whereas the length and the contribution of the positive column (γ' regime) increase. In a planar diode RF discharge three regimes (α, γ, γ') can exist. The α regime is unique to RF discharges. It allows operating conditions at lower pressure and sheath potential than in DC discharges in the same geometry. The transition from the α regime to a situation in which the γ or γ' regimes become dominant is rather sudden and appears as a secondary breakdown [8]. In a rare gas such as He, the transition directly takes place between the α and γ regimes since there is no electron attachment and the plasma-bulk electric field remains very weak. This transition has been well described by numerical modeling in excellent agreement with experimental results [9, 10]. On the contrary, in SiH_4, the transition that had been originally interpreted as a α–γ transition [11], appeared to be a α–γ' transition where plasma-bulk Joule heating suddenly takes over wave riding [12]. A comparison with numerical modeling [13, 14] revealed that the effective-electron attachment coefficient in the γ' regime is 15 times larger than the electron attachment coefficient on SiH_4 molecules. This is attributed to an enhanced attachment onto radicals and powders generated in the plasma. Actually powder particles trapped in the plasma act as electron scavengers and trigger the α–γ' transition. The domains of the α regime and of the γ' regime in a V_{RF}/P diagram for a 13.56-MHz symmetrical planar diode RF discharge with $d = 3.6$ cm and a gas temperature of 200°C are shown in Fig. 4 (see also Section III.D and Fig. 16). When working with constant RF power at the generator, the α–γ' transition results in a sudden drop of V_{RF} as $V_{\alpha\gamma'}$ is approached since the discharge becomes more resistive in the γ' regime. Moreover, when the γ' regime is established and the plasma filled with powders (see Section III.D), the reverse γ'–α transition occurs at a RF potential $V_{\gamma\alpha}$ lower than $V_{\alpha\gamma'}$ (hysteresis as shown in fig. 5)

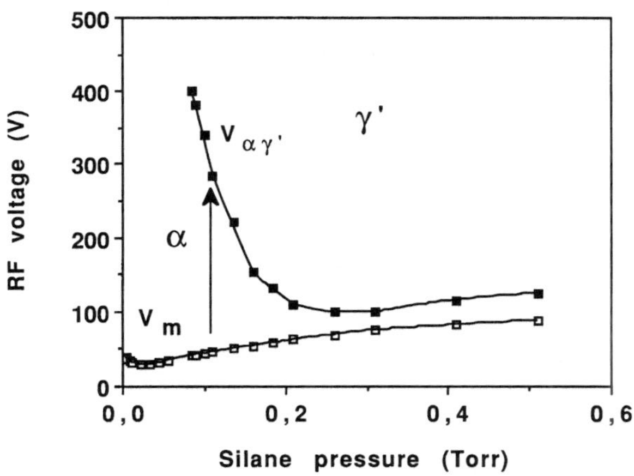

FIGURE 4. Existence domains of the α and γ' regimes in a SiH_4 13.56-MHz RF discharge as a function of pressure and RF potential, for an electrode distance of 3.6 cm and a gas temperature of 200°C; V_m is the minimum RF voltage to maintain the discharge, and $V_{\alpha\gamma'}$ is the α–γ' transition voltage.

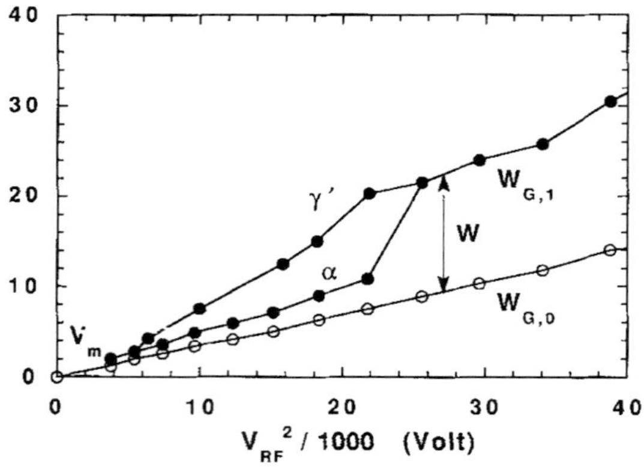

FIGURE 5. Illustration of the substractive method to derive the actual power W dissipated in a rf discharge [13.56-MHz discharge in SiH_4(10%)/He at 0.5 torr and 200°C]. The net power at the generator is measured when the discharge in on $W_{G,1}$, and when the discharge in off $W_{G,0}$, as a function of the RF voltage V_{RF}. $W = W_{G,1} - W_{G,0}$ at a fixed value of V_{RF}. The two curves for $W_{G,1}$ vs. V_{RF}^2 correspond to the α and γ' regimes.

C. Macroscopic Electrical Parameters

1. Orders of Magnitude

The discharge electrical properties are first characterized by external parameters:

- The total power W, from which one derives the power volumic density $w = W/V$ or surface density $w = W/S$, where V and S are the discharge volume and electrode area, respectively. In RF or MW discharges the actual value of the power (watts) fed into the discharge can be significantly smaller than the net power delivered by the generator due to the losses in the matching circuit (see Section II.C.2).
- The voltages V_{DC} (DC discharge) or V_{RF} and the self-bias V_B (RF discharge) from which the sheath potentials can be estimated [see Section II.A and equations (2) and (3)].
- The discharge current I or current density $J = I/S$. In RF discharges the instantaneous current $I(t)$ has a phase shift ϕ with respect to $V(t)$ so that the total peak RF current density J_{RF} on the electrode can be decomposed in a large contribution of displacement current and much smaller conduction current densities of electrons J_e and positive ions J_+.

In usual operating conditions of a-Si:H PECVD, the power (watts) lies typically between a few milliwatts per square centimeter and a few 100 mW/cm^2. In

DC discharges V_{DC} is several hundred volts. For the same power density in RF discharges, V_{RF} tends to decrease with increasing ω, due to the efficiency of the sheath heating mechanism in the α regime. Consequently, the average sheath potential drop V_s is also a decreasing function of ω. J_+ is in the micro- to milliampere per square centimeter ($\mu A/cm^2$ to mA/cm^2) range, whereas at 13.56-MHz J_{RF} is a few milliamperes per square centimeter up to a fraction of amperes per square centimeter (A/cm^2) and the current–voltage phase shift varies from $\cos\phi \approx 0.1$ (α regime) up to 0.5 (γ' regime) [12, 13].

The internal electrical parameters relevant to the discharge physicochemistry are

- The plasma-bulk positive-ion and electron densities n_+ and n_e between 10^8 and 10^{10} cm^{-3}. In SiH_4, the negative-ion density n_- (or negative charge density as powders) can exceed n_e by an order of magnitude [5, 7, 12, 13, 15]. The ratio n_+/N or ionized fraction of the gas ranges from 10^{-7} (low power and high pressure) to 10^{-3} (high power and low pressure).
- The plasma-bulk electron and ion energy distribution and mean energies or temperatures: $T_e \approx 1$ to 4 eV [5, 16], and $T_i \approx 0.05$ eV (close to the gas temperature). However, the electron energy distribution is not Maxwellian in ultra-low-pressure discharges and in the sheath regions, so the concept of electron temperature must be taken with caution.
- The reduced electric field E/P ($E = V/t$), derived from the voltage distribution V and the plasma and sheath thicknesses t and normalized to the pressure P. In SiH_4 gas, DC or RF discharges, the instantaneous value of E/P decreases continuously across the sheaths from 10^3 to 10^4 V cm^{-1} $torr^{-1}$ at the electrode surface. In the plasma bulk, the DC field or effective RF field built up to compensate electron attachment is around 150 V cm^{-1} $torr^{-1}$ [12, 13, 15, 16].
- The positive-ion flux Φ_+ and kinetic energy distribution $f(E_+)$ on the wall. At low pressure (collisionless "drift" transport) the ions acquire, through the sheath voltage V_s, the potential energy $E_+ \approx eV_s$. As the pressure increases, the ion collision mean free path becomes smaller than the sheath thickness ("mobility limited" transsport), resulting in the dispersion of ion energy below eV_s [12]. Since V_s decreases as a function of ω in RF discharges (see Section II.B.2), E_+ should also decrease for a given value of Φ_+.

2. *Diagnostics*

Although this chapter is not devoted to diagnostics, it is worth emphasizing the importance of proper measurement and control of relevant electrical parameters for understanding a PECVD process.

In RF discharges the measurement of V_{RF} can be erroneous if not done as close as possible to the electrode plate because of the Ferranti effect, which induces a rise of V_{RF} along a transmission line terminated by a capacitor [12].

Also the measurement of the net discharge power W is not always straightforward in RF discharges, due to power losses in the matching circuit and electrical feedthroughs. One can measure separately the time-dependent RF voltage and current and integrate the product $V(t) \cdot I(t)$. Such a *direct* procedure and the tedious precautions necessary to achieve proper measurements of $V(t)$ and $I(t)$ and their phase shift are discussed in Böhm and Perrin [12]. An easier *substractive* procedure based on the sole measurements of the net power delivered by the generator $W_G = W_I - W_R$ (incident minus reflected power) and of V_{RF} is discussed in refs. [17, 18]. We have recently confirmed its validity by comparison with the *direct* method. Briefly, it consists of measuring W_G with the discharge on (1) $W_{G,1}$, for a given gas and pressure, and without discharge; and (2) $W_{G,0}$, obtained by pumping down the reactor and slightly returning the matching circuit, both as a function of V_{RF}. By plotting $W_{G,1}$ and $W_{G,0}$ vs. V_{RF}^2, one obtains two curves—approximately straight lines crossing each other at $V_{RF} \approx V_m$, the minimum maintenance voltage. The net power going into the discharge is then obtained by the subtraction of $W_{G,0}$ from $W_{G,1}$, at the same value of V_{RF}: $W(V_{RF}) = W_{G,1}(V_{RF}) - W_{G,0}(V_{RF})$. A schematic illustration of the procedure is shown on Fig. 5. The power transfer efficiency defined as $W/W_{G,1}$ depends on the design of the reactor and matching circuit and on the discharge impedance, which varies with the pressure and power dissipation regime. It can be as low as 0.1 at low pressure when the discharge is strongly capacitive in the α regime and jumps to 0.7–0.8 when the discharge becomes more resistive in the γ or γ' regime [12].

The ion current density J_+ and ion kinetic energy distribution on an electrode can also be simply measured by using a grid electrostatic analyzer placed on the grounded electrode [12]. Eventually, one can measure the sheath thicknesses by using spatially resolved optical emission spectroscopy [12, 14]. The measurements of n_e, n_+, and T_e, by using other electrostatic probes or microwave interferometry or resonance techniques, are somewhat more complicated. Simple analytical models and the knowledge of basic data on electron and ion transport in SiH_4 allow one to relate in a self-consistent way the various quantities [12].

D. Power Distribution and Relaxation

1. *Inelastic Electron–Molecule Collisional Transfer*

The electrical energy coupled to the electrons is first released in electron–molecule inelastic collisions. For molecules such as SiH_4 or H_2, the main inelastic processes are vibrational excitation, dissociative attachment, dissociation in neutral atoms or radicals, and ionization (dissociative ionization for SiH_4). The set of electron collision cross sections for SiH_4 and H_2 is now complete enough to attempt self-consistent numerical electrical modeling of SiH_4–H_2 glow discharges [5, 13, 14–16, 19, 20] and derive the fractional power transferred into each elementary process. For high-pressure SiH_4 RF discharges (positive columns), Capitelli *et al.* [16] have shown that the fractional power dissipated by electron excitation of the (γ_2, γ_4) and (γ_1, γ_3) SiH_4 vibrations (at 0.11 and 0.27 eV) can be up to 2.6 times larger than the fractional power dissipated into the dissociation, ionization, or attachment (>8–9 eV), This ratio evolves with gas composition in SiH_4/H_2 mixtures [16, 19] and decreases at low SiH_4 pressure when the discharge is in the α regime and is not dominated by the positive column [13, 14, 18, 20]. Figure 6 shows the computed spatial distribution of the power density dissipated in each process for a pure SiH_4 13.56-MHz RF discharge in a symmetrical configuration at 55 and 185 mtorr, with the temperature at 200°C, using a "particle in cell" numerical method [13, 20].

2. *Vibrational, Rotational, and Translational Heating of the Gas*

The large fractional power dissipated in vibrational excitation of SiH_4 molecule has led to a controversial statement [21] that the main dissociation route in glow discharges could be a vibrationally induced "pyrolysis":

$$\text{e (slow)} + SiH_4 \rightarrow SiH_4^{v*} \rightarrow SiH_2 + H_2$$

instead of the main electron impact dissociation channels [14, 22, 23]:

$$\text{e (fast)} + SiH_4 \rightarrow SiH_2 + 2\,H \quad \text{or} \quad SiH_3 + H.$$

Two diagnostic tools have given some insight into the vibrational excitation of SiH_4 glow discharges: (1) infrared emission spectroscopy [24] in a high-power SiH_4/H_2 discharge at 0.25 torr, from which rotational and vibrational temperatures 300 K $< T_R <$ 480 K and $T_V \approx$ 850 K were derived; and (2) coherent anti-Stokes Raman scattering (CARS) in pure SiH_4 (0.1 torr) or SiH_4/H_2 (1 torr) discharges [25] revealing an inhomogeneous rotational and vibrational excitation peaking in the plasma bulk (see Fig. 7a; a minimum of CARS intensity corresponds to a maximum in T_R and T_V).

We have developed a model [26] to compute both the transrotational temperature T_R (rotation and translation are equilibrated) and T_V from a detailed balance

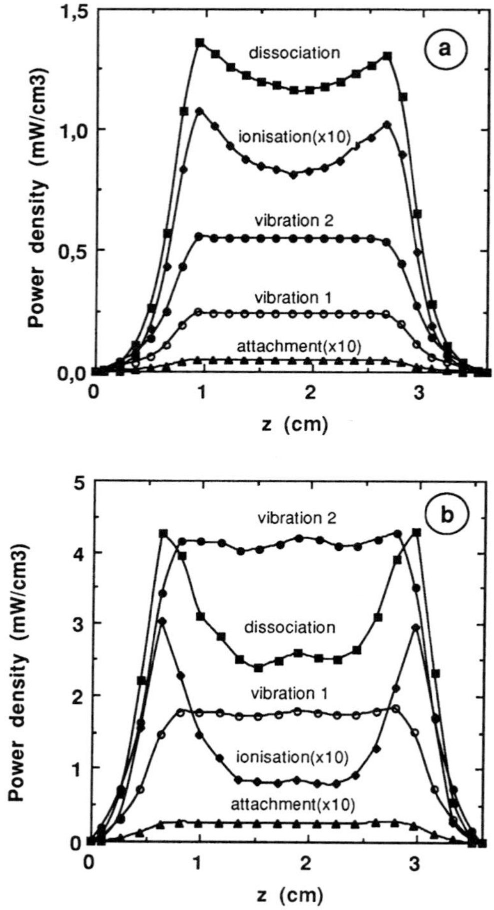

FIGURE 6. Computed spatial distribution of power densities in 13.56-MHz RF discharges in pure SiH_4 for two pressure conditions at 200°C: (a) 55 mtorr; (b) 185 mtorr using a "particle in cell" numerical method [13, 20].

of the power dissipation channels described in Fig. 8, which can determine whether the contribution of vibrationally induced pyrolysis could be of any significance in the SiH_4 dissociation mechanism. The model has been tested in the discharge conditions (geometry, RF power, pressure) given in refs. [12, 13, 24, 25]. The SiH_4 partial pressure was derived from the balance between the flow rate and the SiH_4 decomposition rate corresponding to a-Si:H deposition (the

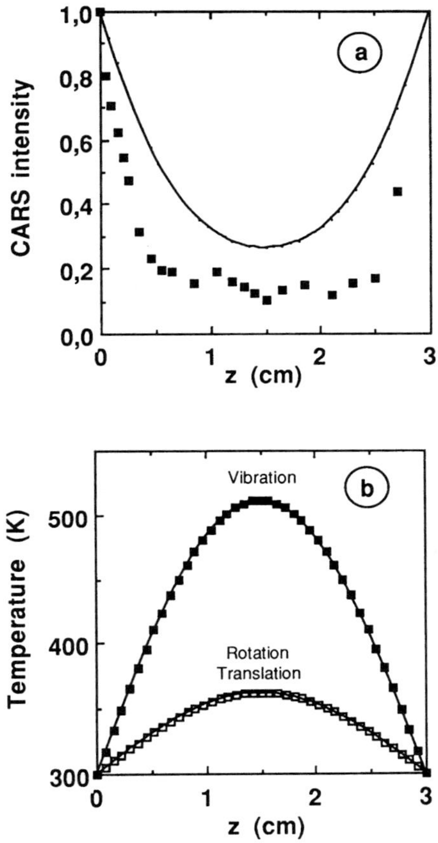

FIGURE 7. (a) Measured CARS intensity profile in a SiH_4/H_2 RF discharge at 1 torr (10 sccm SiH_4, 90 sccm H_2) and 200 mW/cm^3 total power density (data from Hata and Tanaka [25]) and computed CARS profile derived from the power dissipation modeling [26]; (b) computed transrotational and vibrational spatial temperature profiles.

initial SiH_4 fraction in the mixture is depleted under plasma condition and H_2 builds up; see Section III.A). For the IR emission spectroscopy experiment [24] the computed values of $T_R = 370$ K and $T_V = 870$ K are in fair agreement with the measured values. For the CARS experiments, the computed T_R and T_V temperature profiles shown in Fig. 7b were used to derive the expected CARS intensity spatial profile. Here again, a fair agreement is obtained. Then, applying the model to a variety of discharge conditions used for a-Si:H deposition, it can be concluded that the fractional power going into pyrolytic decomposition of SiH_4

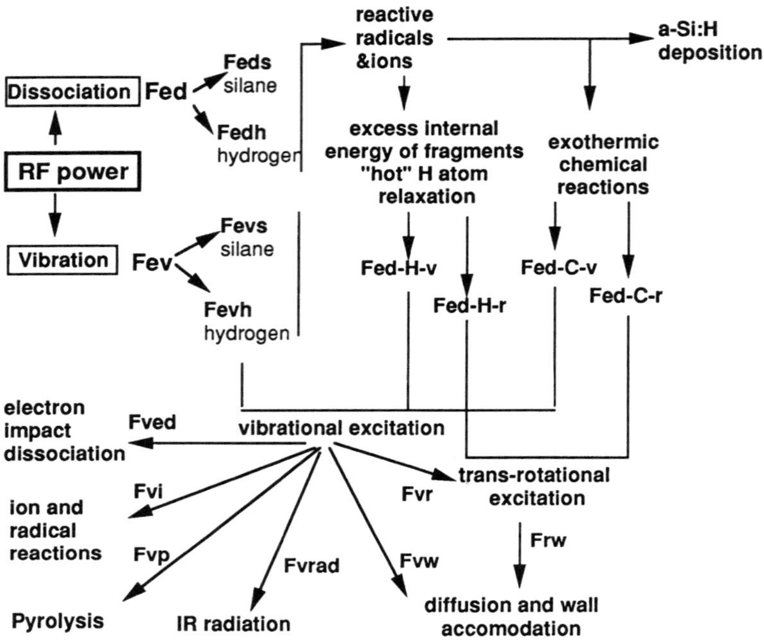

FIGURE 8. Synopsis of power dissipation channels in SiH_4/H_2 RF glow discharges.

never exceeds 1% (for high power density discharges) of the fractional power going into direct electron impact dissociation. This definitively rules out the "pyrolysis" hypothesis raised in ref. [21].

In fact, whatever the power dissipation channels and the chemical reactions, it appears that the global decomposition reaction

$$SiH_4 \rightarrow \text{a-Si:H}_x + (2 - x/2)\,H_2$$

is slightly exothermic. The reaction enthalpy $\Delta H_r = \Delta H_f(\text{a-Si:H}) - \Delta H_f(SiH_4)$ where the ΔH_f values are the formation enthalpies referred to solid crystalline Si and molecular H_2. Since $\Delta H_f(\text{a-Si:H}) \approx \Delta H_f(\text{c-Si})$ and $\Delta H_f(SiH_4) \approx 8.2$ kcal/mol $= 0.36$ eV, $\Delta H_r < 0$. This means that the discharge power that is required to activate the endothermic dissociation of SiH_4 into primary radicals or ions is eventually entirely converted in wall heating or radiative emission by secondary gas-phase and surface reactions leading to the formation of a-Si:H and the release of H_2 (see also Sections III.B and III.C).

III. Material Balance and Gas-Phase and Surface Physicochemistry

A. Material Balance

1. *Gas Consumption and a-Si:H Yield*

The material conversion in a a-Si:H PECVD reactor implies the balance of the molecular flows of reactive source molecules (SiH_4 and H_2) injected into the reactor, with the atomic flow incorporated on the wall as solid deposit (a-Si:H film or powders) and the molecular flows evacuated by the pump (SiH_4, H_2, and also higher order silanes Si_2H_6, Si_3H_8, etc. produced by gas-phase reactions). These flows are designated by F_{0,SiH_4}, F_{0,H_2}, $F_{a\text{-}Si:H}$, F_{1,SiH_4}, F_{1,H_2}, F_{1,Si_2H_6}, and F_{1,Si_3H_8}, respectively. Let us recall the conversion from the usual gas flow rate unit in sccm (standard cubic cm per minute) into molecule/s:

$$F(\text{molecule/s}) = F(\text{sccm})\frac{N_L}{60}, \tag{5}$$

where $N_L = 2.69\ 10^{19}$ molecule/cm^3 is the Loschmidt number. Similarly, the a-Si:H deposition rate, r_d in Angströms per second, can be converted into an atomic Si flow, $F_{a\text{-}Si:H}$, using an estimate of the total deposition area S_d, the molar mass of silicon $M_{Si} = 28$ g/mol, and the specific gravity of a-Si:H $\rho_{a\text{-}Si:H} \approx 2 - 2.3$ g/cm^3 depending on the PECVD conditions (especially the temperature):

$$F_{a\text{-}Si:H}\ (\text{Si atom/s}) = 10^{-8}\ r_d\ (\text{Å/s}) S_d\ (\text{cm}^2)\ \frac{\rho_{a\text{-}Si:H}}{M_{Si}}\ N_A, \tag{6}$$

where $N_A = 6.02\ 10^{23}$ atom/mol is the Avogadro number. The steady-state Si and H atomic balance flow equations are

$$F_{0,SiH_4} = F_{a\text{-}Si:H} + F_{1,SiH_4} + 2F_{1,Si2H_6} + 3F_{1,Si_3H_8} + \cdots, \tag{7}$$

$$2F_{0,SiH_4} + F_{0,H_2} = (x/2)F_{a\text{-}Si:H} + 2F_{1,SiH_4} + 3F_{1,Si_2H_6} + 4F_{1,Si_3H_8} + \cdots, \tag{8}$$

where x is the atomic H content in a-Si:H, $x \approx 0.05$–0.3 depending on PECVD conditions (usually $x \approx 0.1$ at 200°C).

The depleted fraction or consumption efficiency of SiH_4 and the yields of a-Si:H and higher-order silanes are derived by normalization to the injected SiH_4 flow rate:

$$Y_{\text{-SiH}_4} = \frac{F_{0,\text{SiH}_4} - F_{1,\text{SiH}_4}}{F_{0,\text{SiH}_4}}, \quad (9)$$

$$Y_{\text{a-Si:H}} = \frac{F_{\text{a-Si:H}}}{F_{0,\text{SiH}_4}}, \quad (10)$$

$$Y_{\text{Si}_2\text{H}_6} = \frac{F_{1,\text{Si}_2\text{H}_6}}{F_{0,\text{SiH}_4}}, \quad (11)$$

$$Y_{\text{Si}_3\text{H}_8} = \frac{F_{1,\text{Si}_3\text{H}_8}}{F_{0,\text{SiH}_4}}. \quad (12)$$

Eventually, the fractional Si atom conversion efficiencies from SiH_4 are given by

$$\eta_{\text{a-Si:H}} = \frac{Y_{\text{a-Si:H}}}{Y_{-\text{SiH}_4}} = \frac{F_{\text{a-Si:H}}}{F_{0,\text{SiH}_4} - F_{1,\text{SiH}_4}}, \quad (13)$$

$$\eta_{\text{Si}_2\text{H}_6} = \frac{2Y_{\text{Si}_2\text{H}_6}}{Y_{-\text{SiH}_4}} = \frac{2F_{1,\text{Si}_2\text{H}_6}}{F_{0,\text{SiH}_4} - F_{1,\text{SiH}_4}}, \quad (14)$$

$$\eta_{\text{Si}_3\text{H}_8} = \frac{3Y_{\text{Si}_3\text{H}_8}}{Y_{-\text{SiH}_4}} = \frac{3F_{1,\text{Si}_3\text{H}_8}}{F_{0,\text{SiH}_4} - F_{1,\text{SiH}_4}}, \quad (15)$$

and the fractional H-atom conversion efficiency from SiH_4 into H_2 is

$$\eta_{\text{H}_2} = 4 - \text{x}\ \eta_{\text{a-Si:H}} - 3\ \eta_{\text{Si}_2\text{H}_6} - (8/3)\eta_{\text{Si}_3\text{H}_8} + \cdots. \quad (16)$$

Since the cost of SiH_4 gas is high, then the optimization of $Y_{\text{a-Si:H}}$, and hence in $\eta_{\text{a-Si:H}}$, is desired. However, because of the requirements of a-Si:H deposition homogeneity on large areas, gas dynamics have to be taken into account (Section III.A.2). When using pure SiH_4 gas, it is found that the best conditions of gas flow management and optimized a-Si:H films usually limit, $Y_{-\text{SiH}_4}$ to 50%. It should be noted that the fractional Si atom conversion efficiency into a-Si:H, $\eta_{\text{a-Si:H}}$, and higher-order silanes, η_{SinHm}, strongly depend on the pressure and discharge power conditions. As the SiH_4 gas pressure increases, $\eta_{\text{a-Si:H}}$ decreases from 1 to 0.6 for a deposition pressure of 0.25 torr [22]; this is due to secondary gas-phase reactions, which produce higher order silanes. Hence, the quantities $\eta_{\text{Si}_2\text{H}_6} + \eta_{\text{Si}_3\text{H}_8} \approx (1 - \eta_{\text{a-Si:H}})$ can reach 0.4. The corresponding molecular yield of disilane, $Y_{\text{Si}_2\text{H}_6}$ can reach 10–20% of the depleted SiH_4 fraction, whereas $Y_{\text{Si}_3\text{H}_8}$ is in the 0.1–1% range [note that the multiplication factors 2 and 3 of molecular yield $Y_{\text{Si}_2\text{H}_6}$ and $Y_{\text{Si}_3\text{H}_8}$, respectively to derive $\eta_{\text{Si}_2\text{H}_6}$ and $\eta_{\text{Si}_3\text{H}_8}$ in equations (14) and (15)]. Eventually, with typical values of $\eta_{\text{a-Si:H}} \approx 0.7$, $x \approx 0.1$, $\eta_{\text{Si}_2\text{H}_6} \approx 0.27$, and

$\eta_{Si_3H_8} \approx 0.03$ for 0.1–0.2 torr deposition pressure at 200°C, one can derive from equation (16) that $\eta_{H_2} \approx 3$, i.e., one SiH_4 molecule produces 1.5 H_2 molecule; the remaining H atoms are either incorporated in the a-Si:H film or are mostly pumped out as higher-order silanes. However at lower pressures η_{H_2} can increase up to 3.8.

The measurements of Y_{-SiH_4}, $Y_{Si_2H_6}$, and $Y_{Si_3H_8}$ can be performed by sampling the gas through a microleak placed before the pumping outlet of the reactor, and by analyzing the gas composition by mass spectrometry [21, 22, 27–30]. The main difficulties consist in calibrating the mass-spectrometric signals in terms of partial pressures at the sampling point and to properly take into account possible differences in the pumping velocity or the residence time of the various molecules in the reactor (see Section III.A.2) in order to derive the outlet flows F_{1,SiH_4}, F_{1,H_2}, F_{1,Si_2H_6}, and F_{1,Si_3H_8} [28].

In terms of electrical power efficiency, it is interesting to know the electrical energy needed to decompose one SiH_4 molecule. This can be done by dividing the net power (W) fed into the discharge by the consumed SiH_4 flow: $\Delta F_{SiH_4} = F_{0,SiH_4} - F_{1,SiH_4}$, where

$$E_{-SiH_4}\,(\text{eV}) = \frac{(W/e)}{\Delta F_{SiH_4}} \tag{17}$$

with W in watts, ΔF_{SiH_4} in molecules per second, and $e = 1.6\ 10^{-19}$ C. A plot of E_{-SiH_4} vs. W for pure SiH_4 RF discharges at different pressures and injected flow rate F_{0,SiH_4} is shown in Fig. 9. E_{-SiH_4} slightly decreases with increasing power

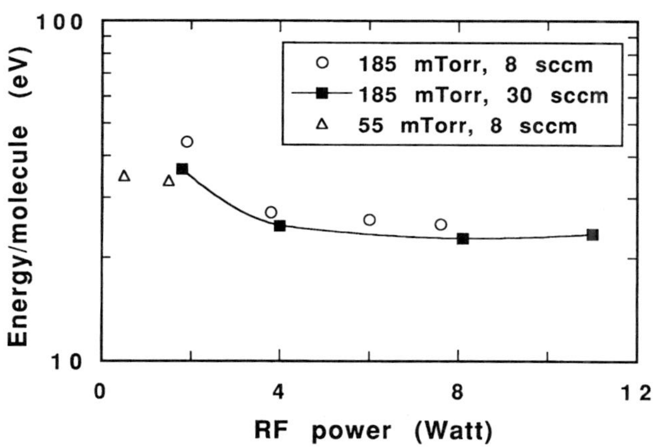

FIGURE 9. Electrical energy consumed per dissociated SiH_4 molecule for three pressures and flow rate conditions as a function of discharge RF power.

but levels off at around 23 eV. Under these conditions where the depleted fraction, $Y_{-\mathrm{SiH}_4}$, remains smaller than 50%, $\Delta F_{\mathrm{SiH}_4}$ is roughly proportional to W. Moreover, there is no apparent change of $E_{-\mathrm{SiH}_4}$ at the discharge transition from the α regime to the γ' regime at the highest pressure of 0.185 torr. However, one might expect a rapid increase of $E_{-\mathrm{SiH}_4}$ at higher depletion where the gas composition becomes dominated by H_2 as reported in a similar study by Gallagher [27]. The value of $E_{-\mathrm{SiH}_4} \approx 23$ eV in RF discharges rich in SiH_4 seems quite high in comparison with the electron impact dissociation threshold around 8–10 eV. This can be accounted for by the large fractional power dissipated in vibrational excitation (see Section II.D), which is eventually redistributed into transrotational gas heating and released on the wall (see Fig. 8).

2. *Gas Dynamics and Partial Pressures*

The partial pressures of SiH_4 and the molecular products of the plasma—H_2, Si_2H_6, and Si_3H_8—are determined by the injected flow rates, the pumping velocity or gas residence time in the reactor, and the dissociation or production rate of each species. In a PECVD reactor, the gas dynamics are controlled either by diffusion or convection and can be characterized by the Péclet number:

$$\mathrm{Pe} = \frac{\tau_\mathrm{D}}{\tau_\mathrm{C}}, \tag{18}$$

where τ_D and τ_C are the characteristic times for diffusion and convection across the discharge volume. For high flow rates and low diffusivities (high pressures) or long distance between the gas injection point and pumping port, $\mathrm{Pe} \gg 1$ and convection dominates. This corresponds to a "plug-flow" reactor type where the partial pressures of the molecules vary along the flow axis. Therefore, at high SiH_4 dissociation rate (high power), the large SiH_4 partial pressure gradient is likely to result in an inhomogeneous a-Si : H deposition. On the contrary, for low flow rates and high diffusivities (low pressure) or distributed gas injection (e.g., by using a shower-head electrode), $\mathrm{Pe} \ll 1$ and diffusion and backmixing of the gas dominate. This corresponds to a well-mixed or continuous stirred tank reactor where the partial pressures are spatially homogeneous.

We further consider the case of low Péclet numbers and assume homogeneous partial pressures in the whole reactor. In a steady-state plasma condition, each molecule X can be attributed a characteristic residence or pumping time, $\tau_{\mathrm{R,X}}$. In the most general case [28] $\tau_{\mathrm{R,X}}$ depends not only on the nature of the molecule but also on the composition of the gas mixture for a fixed pumping aperture. Therefore, a given molecule may have different residence time in plasma-off and plasma-on conditions, which can be represented as $\tau_{R_0,X}$ and $\tau_{R_1,X}$ respectively. Considering only SiH_4 and H_2 gases for simplicity, the molecular flows (Section III.A.1) and densities $N_{0,X}$ and $N_{1,X}$ are related via the steady-state rate equations:

$$\frac{\partial N_{0,SiH_4}}{\partial t} = \frac{F_{0,SiH_4}}{V_R} - \frac{N_{0,SiH_4}}{\tau_{R_0,SiH_4}} = 0, \tag{19}$$

$$\frac{\partial N_{1,SiH_4}}{\partial t} = \frac{F_{0,SiH_4}}{V_R} - \frac{N_{1,SiH_4}}{\tau_{R_1,SiH_4}} - \frac{N_{1,SiH_4}}{\tau_P}\frac{V_P}{V_R} = 0, \tag{20}$$

$$\partial\frac{N_{0,H_2}}{\partial t} = \frac{F_{0,H_2}}{V_R} - \frac{N_{0,H_2}}{\tau_{R_0,H_2}} = 0, \tag{21}$$

$$\frac{\partial N_{1,H_2}}{\partial t} = \frac{F_{0,H_2}}{V_R} - \frac{N_{1,H_2}}{\tau_{R_1,H_2}} - \frac{\eta_{H_2}}{2}\frac{N_{1,SiH_4}}{\tau_P}\frac{V_P}{V_R} = 0, \tag{22}$$

where V_P and V_R are the plasma volume and reactor volume, respectively; τ_P is the characteristic dissociation time of SiH_4 in the plasma; and $3 \leq \eta_{H_2} \leq 3.8$ is the probability of H-atom conversion from SiH_4 into H_2 [see equation (16)]. The injected molecular flows $F_{0,X}$ are usually known from the mass flow controllers, and equations (19) and (21) allow the determination of the $\tau_{R_0,X}$ from the values of partial densities $N_{0,X}$ or partial pressures $P_{0,X} = N_{0,X}/k_B T$ derived by a combination of the total pressure measurement and mass-spectrometry gas analysis. However, in the plasma-on condition, even if the partial densities $N_{1,X}$ can be measured, the actual molecular flows to the pump, $F_{1,X} = N_{1,X}/\tau_{R_1,X}$, are difficult to estimate if the quantity $\tau_{R_1,X}$ varies with the gas composition.

The situation is simplified in two limiting cases. If the reactor is in the *molecular flow* pumping regime [30] (molecular mean free path larger than the pumping port orifice), $\tau_{R,X}$ is proportional to $M_X^{1/2}$ (the square root of the molecular mass) and $\tau_{R_1,X} = \tau_{R_0,X}$. Therefore, H_2 is pumped 4 times faster than SiH_4 and

$$\frac{N_{0,SiH_4}}{N_{0,H_2}} = 4\,\frac{F_{0,SiH_4}}{F_{0,H_2}} \tag{23}$$

$$\frac{N_{1,SiH_4}}{N_{1,H_2}} = 4\,\frac{F_{1,SiH_4}}{F_{1,H_4}} \tag{24}$$

$$F_{0,SiH_4} - F_{1,SiH_4} = \frac{N_{0,SiH_4} - N_{1,SiH_4}}{\tau_{R,SiH_4}}V_R = \frac{N_{1,SiH_4}}{\tau_P}V_P. \tag{25}$$

Moreover, the total pressure should decrease after plasma ignition. Indeed, if one neglects the small partial pressures of higher-order silanes, the total molecular density becomes

$$N_{1,H_2} + N_{1,SiH_4} = N_{0,H_2} + N_{0,SiH_4} \left(\frac{1 + \dfrac{\eta_{H_2}}{8} \dfrac{\tau_{R,SiH_4}}{\tau_P} \dfrac{V_P}{V_R}}{1 + \dfrac{\tau_{R,SiH_4}}{\tau_P} \dfrac{V_P}{V_R}} \right), \tag{26}$$

which is smaller than $(N_{0,H_2} + N_{0,SiH_4})$ since $(\eta_{H_2}/8) < 1$. This is usually observed when the molecular regime is achieved at a moderate pressure by strongly throttling a valve (hence having very small pumping apertures) above a pumping system of large capacity. But when using a turbomolecular pump of small capacity at low pressure with a large pumping aperture, it is well known that the pumping efficiency of light molecules such as H_2 is poor compared to heavy molecules, so that the molecular pumping regime is not achieved and the reverse effect can be observed.

If the reactor is in the *viscous flow* pumping regime (molecular mean free path much smaller than the pumping port orifice), all the molecules are pumped at the same speed: $\tau_{R,SiH_4} = \tau_{R,H_2} = \tau_R$, but τ_{R_1} is different from τ_{R_0} because of the change of gas composition. Then

$$\frac{N_{0,SiH_4}}{N_{0,H_2}} = \frac{F_{0,SiH_4}}{F_{0,H_2}}, \tag{27}$$

$$\frac{N_{1,SiH_4}}{N_{1,H_2}} = \frac{F_{1,SiH_4}}{F_{1,H_2}}, \tag{28}$$

and $F_{1,SiH_4}/F_{0,SiH_4}$ is not equal to $N_{1,SiH_4}/N_{0,SiH_4}$ as in the molecular regime

$$F_{0,SiH_4} - F_{1,SiH_4} = \left(\frac{N_{0,SiH_4}}{\tau_{R_0}} - \frac{N_{1,SiH_4}}{\tau_{R_1}} \right) V_R = \frac{N_{1,SiH_4}}{\tau_P} V_P. \tag{29}$$

The total pressure variation is also more difficult to predict since

$$N_{1,H_2} + N_{1,SiH_4} = \left(\frac{\tau_{R,1}}{\tau_{R,0}} \right) \left[N_{0,H_2} + N_{0,SiH_4} \left(\frac{1 + \dfrac{\eta_{H_2}}{2} \dfrac{\tau_{R_1}}{\tau_P} \dfrac{V_P}{V_R}}{1 + \dfrac{\tau_{R_1}}{\tau_P} \dfrac{V_P}{V_R}} \right) \right], \tag{30}$$

where $(\eta_{H_2}/2) > 1$ would result in a pressure increase but (τ_{R_1}/τ_{R_0}) decreases because of the changing gas composition from a SiH_4-dominated mixture to a H_2-dominated mixture.

An additional complication to these simple rate equations comes from the fact that PECVD reactors are usually not isothermal. If the partial pressures are homogeneous between different regions of a well-mixed reactor, there are gradients of molecular densities since $N_X = P_X/k_B T$, so that rate equations have to be modified accordingly. Indeed, a rigorous treatment in the general case of a non-isothermal and/or plug-flow reactor requires solving the Navier–Stokes equations by numerical techniques as in ref. [31]. Nevertheless, a first-hand approach considering the pumping regime is of primary importance for transferring a PECVD process from one reactor to another. Indeed, keeping the total pressure and the power density constant, and the SiH_4 and H_2 flow rates scaled to the plasma volume, does not guaranty that the partial pressures of SiH_4 and H_2 are reproduced if the pumping system is not adapted. The only solution consists in checking the partial pressures by mass spectrometry.

In addition, as we shall discuss in Section III.D, the hydrodynamics and the temperature gradients of a PECVD reactor play a considerable role in the control of the evacuation of powders. Any flow vortex or cold spot will tend to be a zone of powder accumulation.

B. Gas-Phase Chemistry and Transport to the Walls

Here, we review only the most important elementary mechanisms in order to outline the main ingredients of the gas-phase chemistry and the dominant precursors to a-Si:H deposition, following the synopsis of Fig. 1.

Since the "pyrolysis" hypothesis can be ruled out (see Section II.D.2), the initial step is the fast-electron impact dissociation of SiH_4 into primary ions and radicals, namely, $SiH^+_{m\leq3}$, H_2^+, and H^+ positive ions, $SiH_{m\leq3}$ and H radicals, and $SiH^-_{m\leq3}$ and H^- negative ions. Although the density of negative ions in the plasma can eventually be rather high because of electrostatic confinement into the plasma volume, the electron attachment rate is much smaller than the rate of dissociation into positive ions and neutral radicals. A schematic representation of the electron impact fragmentation pattern of SiH_4 into $SiH^+_{m\leq3}$ and $SiH_{m\leq3}$ is shown in Fig. 10. Note that the main dissociation channel at low energy is [23] given by

$$e + SiH_4 \rightarrow e + SiH_2 + 2H$$

and not

$$e + SiH_4 \rightarrow e + SiH_2 + H_2.$$

In ultra-low-pressure plasmas ($P \leq 1$ mtorr), such as multipole DC discharges [30, 32, 33] or ECR discharges (see Sections II.B.4 and IV.B), the electron energy distribution is non-Maxwellian [30, 33] and fast electrons can reach energies much higher than the ionization threshold. In such a case, the fragmentation into SiH_m^+ can represent 40–50% of the primary silicon-containing fragments. More-

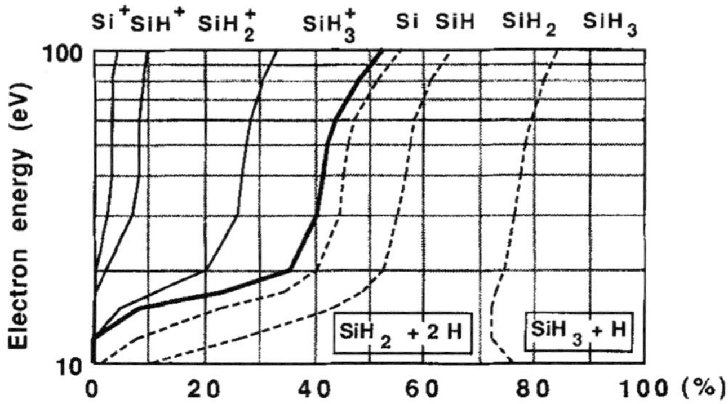

FIGURE 10. Electron impact fragmentation pattern of SiH_4 derived from Doyle *et al.* [22], Perrin *et al.* [30], Perrin and Schmitt [33] and Schmitt [34].

over, since the ion and radical mean free path is of the order of the discharge dimension (1–20 cm) the primary SiH_m^+, SiH_m, H, H^+, and H_2^+ species will mostly reach the walls without secondary gas-phase reactions. The plasma chemistry is therefore not selective but one can take advantage of the incident energy of the positive ions accelerated through the sheaths (see Sections II.A and II.C).

In higher-pressure discharges (0.1-torr range), the electron energy distribution is thermalized and electron impact dissociation close to the ionization threshold (8–15 eV) dominates. Consequently the primary dissociation favors dissociation into neutral fragments, and as shown in Fig. 10, the dominant primary species are likely to be SiH_2, SiH_3, and H [22, 23]. Moreover, secondary reactions take place. Among the primary Si-containing radicals, SiH_3 does not react with SiH_4 except by the H-transfer reaction

$$SiH_3 + SiH_4 \rightarrow SiH_4 + SiH_3,$$

which does not modify the chemistry, whereas SiH_2, SiH, and Si rapidly react with SiH_4 to give

$$SiH_2 + SiH_4 \rightarrow Si_2H_6 \text{ and } Si_2H_4 + H_2,$$

$$SiH + SiH_4 \rightarrow Si_2H_3 + H_2,$$

$$Si + SiH_4 \rightarrow Si_2H_2 + H_2,$$

with rate constants of $k \approx 10^{-12}–10^{-10}$ cm^3 s^{-1} mol^{-1} leading to subsequent chain reactions and production of higher-order silanes (Si_2H_6, Si_3H_8). Moreover, the reactions of H atoms increase drastically the overall production rate of SiH_3

$$H + SiH_4 \rightarrow SiH_3 + H_2,$$

which becomes the dominant species contributing to film growth as the SiH_4 pressure increases, as confirmed by IR absorption spectroscopy [35]. However, at high power density, the density of SiH_3, or analogous radicals such as Si_2H_5, becomes self-limited by the fast radical–radical disproportionation reaction ($k \approx 10^{-10}\,cm^3\,s^{-1}\,mol^{-1}$):

$$SiH_3 + SiH_3 \rightarrow Si_2H_6^{**} \rightarrow SiH_2 + SiH_4,$$

where $Si_2H_6^{**}$ carries ~3.5 eV internal energy and cannot be stabilized in usual pressure conditions of PECVD.

Although they are minority species with respect to radicals, positive ions react with SiH_4 with larger rate constants ($k \approx 10^{-10}$–$10^{-9}\,cm^3\,s^{-1}\,mol^{-1}$), leading either to oligomerization or proton exchange, resulting in the production of additional SiH_3 radicals

$$SiH_2 + SiH_4 \rightarrow Si_2H_4^+ + H_2,$$

$$SiH_2^+ + SiH_4 \rightarrow SiH_3^+ + SiH_3.$$

Actually, the positive-ion chemistry in the central plasma region of the discharge, where the ion transport is governed by ambipolar diffusion in the small plasma-bulk DC electric field, is quite different from the ion chemistry occurring in the sheaths, where the large ion drift velocity permits endothermic reactions. Therefore, the composition of the ion flux arriving on the electrodes depends not only on the SiH_4 pressure and the discharge power density but also on the potential distribution across the discharge and the sheath electric field. Figure 11 shows the distribution of the $Si_nH_m^+$ species as a function of the number n of Si atoms on the electrodes for three types of discharge [27]. In particular, on the cathode of a DC discharge, the positive ions appear to be less polymerized than on the electrode of an RF discharge where the sheath voltage is smaller. Moreover, it has been shown that the secondary reactions of positive ions are the main cause of SiH_4 dissociation in the cathodic sheath of a DC discharge [36], whereas the primary electron impact dissociation of SiH_4 dominates in the positive column region, as well as in a symmetric planar diode RF discharge [23].

Negative ions, SiH_m^-, although produced at much smaller rates than positive ions, are trapped in the plasma, due to the potential barriers of the sheath (see Fig. 1); because of this longer residence time, limited only by collisional detachment, photodetachment, or mutual anion–cation recombination, negative ions can accumulate in the plasma bulk and reach densities of the same order of magnitude as positive-ion densities [5, 7, 12, 15, 37]. Moreover, even if their reaction rate

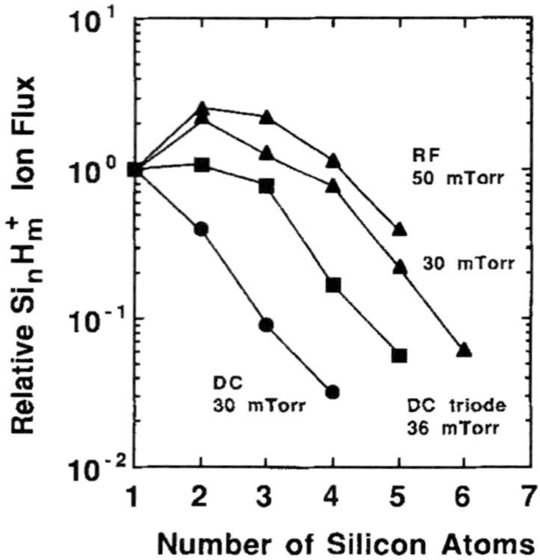

FIGURE 11. Distribution of positive ion $Si_nH_n^+$ flux vs. the number n of Si atoms on the electrodes of three types of SiH_4 discharge: cathode in a DC discharge (circles), cathode covered by a grid to slow down the ions in a triode DC discharge (squares), and RF discharge (triangles). (From Gallagher [27].)

constant is smaller, negative ions polymerize faster than positive ions, as shown in Fig. 12 by the pressure dependence of the fractions of n-Si cations and anions measured by mass spectrometry in a multipole DC discharge [38]. Recently, negative-ion mass spectrometry in RF SiH_4 discharge conditions has confirmed the faster polymerization of negative ions [39]. They are the best candidates to explain the initiation of powder formation in SiH_4 discharges (see Section III.D).

The general rate equation for production and loss by reactions and diffusion to the wall can be expressed as [19]

$$\frac{\partial N_i\,(x,\,t)}{\partial t} = \sum_j k_{ej,i}\, n_e\, N_j + \sum_{j,l} k_{jl,i}\, N_j\, N_l - N_i \left(\sum_j k_{ei,j}\, n_e + \sum_{j,l} k_{il,j}\, N_l \right) + \nabla . D_i\, \nabla N_i - \nabla . \mu_i\, E N_i, \tag{31}$$

where all quantities are functions of time and position and $k_{ej,i}$ is the rate constant for production of the i species by electron impact on the j species (derived by convolution of electron collision cross sections and electron energy distribution),

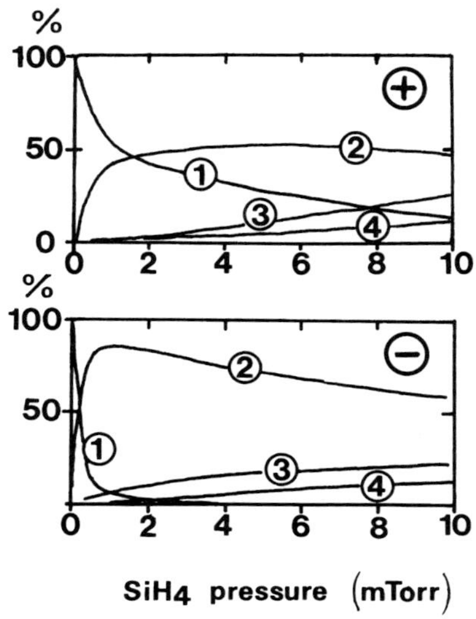

FIGURE 12. Evolution of the fractions of n-Si cations and anions vs. the SiH_4 pressure measured by mass spectrometry in a multipole DC discharge [38].

$k_{jl,i}$ is the rate constant for the reaction of j and l species producing the i species. D_i is the diffusion constant, and μ_i is the mobility for charged species in the local electric field E. The appropriate boundary condition for diffusive transport of radicals to the wall is given by

$$D_i \left|\frac{\partial N_i}{\partial x}\right|_{\text{wall}} = N_i \frac{v_i}{4}\left(\frac{1 - r_i}{1 + r_i}\right) = N_i \frac{v_i}{2} \frac{\beta_i}{(2 - \beta_i)}, \tag{32}$$

where $v_i = (8k_BT/\pi m_i)^{1/2}$ is the radical thermal velocity; r_i, the reflection coefficient; and $\beta_i = 1 - r_i$, the surface reaction or loss probability (see Section III.C.1).

More or less sophisticated modelling of the plasma chemistry and/or Si mass transfer on the walls in different discharge configurations can be found in refs. [5, 19, 23, 29, 31, 36, 37, 40, 41). Figure 13 shows the typical evolution of the ion contribution (Φ_+/Φ_0) to the a-Si:H film growth and of the main deposition precursors as a function of SiH_4 partial pressure. All effects associated with the cooling of the electron energy distribution and the development of secondary reactions such as the reduction of (Φ_+/Φ_0), the increase of the SiH_3 density, and the development of gas-phase polymerization up to powder formation (associated with the transition to the γ' discharge regime; see Section III.D), are shifted to higher pres-

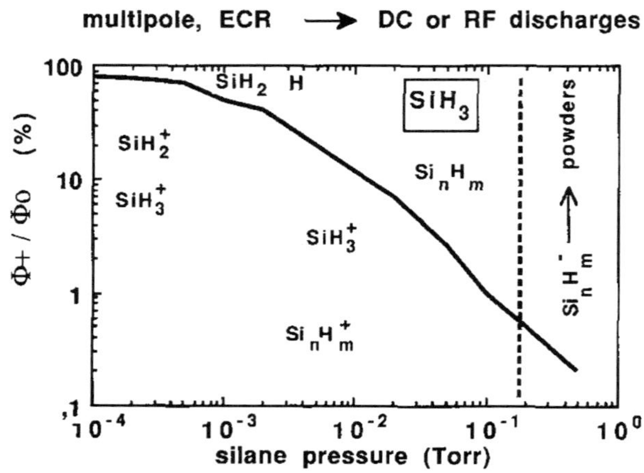

FIGURE 13. Evolution of the main reactive species in the plasma and of the ion/radical contribution to a-Si:H deposition as a function of SiH_4 partial pressure.

sure when decreasing the interelectrode space. An increase of the temperature at a constant pressure decreases the molecular density and enhances the diffusion to the wall so that most secondary reactions are also delayed at a higher pressure, except thermally activated reactions such as the H insertion into SiH_4 [4]. Eventually, an increase of the discharge power density not only depletes the SiH_4 concentration but also heats up the electron energy distribution. This enhances ionization vs. dissociation so that (Φ_+/Φ_0) increases. Moreover, the SiH_3 density saturates as a result of biradical recombination, gas-phase polymerization, and powder formation, which are accelerated by the concomitance of radical–radical, electron–ion, and anion–cation recombination reactions.

C. Surface Reactions and a-Si:H Film Growth

1. *Selection of the "Good" Radicals*

Let us first consider the case where a-Si:H deposition occurs on a wall where the sheath voltage is low, and hence the maximum ion energy E_+ is minimized (typically $\leq$10–20 eV). This can be obtained either on the anode of a planar diode DC discharge, or on the grounded electrode of an asymmetrical RF discharge with a large negative self-bias V_B on the RF-powered electrode [42, 43] (see Section II.A and Fig. 2), or in any discharge structure where the plasma potential is sufficiently low such as in a multipole discharge [44]. Then, in the absence of significant kinetic energy delivered to the surface (see Section III.C.2 for the effect of ion bombardment) the a-Si:H film growth process is very sensitive to the chemical nature of the incoming species (positive ions and radicals). Depending on the total pressure and SiH_4 partial pressure, the interlectrode distance, the gas temperature, or the power density and discharge regime, the plasma chemistry is subjected to large variations (see Fig. 13). In particular, at low SiH_4 partial pressure the radical flux involves mostly SiH_2 and the ion contribution to film growth can exceed 50%, whereas at moderate pressure—i.e., when the product of the SiH_4 pressure and the interlectrode distance $P_{SiH_4}.d$ lies typically between 0.1 and 0.5 torr/cm—the ion contribution (Φ_+/Φ_0) to a-Si:H deposition is a few percent or less, and SiH_3 becomes the dominant radical.

In these specific conditions of minimum ion bombardment many correlations of the a-Si:H microstructure with discharge parameters (e.g., refs. [42–44]) can be simply interpreted in terms of differences of surface reaction kinetics among the depositing precursors.

In order to understand these phenomena it is of primary importance to know what the surface of a growing a-Si:H film looks like. At the usual deposition temperature (≤300°C), the steady-state growing film surface is covered by hydrogen as revealed by IR ellipsometry [45] or surface sputtering [46]. A pictorial view of a a-Si:H growing film id shown in Fig. 14. In the *growth zone* (a few monolayers), thermally activated ($E_a \approx 0.1$ eV) crosslinking of neighboring SiH bonds and H_2 desorption take place, leading to the *stabilized a-Si:H network* underneath. But the surface reaction kinetics of incoming species are determined largely by the hydrogen-rich *surface* composed of the first monolayer of chemisorbed species.

In Section III.B we introduced the concept of *surface reaction probability*, $0 \leq \beta \leq 1$, involving all loss processes of a radical on the surface. Since $\beta = 1 - r$, where r is the reflection coefficient, a low value of β allows a *conformal coverage*

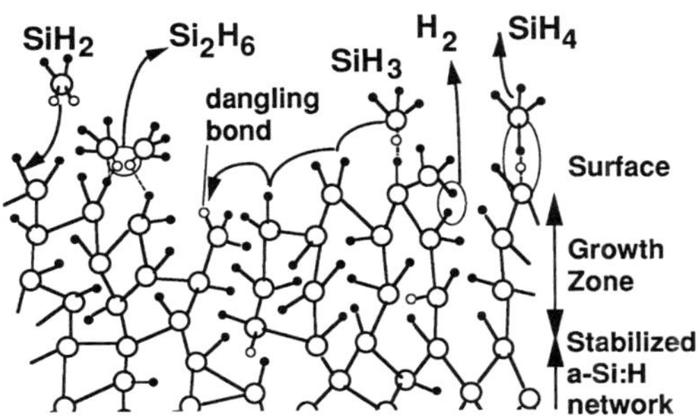

FIGURE 14. Pictorial view of a growing a-Si:H film with SiH_3 and SiH_2 radicals.

of steps or trenches on the substrate at a scale of micrometers to millimeters because of multireflections on shaded areas [43]. On the contrary, for $\beta \approx 1$ coverage is limited by the solid angle of collection of incident species.

However the value of β does not tell us why the a-Si:H film microstructure can be dense or porous and the surface smooth or rough at a scale of 10–100 Å [43, 44]. In fact, the film microstructure is entirely determined by the *surface diffusion length* of adsorbed species. A high *surface mobility* is necessary to reduce the surface roughness and porosity due to *self-shadowing* of incident species at a scale of 10–100 Å [47].

In fact, for a Si-containing radical, β consists of the sum of two probabilities:

- The *sticking probability* s or the probability that the Si atom remains chemisorbed and eventually incorporated in the a-Si:H film.
- The *recombination probability* γ or the probability that the Si atom is desorbed as a saturated volatile species (SiH_4 or Si_2H_6):

$$\beta = s + \gamma \tag{33}$$

Both s and γ and the surface mobility of a given species can be related to a kinetic description of the elementary microscopic processes on the H-covered surface of a-Si:H.

SiH_2 (or SiH or Si) radicals as well as low-energy $Si_nH_m^+$ ions, which dominate under low SiH_4 partial-pressure conditions (see Fig. 13) are very reactive in the gas phase with the H-saturated molecule SiH_4. Therefore, they should easily insert in SiH surface bonds, resulting in low surface mobility and $\beta \approx 1$. Indeed, we determined an effective β value of ≈ 0.7 in low-pressure multipole discharges [34], and a specific experiment gave $\beta \geq 0.94$ for SiH [48]. Consequently, a-Si:H film exhibits porosity, surface roughness, and columnar morphology, due to the self-shadowing effect, when incident ions are not accelerated [44]. Then, the only way to enhance surface mobility and reduce microstructural inhomogeneities is to accelerate the ions (cf. Section III.C.2).

On the contrary, SiH_3 does not insert in SiH bonds and can physisorb only on a H-covered surface site and diffuse on neighboring until it chemisorbs on a dangling bond, or recombines as SiH_4 by H abstraction, or as Si_2H_6 with another physisorbed SiH_3 (see Fig. 14). The low value of $\beta \approx 0.26 \pm 0.05$ determined for a selective source of SiH_3 [49, 50] provides good conformal coverage, and the surface diffusion explains the dense microstructure and the smooth surfaces obtained in these conditions. Moreover, the value of $s \approx 0.4\,\beta$ is consistent with a kinetic model [49–51] showing that at least two SiH_3 have to react for one Si incorporation, according to the overall reaction

$$2\ SiH_3 \rightarrow SiH_4 + \text{a-Si:H}_x + (1 - x/2)\ H_2,$$

in which one SiH_3 abstracts a H atom to create a dangling bond, on which the other SiH_3 chemisorbs.

Other candidates having low surface reactivity and high mobility may be Si_2H_5 or the stable configuration of Si_2H_4: disilene $H_2Si{=}SiH_2$. In a pure SiH_4 discharge where SiH_3 is dominant but mixed with more reactive species, Doughty *et al.* [52] determined a slightly higher "effective" β value of 0.37 ± 0.05, but they also concluded that the a-Si:H film quality obtained in this condition is related essentially to the ability of the film producing radicals to diffuse on the surface. In this kinetic description, the effect of temperature is twofold: (1) activation of the surface mobility and increase of the diffusion length and (2) activation of the crosslinking and exodiffusion of H_2 from the growth zone.

The surface reactions of Si containing radicals may also be drastically affected by catalytic effects due to some impurities or to gases introduced at parts per million (ppm) to percent levels in the plasma, such as diborane B_2H_6 or phoshine PH_3 used for depositing B-doped or P-doped a-Si:H. A kinetic study of the catalytic effect of B_2H_6 is reported in Perrin *et al.* [49].

2. *The Effect of Ion Bombardment*

If the adsorbing species are intrinsically too reactive to allow surface diffusion, the surface mobility can be enhanced by a physical process involving momentum transfer from incident ions (or fast neutrals) arriving with enough kinetic energy on the surface. The ions are either reactive ($Si_nH_m^+$) or inert (rare-gas ions). However, the nominal energy and the mass of the incident ion govern to a large extent the efficiency of the collisional momentum transfer and the penetration depth of the ion in the growth zone. Therefore, ion bombardment may have either beneficial or deterious effects on the film growth and in the final a-Si:H film properties.

Light and slow ions mostly excite vibrationally SiH surface bonds. This can enhance the diffusion or desorption of weakly physisorbed species such as SiH_3 but not dislodge a reactive radical such as SiH_2 which has just chemisorbed by insertion into a SiH surface bond. Thus the ion mass and/or energy must be large enough to induce "shot-peening" of the nonmobile chemisorbed species of the first monolayer and overcome the "self-shadowing" effect, which is responsible for the development of surface roughness and columnar growth morphology. On the other hand, as the energy increases, the ion penetrates deeper into the growth zone and can induce collision cascades or thermal spikes, which may create local defects, and dangling bonds remain trapped in the stabilized a-Si:H network. Hence, the resulting film might well be mirror-like and dense but with poor electronic properties and highly stressed.

In the abundant literature reporting correlations between a-Si:H film properties and discharge parameters, the effect of ion bombardment is often invoked, but very often the discussions of the observed phenomena may be misleading. Indeed, if one varies the negative bias on the electrode on which the substrate is situated in a RF or DC planar diode discharge, the whole potential distribution across the plasma can be modified so that the power density distribution and the mechanisms of fast electron generation are perturbed (see Section II.A). Consequently, not only the ion energy but also the plasma chemistry and the chemical composition of reactive species impinging on the substrate may be significantly modified. Therefore, the sole effect of the ion bombardment cannot be clearly assessed. Another example is the study of the effect of the rare gas (Ar, Xe, He) or H_2 dilution of SiH_4 on the a-Si:H properties, which has sometimes been correlated with a variation of ion bombardment on the substrate [53, 54]. However, it has been shown that Ar dilution (and probably Xe dilution) tends to decrease the production of SiH_3 radicals in favor of $SiH_{m<3}$ radicals [55, 56] and favors powder generation in the plasma [57], whereas He dilution or H_2 has the reverse effect [58]. Moreover, in the case of H_2 dilution, the main effect is to increase atomic H production, which has a "chemical" effect on the surface reactions (see Section III.C.3), whereas the collisional momentum transfer of the light H^+, H_2^+, or H_3^+ ions is presumably less efficient because of the mass difference between H and Si.

In some cases, however, the ion energy distribution can be controlled without modification of the plasma properties [44, 59], and hence with constant radical and ion flux. For low-pressure plasmas where the reactive ion contribution (Φ_+/Φ_0) is greater than $\approx$10%, the correlations between film properties and increasing ion energy up to 150 eV clearly show the following:

1. The densification of the film microstructure, the disappearance of the columnar growth morphology, and a reduction of surface roughness.
2. The modification of the H incorporation in the a-Si:H bulk with a reduction of the $=SiH_2$ polymer-like bonding modes in favor of $\equiv SiH$ groups.
3. An increasing compressive stress.

All these effects tend to saturate for ion energy exceeding 80–100 eV and (Φ_+/Φ_0) $\approx$ 20%.

Complementary studies of the optoelectronic film properties [59, 60] show that these microstructural modifications are accompanied by an improvement of the film photoconductivity and a reduction of the bulk-defect density as long as the ion energy does not exceed $\approx$75 eV [60]. This puts an upper energy limit to the beneficial effect of ion bombardment.

The best compromise is therefore to combine large ion flux and moderate energy ($\approx$20–50 eV). In this condition, ion bombardment is an efficient way to maintain the a-Si:H film quality when increasing the deposition rate, and hence the discharge power density. However, this can be better achieved in ultra-low-pressure (milliTorr range) magnetically confined discharges such as multipole or ECR discharges (see Section IV.B), where the power density can be increased without increasing the sheath voltage in front of the substrate electrode, in contrast to the self-sustained planar diode DC or RF discharges at moderate pressures where (Φ_+/Φ_0) is only a few percent and the sheath voltage increases with the power density. Increasing the RF frequency towards MW range is also a way to allow larger power density at lower sheath voltages due to the increasing efficiency of the "wave-riding" mechanism in the α regime (see Section II.B.2).

3. *The Effect of Atomic Hydrogen, Rare-Gas Metastables, and UV*

An alternative way to release energy to the surface is the electronic energy transfer via the surface reaction enthalpy (recombination or chemisorption), or the deexcitation of a metastable (e.g., Ar or He metastable), or a UV irradiation. This excess energy can increase the "effective temperature" of the surface and enhance the surface mobility and structural rearrangement.

For example, microcrystalline (μc)-Si:H films are obtained in H_2-diluted SiH_4 discharges at high power density where H atoms play an important role both in the gas phase by favoring SiH_3 radical production and on the surface. The mechanisms of μc-Si:H growth have been studied by alternating SiH_4 and H_2 discharges [61–63]. In spite of an apparent controversy [61, 62], it seems clear that H atoms are acting in three complementary ways:

1. H atoms *etch* the polymeric loosely bound phase corresponding to the growth zone in a-Si:H deposition, leaving only a dense Si surface covered by H atoms [63].
2. H atoms induce *chemical annealing* [62]. Indeed, the chemical reactions of H atoms with SiH surface bonds release a considerable energy per deposited Si atom. For a high-H atom flux, the overall reaction is given by

$$2n\ \mathrm{H} + \mathrm{SiH_3} \rightarrow \mu\mathrm{c\text{-}Si\!:\!H}_x + [n+(3-x)/2]\ \mathrm{H_2},$$

   with $\Delta H \approx \Delta H_f(\mathrm{SiH_3}) + n\,D(\mathrm{H\text{-}H}) \approx 2 + n \times 4.5$ eV. If n is large enough, this energy is much higher than for a-Si:H growth with only SiH_3 radicals (see Section III.C.1), which releases only $\Delta H \approx 3.7$ eV, so that the "effective" surface temperature is high enough to induce nucleation of microcrystallites.
3. H atoms saturate the surface of the growing microcrystallites, allowing surface diffusion of SiH_3 radicals and chemisorption in epitaxial conditions. This

H coverage (of different nature than the polymeric-like H coverage corresponding to a-Si : H deposition) is indeed a necessary condition for μc-Si : H growth [61].

The effect of chemical annealing on a-Si : H film growth has also been demonstrated by using sources of Ar or He metastable species [64]. Concerning the effect of UV-light irradiation during a-Si : H film growth, two recent studies [65, 66] have reported a general improvement of the film properties but with some contradicting results concerning the photoconductivity and density of defects.

D. POWDERS

As mentioned in Section II.A.5, the transition from the α to the γ' regime in a SiH_4 discharge is associated with the generation of negatively charged powders that accumulate in the plasma and are electrostatically trapped by the potential barriers located in the sheaths. The growth of powders, their spatial distribution in the plasma, and their kinetic behavior have been studied by combining laser Mie scattering, optical emission, and electrical diagnostics [57, 67, 68]. Figure 15 shows the temporal evolution of Mie signal and Ar line emission in a Ar diluted SiH_4 discharge. A refined analysis of the evolution of discharge electrical properties, powder volumic density and size distribution [57], and the effect of different initial discharge parameters (RF voltage, partial pressure or partial SiH_4 flow,

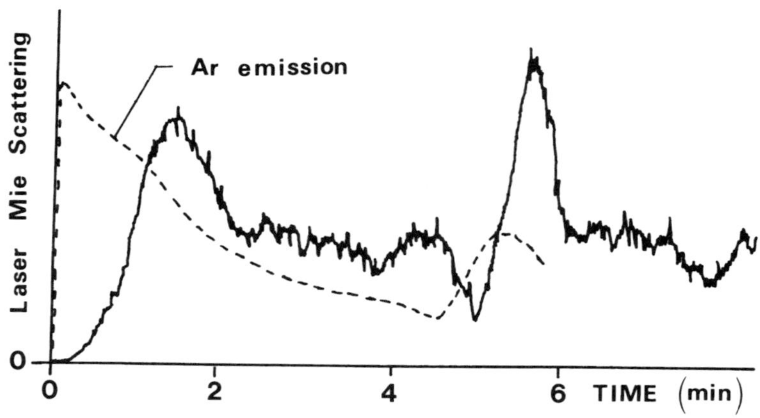

FIGURE 15. Temporal evolution of the Mie scattering intensity and Ar line emission after plasma ignition in a 4% SiH_4/Ar RF discharge at 0.12 torr, room temperature and for an initial RF voltage of 300 V. (After Bouchoule *et al.* [57].)

gas temperature) reveals that the powder formation proceeds in successive phases after plasma ignition. A general scenario involves five steps of respective durations defined as t_0, t_1, t_2, t_3, and t_4 [69]. The initiation of this powder formation is attributed to electron attachment process and negative ions formation.

1. For $0 < t < t_0$ with $t_0 \approx$ ms: The primary radicals, positive ions, and negative ions created by electron–SiH_4 collisions diffuse to the plasma–sheath boundary, but there are clear indications that an enhanced attachment occurs on the new SiH_x radicals [70] and a cumulative and as yet unclarified process of ion–molecule oligomerization, mutual anion–cation recombination, and reattachment on neutral oligomers is initiated in the plasma.
2. For $t_0 < t < t_1$ with $0 < t_1 < \infty$ and depending on initial discharge conditions: The growth of negative ions competes with detachment processes and diffusion as neutral oligomers to the walls, but the negatively charged species are still mostly singly charged. Their density is of the order of the positive ion density (typically $n_- \approx 10^9$ cm^{-3}). However, the discharge power dissipation regime is not yet modified; the RF discharge is still in the α regime, and powders cannot be detected by the Mie scattering techniques. This phase can be extremely short at high pressure and high power density or infinitely long at low pressure and low power density. If the diffusion of neutral radicals is faster than the time for reattachment, the negative ion oligomerization is prevented and powder growth will not develop.
3. For $t_1 < t < t_2$ with $t_2 \approx 1$–5 s: The α–γ' transition in the electrical discharge properties starts and develops in a time on the order of the gas residence time, τ_R. This fast transition is associated with an aggregation of small clusters (≤20 Å) generated in the previous phase into multiply charged powders and completely modifies the fast-electron energy distribution. The particle number density drops by one to two orders of magnitude (from 10^9 to 10^7 cm^{-3}) as their radius increases by aggregation showing that the total mass of all the particles remains constant within this short phase. The particle sizes are remarkably uniform, and the number of electrons per particle increases with the particle diameter and can reach hundreds of electrons. Therefore, each particle becomes surrounded by a space-charge sheath of positive ions. Because of the combined actions of various forces involving ion drag [71], thermophoresis [72, 73] and polarizability [57], the particles begin to drift toward the sheath edges, where they accumulate.
4. For $t_2 < t < t_3$ with t_3 in the order of several seconds to several minutes: The particle number density established during the phase t_2 remains roughly constant, but the powders continue their growth by condensation and surface reactions of radicals similar to the growth mechanism of a-Si:H films on the electrodes. This phase corresponds to an homogeneous nucleation, and the size growth rate is much smaller than during the aggregation period.

5. For $t_2 < t < t_4$, where t_4 is a pseudoperiod of the order of several seconds to several minutes: The powders accumulated at the sheath edge induce perturbations of the sheath field. The gas convection creates some leakage channels across the sheaths along the lines of maximum flow velocity and main temperature gradients (thermophoresis). Powders are expelled from the plasma, sometimes by sudden bursts, thus explaining the oscillatory behavior of the Mie scattering signal in Fig. 15. This leads to a dispersion of powder size [57].

The total transition time to reach the γ' regime involving the first three phases $t_0 + t_1 + t_2$ of powder formation is plotted as a function of the initial RF voltage in Fig. 16 for a pure SiH_4 discharge in the conditions of Fig. 4. As the initial V_{RF} reaches the value of $V_{\alpha\gamma'}$ in Fig. 4, t_1 becomes negligible and the transition is immediate.

Powder accumulation in the plasma can be inhibited by increasing the flow rate, or the temperature, or by square-wave modulation of the discharge at low frequency (10^2–10^3 Hz) (see Fig. 16) in order to allow the smallest negatively charged species to be swept out regularly, hence preventing further growth [70, 74]. However, since the latter procedure prevents the discharge to enter in the γ' regime, the discharge impedance remains essentially capacitive, and it is difficult to raise the power density in order to increase the a-Si:H deposition rate without increasing the RF voltage. Consequently, the ion bombardment energy on the growing a-Si:H films increases. In fact, it is often necessary to accept the presence

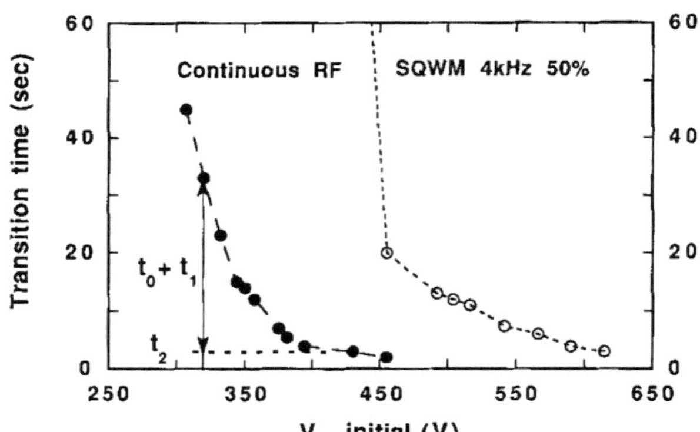

FIGURE 16. Total transition time from the α regime to the γ' regime in a SiH_4 RF discharge at 0.1 torr as a function of the initial RF voltage V_{RF} for continuous discharge and square-wave modulated discharge at 4 Hz and 50% duty cycle. The total time corresponds to the addition of the three phases ($t_0 + t_1 + t_2$) of the particle growth in the plasma (see text).

of powders in the plasma for a fast deposition rate. The only problem is then to control the leakage of powders to avoid their deposition on the substrate area.

IV. Concepts of Reactors for a-Si:H Deposition

In this section we first review the different concepts of discharges used for PECVD of a-Si:H from SiH_4. The chosen classification is based primarily on the range of operating pressure, which is one of the main parameters governing the electron energy distribution in the plasma, the SiH_4 plasma chemistry, and the ion contribution to the film growth Φ_+/Φ_0 (see Fig. 13). Since the pressure range is restricted by the onset of powder formation, we distinguish the medium pressure range (0.01 – 1 torr), where radical chemistry dominates and $\Phi_+/\Phi_0 < 10\%$ from the low pressure range (1 to 10 mtorr), where $\Phi_+/\Phi_0 > 10\%$ and the deposition is necessarily assisted by ion bombardment. Discharges operating at medium pressure under DC, RF, VHF (very high frequency), or MW excitation with capacitive or inductive coupling and various structures of plasma confinement and discharge configurations are examined in Section IV.A. Low-pressure discharges generated in magnetically confined structures with fast electron injection are described in Section IV.B. Then in Section IV.C we consider together the various concepts of PECVD in the so-called triode or remote plasma configurations, which are intended to provide selective gas-phase and/or surface chemistry.

The next sections cover the non-PECVD techniques that have been developed to deposit a-Si:H. Most of them still use SiH_4 as a Si source but are based on different initial dissociation mechanisms: photolysis with IR or UV sources (Section IV.D), or pyrolysis in homo-CVD or hot-filament CVD (Section IV.E). Eventually the specific case of reactive sputtering using a solid Si source is analyzed in Section IV.F. In each case, the key elements of the gas phase or surface physicochemistry are outlined to make relevant comparisons and analogies with PECVD.

The respective advantages and limitations of each reactor type for technological applications are discussed.

A. PECVD at Medium Pressure

1. *Gas Distribution*

At medium pressure, the gas distribution in the reactor may be a determining factor of the a-Si:H film deposition homogeneity on the substrate, especially for large-area PECVD. As mentioned in Section III.A the gas flow across the dis-

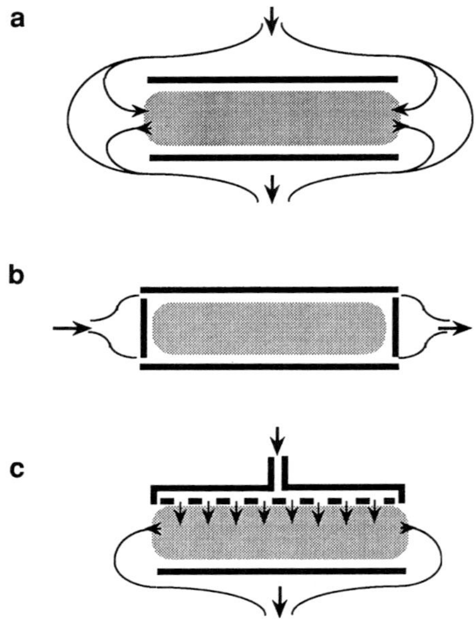

FIGURE 17. Three configurations of gas injection in a PECVD reactor: (a) lateral diffusion; (b) laminar plug-flow; (c) shower-head injection through electrode.

charge may be limited by convection of diffusion, and an estimation of the Péclet number allows us to determine what is the limiting flow regime. Different types of gas distribution are presented in Fig. 17. Obviously, the laminar plug-flow (Fig. 17b) is the most restrictive one. The symmetrical lateral diffusion (Fig. 17a) is better, but the shower-head injection through an electrode (Fig. 17c) is the best. The direct injection of the gas into the discharge zone has another advantage if the discharge is confined since the slight overpressure in the discharge volume tends to prevent backdiffusion of chemical impurities coming from outgassing of the reactor wall. This has led to the "plasma-box" concept [75, 76], in which oxygen contamination in the film could be reduced down to 5×10^{18} atom/cm^3, whereas the contamination level is usually larger than 10^{19} atom/cm^3 in conventional "open" discharge configurations.

2. *DC Discharges*

The classical DC diode planar discharge structure is described schematically in Fig. 18a. Films deposited on the anode or on the cathode have very different mi-

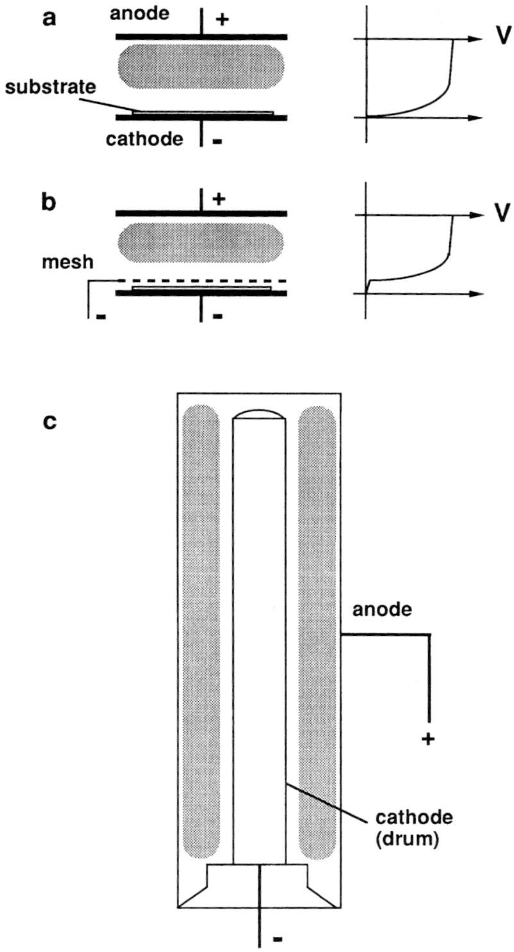

FIGURE 18. Configurations of DC discharges for a-Si:H PECVD: (a) planar diode; (b) DC proximity triode; (c) cylindrical geometry for coating drums for xerography.

crostructural properties as a result of the difference in sheath voltage and ion bombardment. Anodic film are usually porous and inappropriate for optoelectronic applications (most likely because of powder deposition since there is no confinement of negatively charged particulates at the anodic sheath edge), whereas cathodic films submitted to energetic ions are denser but usually stressed with electronic defects. A method to reduce the ion bombardment energy consists in using a screen or mesh electrode place above the substrate (Fig. 18b) and polarized at the same potential as the cathode. Then the discharge is confined between the anode and the mesh, which acts as the new cathode. During their transport in the zero-field region between the mesh and the substrate electrode, ions are slowed down by gas-phase collisions. Moreover, the most reactive species of the radical flux are also filtered (see the triode concept discussed in Section IV.C.1), which favors the most mobile species and the growth of a-Si:H films of good optoelectronic properties. This concept of "DC-proximity" discharge has been developed by a few groups [27, 77] but is not widely used. Actually, there are two limitations to the development of DC discharge reactors: (1) the a-Si:H film resistivity induces a variation of the effective sheath voltage as the film thickness increases, and (2) it is difficult to scale up the mesh electrode to a large area because of thermal expansion during deposition at temperature of 200–300°C.

Cylindrical DC discharge geometry have also been used for a-Si:H deposition on drums of Xerox copiers [78] as illustrated in Fig. 18c. In this case the drum is the cathode, and there is no protecting mesh electrode. However, the a-Si:H materials requirements for xerography are less stringent than for solar cell applications (see Chapter 6). On the contrary, the main issue is to obtain fast deposition rate.

3. RF and VHF Discharges

1. Capacitive coupling. Capacitively coupled planar diode RF glow discharges are by far the most widely used in a-Si:H PECVD. The most common frequency is 13.56 MHz, but two groups of investigators tried to increase the RF frequency toward VHF up to 300 MHz [79–81]. Both groups reported an increasing a-Si:H deposition rate when working at constant SiH_4 pressure and constant RF power *at the generator* (termed W_G in Section II.C.2). In one case the deposition rate reached a maximum around 70 MHz [79, 80], whereas in the other case the deposition rate increased asymptotically up to 110 MHz [81]. These phenomena are essentially related to the variation of the power transfer efficiency η from the generator to the discharge (see Section II.C.2). Figure 19 represents a typical electrical network involving the RF generator, the matching box, and the reactor. Parasitic resistive losses occur in the matching box or in the reactor. Moreover, the loops in the RF current distribution, especially in the grounded part of the circuit (see Fig. 19), correspond to inductances that are often neglected in the reactor equivalent circuit. As shown in Fig. 19, the reactor impedance without plasma can be described schematically by a capacitance C_0, a resistance R_0, and an inductance L_0. When the discharge is established, another electrical path is created that also involves a capacitance C_1 (representing the sheaths), a resistance R_1 (the

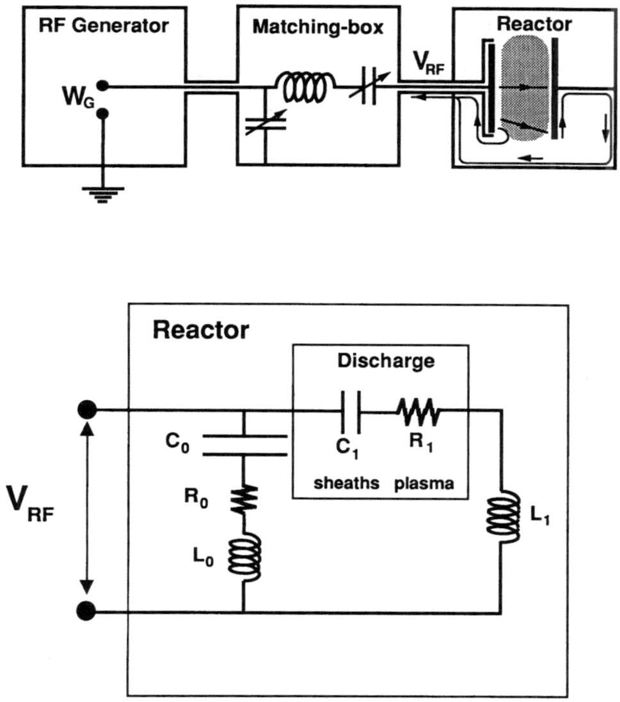

FIGURE 19. Electrical network in an RF discharge with a matching box between the generator and the reactor and equivalent circuits of the reactor and the discharge.

plasma bulk), and an inductance L_1 (due to the discharge current loop on the grounded part). Within these simple approximations, and if one neglects resistive losses in the matching box, the power dissipated at a given value of V_{RF} in the discharge circuit and in the reactor circuit can be expressed as a function of ω, C_0, C_1, L_0, L_1, R_0, R_1. Then an expression of the power transfer efficiency η can be easily derived, showing that η is an increasing function of ω at low frequency for $L_1C_1\omega^2$ and $L_0C_0\omega^2 \ll 1$, *if* $R_1C_1 > R_0C_0$. Nevertheless, depending on the respective values of L_1 and L_0, η reaches a maximum at higher frequency. The measured variation of η with ω in the reactor of the Neuchâtel group [80], using the procedure presented in Section II.C.2, is shown in Fig. 20 and exhibits a maximum around 50–60 MHz. This partly explains the increase of the a-Si:H deposition rate from 2–3 Å/s at 13.56 MHz up to ≈20 Å/s at 70 MHz followed by a decrease at higher frequencies. However, when corrected for the variation of η, the a-Si:H deposition rate appears as continuously increasing with ω for a fixed

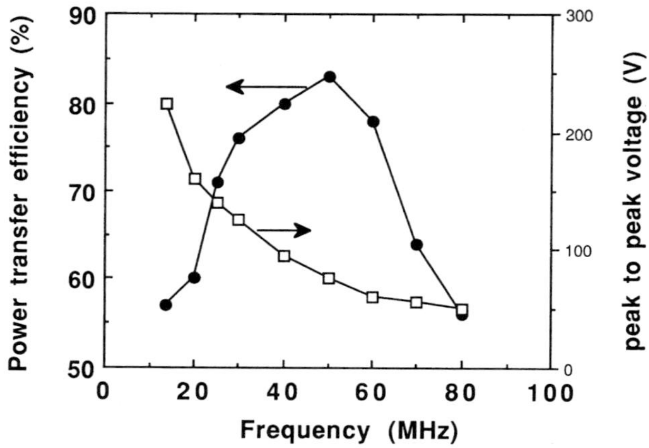

FIGURE 20. Power transfer efficiency and peak-to-peak voltage as a function of RF frequency for SiH_4 discharges at a fixed pressure and RF power at the generator. (From Finger *et al.* [80].)

power dissipated in the discharge. This latter effect is associated with the increase of the high-energy tail of the electron energy distribution, which tends to reduce the fractional power going into vibrational excitation instead of dissociation of the molecule (cf. Section II.D and Fig. 8). In other words, increasing the RF frequency results in a more efficient coupling of the electrical power into SiH_4 dissociation.

Moreover, as shown in Fig. 20, V_{RF} decreases continuously as ω increases, which means that the maximum ion energy on the substrate is reduced, although the ion contribution to the film growth Φ_+/Φ_0 is expected to increase because of the growth of the high-energy tail of the electron energy distribution function. This low-energy ion bombardment appears as the key to understand the maintenance of optimal a-Si:H film microstructural and optoelectronic properties when increasing the deposition rate in VHF discharges, whereas these properties usually degrade in conventional 13.56-MHz RF discharges where the ion impact energy becomes too high as the power increases.

The apparent success of VHF discharges is however counterbalanced by limitations in industrial applications. First, the VHF frequency range ($\approx$100 MHz) is attributed to radio and telecommunications with severe controls against parasitic emitters. Second, as ω increases, the problems of homogeneity of discharge power dissipation and a-Si:H deposition on large-area planar diode system, which are already encountered at 13.56 MHz, become much more difficult to solve, especially because of the series inductance in the electrode plates. A possible solution is the use of guided-microwave applicators (see Section IV.A.4).

Coming back to the classical 13.56-MHz discharges, we want to address the problem of the control of the discharge assymetry, which is revealed by the negative self-bias V_B on the RF-powered electrode. As discussed in Section II.A, the discharge assymetry is related to the effective-area ratio of the RF and grounded electrodes. In laboratory-scale PECVD reactors the effective area of the grounded electrode is much larger than the RF electrode area, due to the lateral expansion of the plasma interacting with the grounded walls of the reactor as shown in Fig. 21a. Consequently, V_B is relatively large and the sheath voltage is much larger on the RF electrode than on the grounded electrode. The assymetry can be significantly reduced by confining the discharge with a grounded wall [76] or mesh [82] surrounding the plasma volume as shown in Fig. 21b. This confined discharge structure also allows higher-power density in the discharge, and hence higher deposition rates [82]. The assymetry is further reduced when scaling up the elec-

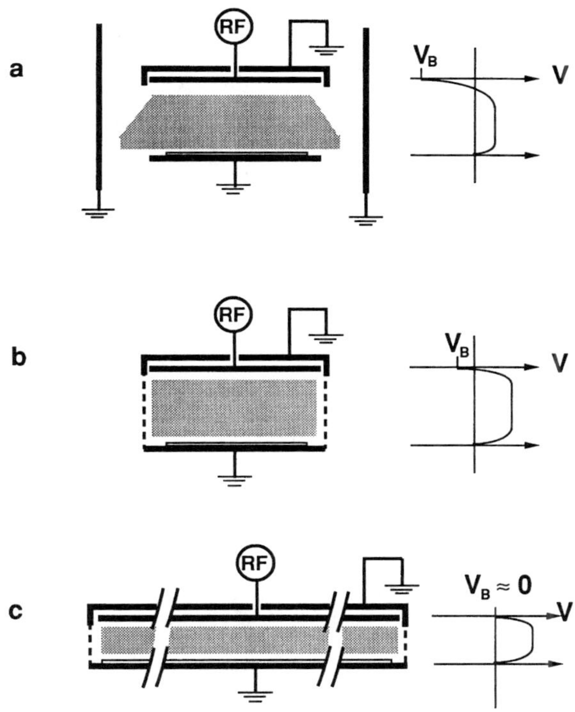

FIGURE 21. Planar diode RF discharges and effect of electrode asymmetry on the bias voltage and sheath potential distribution: (a) small electrode area and interactions with grounded walls of the reactor vessel in an open geometry; (b) confinement of the discharge by a grounded mesh; (c) large-area industrial reactor minimizing edge effects and tending toward electrical symmetry.

trode area while keeping a constant interelectrode distance since the edge effects of the plasma become relatively unimportant and V_B tends to be negligibly small with respect to V_{RF}, as shown in Fig. 21c. This point is particularly important when one wants to transfer a PECVD process from a small-size reactor to a large-size reactor, since the whole discharge physicochemistry depends on the potential distribution across the discharge.

2. Other configurations. The capacitively coupled planar-diode RF discharge configuration is the most common one, but other configurations and RF coupling modes can be used. Figure 22a shows an example of concentric electrode configuration [83] in a tubular reactor where the discharge is still capacitively coupled,

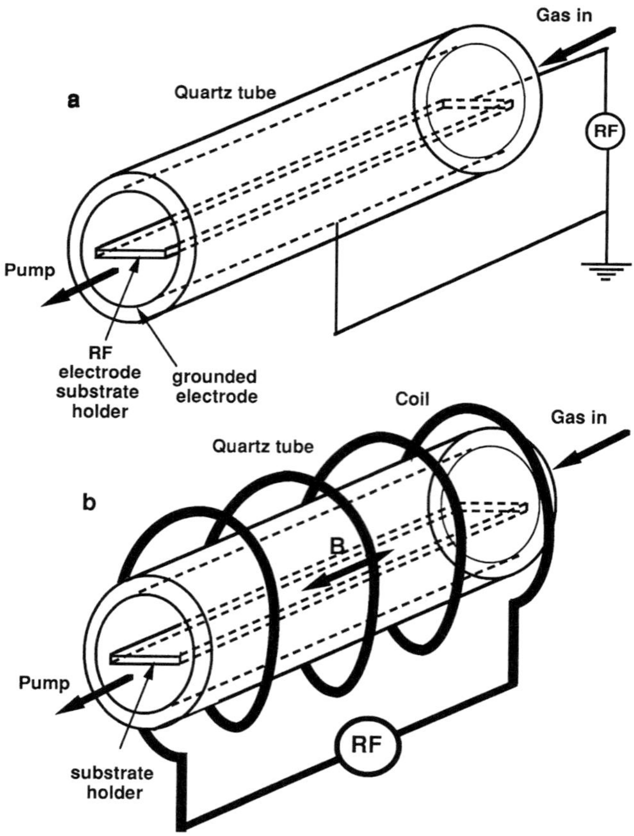

FIGURE 22. Tabular RF discharges for PECVD: (a) capacitive coupling with concentric electrodes (from Chatham *et al.* [81]); (b) inductive coupling with an external coil.

whereas Fig. 22b shows an inductively coupled RF discharge configuration using a coil surrounding the tubular reactor (a recent version of an inductively coupled RF discharge with shield electrodes is given in Yokota *et al.* [84]). Both configurations might be interesting for laboratory-scale experiments but are clearly inappropriate for large-area PECVD.

The combination of a capacitive RF coupling mode and an external magnetic field can be interesting to confine the plasma and to increase the power density in the discharge. The effects of an axial magnetic field (generated by an electromagnet) perpendicular to the electrode planes on the SiH_4 glow discharge behavior and on the a-Si:H film properties have been studied quite early [85]. The controlled plasma magnetron (CPM) structure [86] (see Fig. 23) uses an arrangement of an electromagnet behind the RF electrode plate, which results in a local increase of the electron and ion density as for magnetron sputtering (see Section IV.F). In this case, the aim of enhancing the electron density is a faster decomposition of SiH_4, and hence a larger a-Si:H deposition rate. In addition, the ion bombardment energy on the substrate can be independently controlled by confining the plasma by a mesh and properly biasing the substrate electrode.

4. MW Discharges

When increasing the frequency up to MW (2.45 GHz), the conventional planar-diode capacitive coupling is no more achievable and the MW power has to be fed

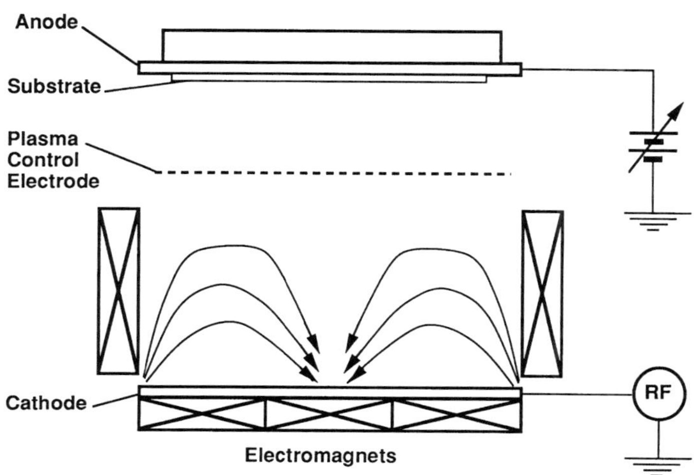

FIGURE 23. The controlled plasma magnetron (CPM) configuration for PECVD. (From Ohnishi *et al.* [86].)

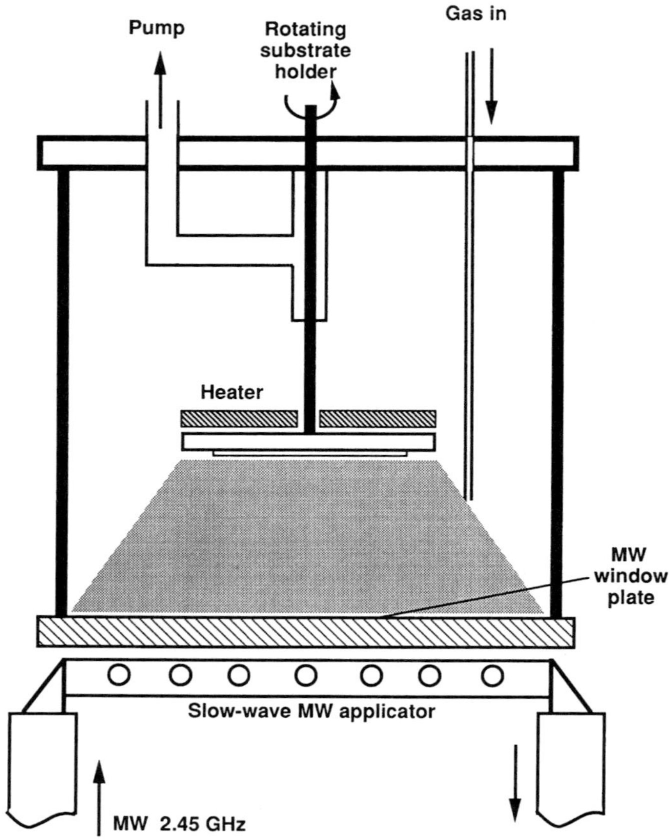

FIGURE 24. The large-microwave plasma (LMP) with a slow-wave microwave applicator. (From Wertheimer *et al.* [87, 88].)

to the discharge via guided-wave structures. MW plasma excitation can be performed at very low pressure in ECR discharges (see Section IV.B), but also at medium pressure. In the latter case, the need for large-area homogeneous deposition has led to the concept of various kinds of microwave applicators or surface-wave launchers [87]. The "large microwave plasma" (LMP) [88] is presented in Fig. 24. The "slow-wave MW applicator" consists of an array of parallel rods. High a-Si:H deposition rates can be achieved (typically ≈20–30 Å/s) due to the more efficient coupling of the electrical power to the high-energy tail of the electron energy distribution function, as already mentioned for VHF discharges (see Section IV.A.3.1).

B. PECVD AT LOW PRESSURE

Conventional DC or RF discharges in diode configurations cannot be maintained below a minimum value of the product *Pd,* which corresponds to a minimum pressure of 10–100 mtorr. Moreover, as the pressure decreases toward this limit, the minimum DC or RF maintenance potentials rise rapidly, which induces a high-energy ion bombardment on the electrodes and deterious effects on the a-Si:H film properties. These restrictions make impossible the combination of a large ion/radical flux ratio Φ_+/Φ_0 with a moderate ion energy (a few tens of electronvolts), which is one of the routes for making good a-Si:H films (see Section III.C.2). However, plasmas can be sustained at very low pressure (typically $\leq$1 mtorr) without imposing high sheath potentials in some discharge configurations where a magnetic confinement is associated with a source of fast primary electrons. We present two sources of this kind: the multipole DC discharge and the ECR discharge. Other configurations might be used, but up to now they have not been applied to a-Si:H PECVD.

1. Multipole

In the DC multipole configuration [32, 33, 38], the confinement is ensured by an arrangement of permanent magnets forming magnetic cusps along the generatrices of a cylindrical anode (see Fig. 25). Fast electrons are generated by thermoionic emission from a hot tungsten wire negatively biased with respect to the walls. When the plasma is established, these electrons gain their energy through the sheath surrounding the hot cathode and can undergo several reflections on the magnetic mirrors before being lost to the anode wall. This increases their geometric mean free path and allows a high ionization efficiency and a high electron density at very low pressure. The plasma potential with respect to the walls varies from a few electronvolts to 10 eV. Consequently, one can achieve a-Si:H deposition conditions involving high Φ_+/Φ_0 at low ion energy. The ion energy on the substrate can be independently varied by a DC bias. This configuration seems limited to small sample areas but there are possibilities to develop the magnetic cusp structures and the available deposition zone at least along one dimension. The hot tungsten wires also induce SiH_4 pyrolysis, as in the so-called hot-filament CVD method (see Section IV.E.2).

2. ECR

Reactors based on electron cyclotron resonance (ECR) (see Section II.B.4) are more widely used than DC multipoles [89–91]. The reactor configuration shown

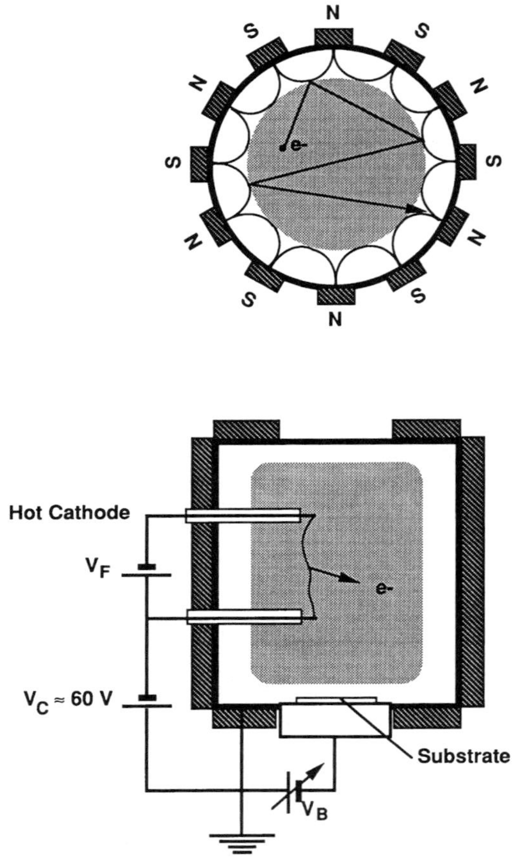

FIGURE 25. The multipole DC discharge configuration with a hot cathode emitting primary electrons and an independent substrate bias for the control of the ion bombardment energy [32, 33, 38].

in Fig. 26 consists of a 2.45-GHz MW guide and antenna coupled through a dielectric window to a cylindrical cavity surrounded by electromagnets generating an axial magnetic field. The electron resonance condition is achieved for a field $B \approx 880$ G at low pressure (10^{-4}–10^{-3} torr) and the plasma expands outside the magnetic cavity. Argon or H_2 are injected inside the magnetic cavity, whereas SiH_4 is injected in the plasma expansion zone just above the substrate, which can be DC biased with respect to the plasma in order to control the ion energy. The DC bias can be achieved by applying a DC voltage or by capacitive coupling of an RF voltage through insulating substrates. The only drawback of this reactor is

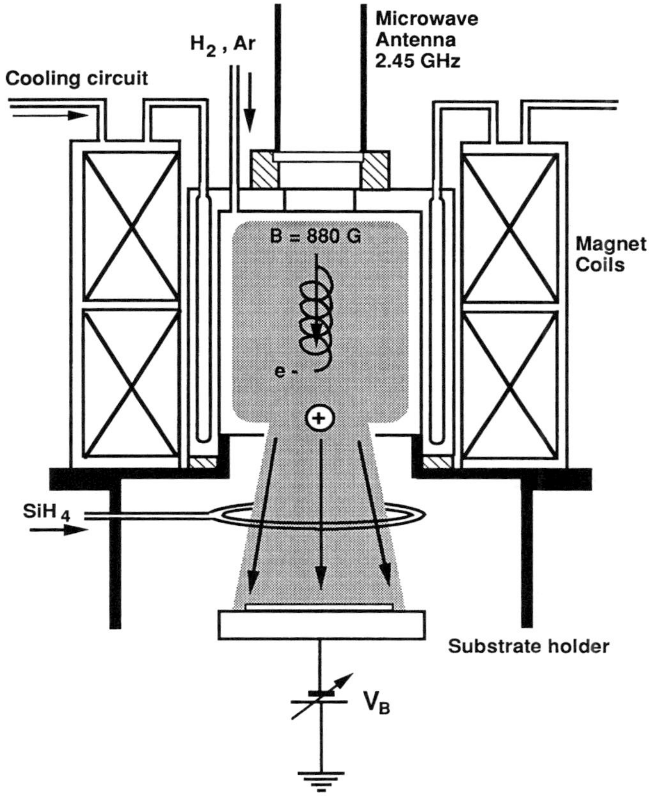

FIGURE 26. Example of ECR reactor used for a-Si:H PECVD. (From Shing *et al.* [90].)

the perturbation of the MW coupling due to the deposition of a semiconducting a-Si:H film on the dielectric window.

C. PECVD in Triode or Remote Plasma Configurations

These reactors involve many different concepts and configurations with one common aspect, which is the physical separation of the production zone of reactive species in a plasma from the deposition on the substrate. The transport of reactive species from the plasma to the substrate can be simply controlled by diffusion as in the triode grid configuration or by the gas flow.

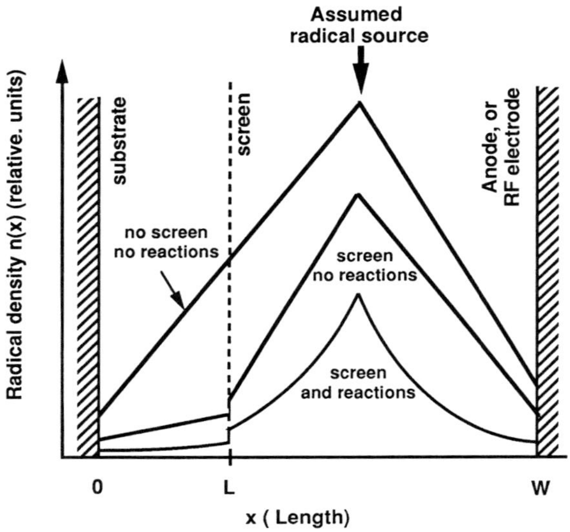

FIGURE 27. Computed radical density profiles in a triode reactor with a mesh or screen electrode showing the effect of radical sticking on the screen and gas-phase secondary reactions on the flux arriving on the substrate. (From Gallagher [94].)

1. Triode Grid Configuration

The planar-triode configuration consists in inserting a mesh-grid electrode between the substrate electrode and the other electrode. The substrate electrode and the mesh grid are at the same potential, or the electric field between them is too low to ignite a plasma, so that the plasma is confined between the mesh grid and the other electrode. In a DC discharge, the mesh grid acts as the cathode (see the DC-proximity discharge configuration in Fig. 18b) and reduces the ion bombardment energy on the substrate. However, in both DC and RF triodes, the grid also acts as a screen for reactive radicals [94] as illustrated in Fig. 27. For example, in SiH_4 discharges, radicals such as SiH_2 that rapidly react with SiH_4 and have a large sticking probability will be lost more rapidly, by sticking to the grid and by gas-phase reactions during their diffusive transport through the grid and toward the substrate, than do the nonreactive radicals such as SiH_3. This selective filtering of SiH_2 radicals has been exploited in RF triode discharges to achieve a-Si:H deposition dominated by the SiH_3 radicals [49, 50, 92]. The improvement of the film microstructure due to the selection of radicals having the larger surface mobility (see Section III.C.1) has also been demonstrated [93].

2. *Flowing Afterglow Configurations*

In flowing afterglow configurations such as those shown in Fig. 28, the transport of reactive species from the plasma source to the substrate is dominated by con-

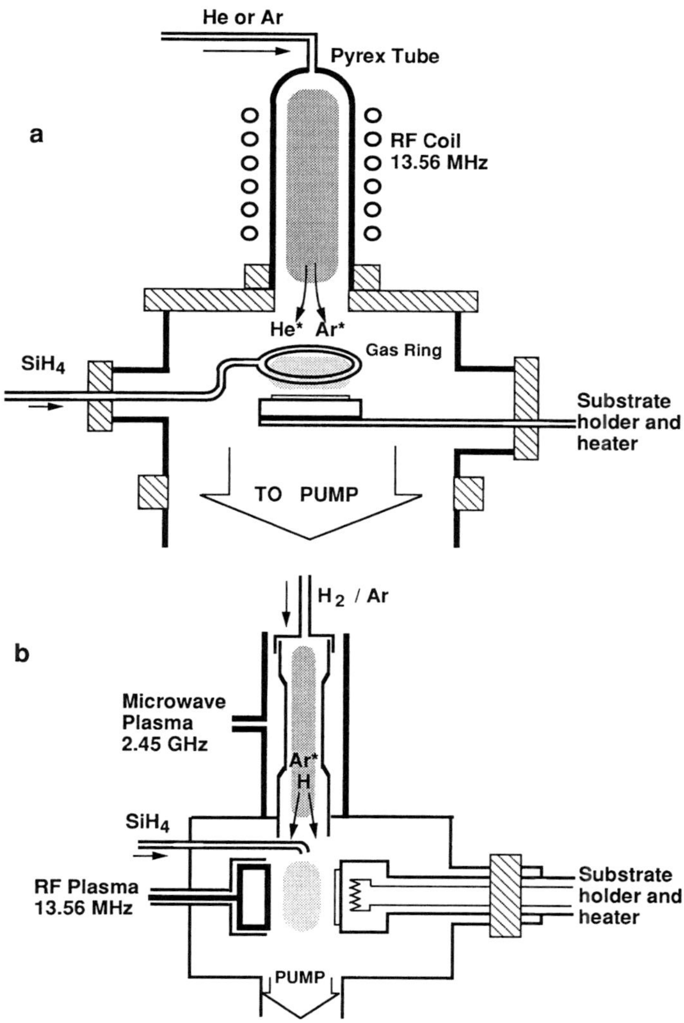

FIGURE 28. Examples of remote plasma PECVD reactors: (a) from Lucovsky *et al.* [95]; (b) from Shirai *et al.* [62].

vection rather than diffusion. This is achieved by flowing the gas along the axis of a tubular plasma source that can be an inductively coupled RF discharge [95, 96] or a MW discharge [62]. At the outlet of the tube where the plasma is extinguished, only the long-lived reactive species can drift far enough to react in the region where the substrate is located. This arrangement is quite versatile since it permits one to create primary-reactive-species-like metastables from Ar or He or atomic H from H_2 plasma sources and to make them react with SiH_4 downstream just above the substrate. Alternatively, the source of H or rare-gas metastables can be used to make a surface treatment or "chemical annealing" of an a-Si:H film deposited from an auxiliary SiH_4 discharge as in Fig. 28b.

3. Plasma Jets

Plasma jets are based on the expansion of a high-pressure thermal plasma created in a small-diameter tube or nozzle in a lower-pressure chamber where the substrate is located. The deposition of a-Si:H by plasma jets has been studied by two groups [97, 98]. In the first case [97] the thermal plasma is created by a RF discharge in a nozzle, and in the second case by a DC cascaded arc [98]. In the DC cascaded arc configuration (see Fig. 29), the plasma jet is supersonic in the first few centimeters after the arc, then subsonic after a shock. The injected gas composition involves 1–10% of SiH_4 in Ar and H_2. In the subsonic flow, the plasma temperature and density gradually decrease and the flux of reactive species on the substrate involves monoatomic Si^+ ions and H atoms created by charge exchange and recombination reactions of Ar^+ with SiH_4 and H_2. The ion kinetic energy on the substrate is on the order of the plasma temperature ($\approx$1 eV). In addition, there is an important contribution of SiH_x radicals arising from the recirculation of SiH_4 molecules created by H-atom etching of the a-Si:H film.

The local a-Si:H deposition rate achievable in plasma jets can reach several 10 nm/s. However, the films exhibit columnar morphology and a propensity to oxidize, which makes them improper for optoelectronic applications up to now.

D. Photo-CVD

The photochemical decomposition of SiH_4 or higher-order silanes (Si_2H_6 or Si_3H_8) is an alternative way of depositing a-Si:H. Hanabusa [99] has presented a review of the various methods of photo-CVD to deposit a variety of materials. We briefly analyze the methods applied to a-Si:H. A main distinction can be established between UV photo-CVD, which involve electronic excitation of the parent molecules; and IR photo-CVD, which involves multiphoton vibrational excitation.

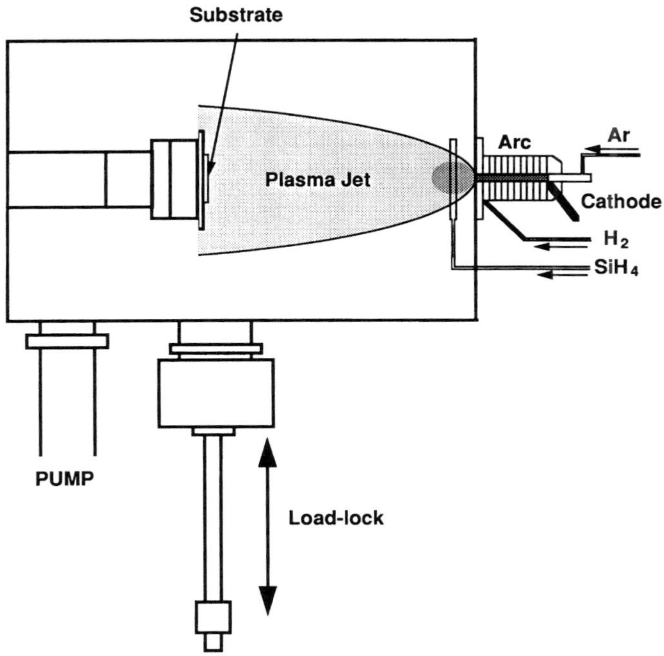

FIGURE 29. Plasma jet for a-Si:H deposition created by a DC cascaded arc. (From Meeusen *et al.* [98].)

1. *UV Photo-CVD*

A variety of UV photon sources have been used for SiH_4 photo-CVD: low-pressure Hg discharge lamps with two narrow resonance lines at 254 and 185 nm, high-pressure discharge lamps using H_2 or D_2 or rare gases such as Xe at 147 nm, and excimer lasers. The silane photoabsorption spectra show broadened features corresponding to successive dissociative electronically excited states.

Two different processes have been exploited:

1. *Hg photosensitization* at 254 nm, using low-pressure Hg lamps, which involves two steps: resonant excitation of Hg atoms premixed with SiH_4, and electrical energy transfer by collision of the excited Hg atom with SiH_4 such that

$$h\nu \text{ (254 nm)} + \text{Hg} \rightarrow \text{Hg}^*, \qquad \text{Hg}^* + \text{SiH}_4 \rightarrow \text{Hg} + \text{H} + \text{SiH}_3.$$

This process, which selectively produces SiH_3 radicals, has been studied and modeled by many groups [100–103] and is known to produce a-Si:H fims of excellent optoelectronic properties. The usual SiH_4 working pressure is around 1 torr, but the feasibility of atmospheric pressure Hg photo-CVD has been demonstrated [104].

2. *Direct photolysis,* where the UV photon can be absorbed by the silane molecule as

$$h\nu + Si_nH_{2n+2} \rightarrow \text{products}.$$

The photolysis threshold is at $\lambda \approx 160$ nm ($h\nu \approx 7.8$ eV) for SiH_4 and shifts to longer wavelengths (or lower photon energies) for higher-order silanes. Therefore, SiH_4 molecules can be photolyzed with a Xe or a D_2 lamp but not with a Hg lamp. However, the 185-nm radiation of the Hg lamp can photolyze higher-order silanes. Examples of direct photo-CVD from Si_2H_6 and Si_3H_8 are given in refs. [105, 106].

The principle of a photo-CVD reactor is illustrated in Fig. 30. The main problems to solve are the control of the lamp temperature, which determines the stability of the UV intensity and the spectrum of the lamp emission; the formation of ozone for $\lambda \leq 200$ nm, which can be avoided by flushing an inert gas around the lamp; and the deposition of a-Si:H on the window (fused silica for the 254- and 185-nm radiations of Hg lamps, or MgF_2 for shorter wavelengths), which results in a loss of transparency. The latter problem can be reduced by some technological "tricks" such as

(a) Flushing an inert rare gas along the window to create a gas curtain and induce a convective flow opposed to the SiH_m radical diffusion.

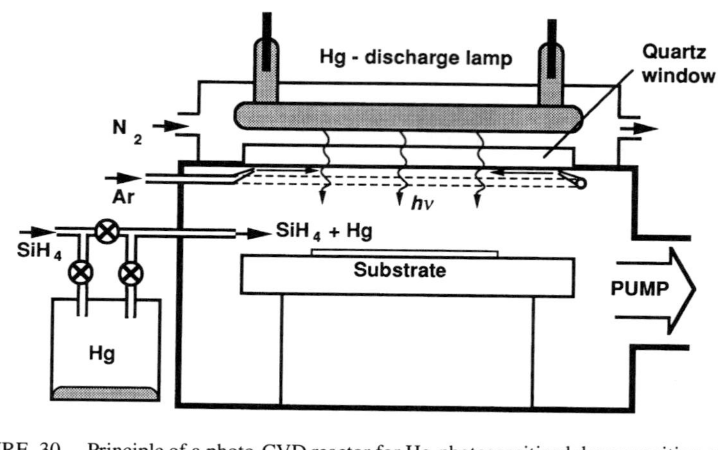

FIGURE 30. Principle of a photo-CVD reactor for Hg-photosensitized decomposition of SiH_4.

(b) Coating the window with a fluorinated oil or grease (e.g., Fomblin) on which the SiH_m radicals have a very low sticking probability [100, 102].
(c) Inserting a rolling Teflon film [103] between the window and the zone of injection of SiH_4.

However, none of these methods is entirely satisfactory, which excludes photo-CVD reactors from industrial applications in a-Si:H technology.

UV photo-CVD in H_2-diluted gas mixtures can lead to epitaxial growth of Si on Si wafers [107, 108], due to the effect of H atoms on the surface kinetics (see Section III.C.3).

With excimer lasers, the photolytic processes are more complex: single-photon or two-photon processes, gas-phase photolysis with a laser beam parallel to the substrate [109], or surface photolysis with a perpendicular laser beam. However, laser photo-CVD is not well adapted for large-area deposition.

2. *IR Photo-CVD*

IR photo-CVD with a CO_2 laser parallel to the substrate has been studied by two groups [110, 111]. The photolytic mechanism proceeds by multiphoton vibrational excitation toward the dissociation limit of the ground electronic state of SiH_4 with

$$n\ h\nu\ (\approx 0.12\ \text{eV}) + SiH_4 \rightarrow SiH_4^{v*} \rightarrow SiH_2 + H_2.$$

The fact that SiH_2 is the primary radical induces chain reactions and the formation of higher-order silane molecules and radicals. In that respect the gas-phase chemistry and the resulting a-Si:H film properties are analogous to the thermal pyrolysis in HOMO-CVD. The vibrational absorption spectrum of a-Si:H films shows a dominance of SiH_2 groups as opposed to the best a-Si:H films obtained by PECVD or UV photo-CVD.

E. Homo CVD and Hot-Filament CVD

1. *Homo CVD*

It is well known that the pyrolytic decomposition of SiH_4 occurs at a temperature much higher than the usual substrate temperature range (≤350°C) compatible with a-Si:H deposition and results in polycrystalline Si film growth. However, by maintaining the substrate at a cooler temperature than the gas-phase temperature, the exodiffusion of H_2 from the growing film surface is restricted and the film

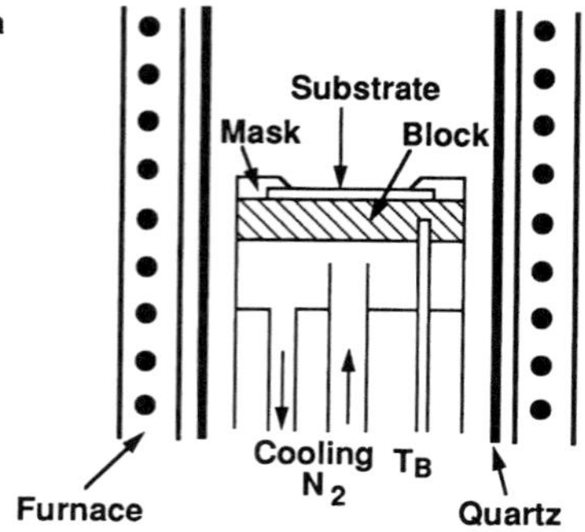

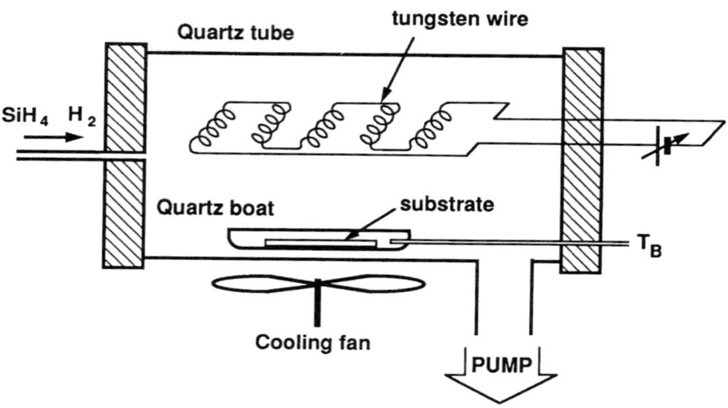

FIGURE 31. (a) Homo–CVD reactor (from Scott *et al.* [112]; (b) hot-filament CVD reactor (from Doyle *et al.* [114]).

remains amorphous. The principle of homo-CVD [112] is illustrated in Fig. 31a. The furnace is heated up to 800°C while the substrate is maintained below 350°C. The gas-phase chemistry is initiated by SiH_4 decomposition into $SiH_2 + H_2$ followed by the formation of higher-order silane molecules and species; however, the strong temperature gradient in front of the substrate is likely to influence the flux of reactive species and the surface reactions. Homo-CVD a-Si:H film properties are distinct but not better than those of conventional PECVD a-Si:H films. The main drawback of the method is the much faster deposition on the furnace walls, which requires frequent cleaning to avoid flaking.

2. *Hot-Filament CVD*

Hot-filament CVD [113, 114] is based on the decomposition of SiH_4 molecules on a hot tungsten filament at 1100–1300°C placed above the substrate maintained at much lower temperature (see Fig. 31b). SiH_4 is diluted in H_2 at a total pressure of a few torrs. The SiH_4 and H_2 decomposition on the hot filament proceeds by Si—H bond and H—H breaking, leading to emission of H, Si, and SiH_3 [114], which then diffuse to the substrate. In addition, H atoms further react with SiH_4 and produce additional SiH_3, or impinge on the growing film, where they control the H surface coverage; hence the surface mobility. In that respect, hot-filament CVD is quite different from Homo-CVD and has more analogy with medium-pressure PECVD or Hg photo-CVD where the SiH_3 radical is the dominant radical responsible for a-Si:H deposition. The a-Si:H films properties are rather good. The main technological difficulty for industrial applications lies in the fragility of the hot-tungsten filament and its development on a large area.

F. Reactive Sputtering

The possibility of depositing a-Si:H films without SiH_4 was demonstrated previously. In fact, one can bring separately Si and H atoms to the growing films by combining a solid Si source emitting Si atoms and an H_2 plasma providing H atoms and H^+, H_2^+, or H_3^+ reactive ions. In principle, the Si atoms can be emitted by evaporation or by ion-induced sputtering from the Si target. In the latter case one can combine Si sputtering and H_2 dissociation and ionization in the same discharge by using a gas mixture of Ar and H_2. This method of reactive sputtering is quite attractive since it offers a great flexibility for the control of the a-Si:H film growth rate, microstructure, and H content, due to the possibility of varying the H_2/Ar partial-pressure ratio in addition to the discharge power, electrical bias on the substrate, and substrate temperature.

Two main discharge configurations have been used:

1. The RF planar diode structure with an asymmetrical potential distribution (see Section II.A). The sheath potential drop has to be much larger on the Si target electrode (referred to as the *cathode*) than on the substrate electrode in order to ensure a high sputtering efficiency on the Si target without resputtering or ion-induced damage on the growing a-Si:H film. The RF diode reactive sputtering method and the properties of sputtered a-Si:H films have been studied extensively [115, 116]. In spite of excellent achievements, it has been recognized that the films usually have poorer optoelectronic properties than do the

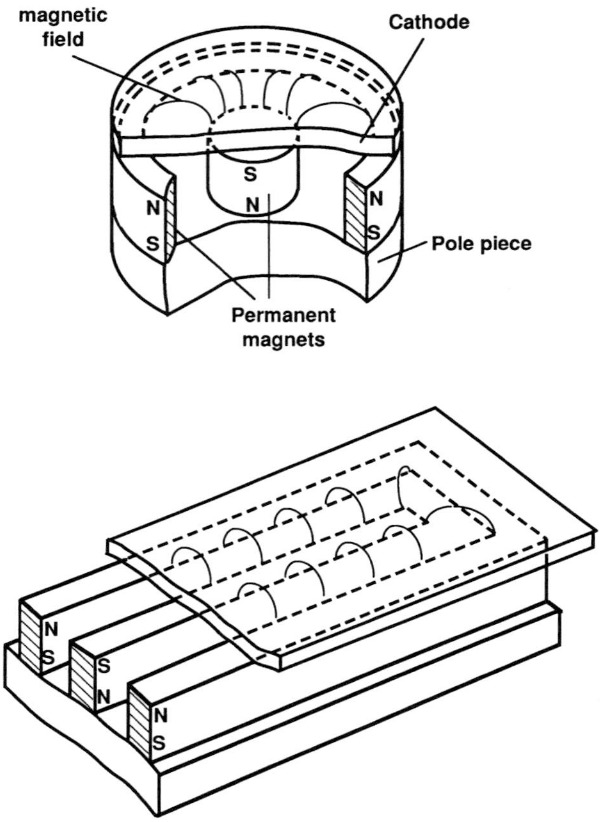

FIGURE 32. Circular and rectangular planar magnetron sputtering sources.

best PECVD a-Si:H films, due to the inclusion of some Ar atoms and to ion-induced damages.

2. The magnetron sputtering configuration, generally used for depositing films of metals and dielectric or transparent conductors in many industrial applications, has been more recently examined for a-Si:H deposition [117, 118]. Its main advantage over RF diode sputtering is the confinement of the plasma close to the target by an arrangement of permanent magnets (see Fig. 32). The electron and ion density is much larger close to the target than above the substrate, which enhances the sputtering efficiency while reducing the ion bombardment energy on the growing film. Moreover, the discharge can be operated at lower pressure than in RF diode sputtering.

The growth mechanisms of sputtered a-Si:H films are expected to differ noticeably from those of PECVD. Indeed, various spectroscopic and mass-spectrometric diagnostics in RF diode sputtering or DC magnetron reactors have shown that the main Si-containing species incoming on the substrate are Si and SiH radicals and Si^+ and SiH^+ ions, whereas SiH_2 and SiH_3 radicals or ions appear as the H_2 pressure increases. The origin of hydrogenated SiH_m radicals is threefold:

1. Gas-phase reactions of sputtered Si atoms with H_2:

$$\mathrm{Si} + \mathrm{H_2} \rightarrow \mathrm{SiH} + \mathrm{H}.$$

2. Direct sputtering from chemisorbed radicals on the target

$$\mathrm{H} + \mathrm{SiH}_m(\text{surface}) \rightarrow \mathrm{SiH}_{m+1}(\text{surface}) \qquad (0 \leq m \leq 2),$$
$$\mathrm{Ar^+}(\text{fast}) + \mathrm{SiH}_m(\text{surface}) \rightarrow \mathrm{SiH}_m(\text{gas}).$$

3. Electron impact dissociation of SiH_4 molecules chemically etched from the target:

$$\mathrm{H} + \mathrm{SiH_3}(\text{surface}) \rightarrow \mathrm{SiH_4}(\text{gas}),$$

$$\mathrm{e^-} + \mathrm{SiH_4} \rightarrow \mathrm{SiH_3} + \mathrm{H} \quad \text{or} \quad \mathrm{SiH_2} + 2\,\mathrm{H}.$$

Since the dominant incoming radicals Si and SiH have sticking probabilities close to unity, their surface mobility is intrinsically low and would lead to self-shadowing and film roughness and porosity (see Section III.C.1) if there were no additional energy transfer on the surface. Actually the good microstructural properties of sputtered a-Si:H films are explained by the contribution of momentum transfer (see Section III.C.2) of moderate energy Ar^+ ions or Ar atoms reflected from the target and by the physicochemical effects of H atoms (see Section III.C.3). In that respect the growth mechanisms of sputtered atoms have many analogies with very low pressure PECVD conditions in multipole or ECR discharges.

V. Summary and Conclusions

The reactor design for a-Si:H deposition has been analyzed first in terms of electrical power dissipation, electron acceleration mechanisms, and discharge regimes (α, γ and γ'). Then we examined the material balance and SiH_4 conversion efficiency into a-Si:H and the main ingredients of the gas-phase and surface physicochemistry controlling the film growth and the final a-Si:H properties. After reviewing the various concepts of reactors that have been developed over the last two decades, we can summarize the respective advantages and drawbacks of the different routes to obtain a-Si:H films of optimal optoelectronic properties.

The first route is the selection of radicals having a high surface mobility on the H-covered growing film surface such as SiH_3, Si_2H_5, or $H_2Si—SiH_2$ without ion bombardment. This requires a selective SiH_4 decomposition and gas-phase chemistry, which is obtained either in nonplasma methods such as Hg photo-CVD or hot-filament-CVD or in low-power and medium-pressure (≈0.1-torr range) PECVD in DC, RF, VHF or MW discharges, or in triode grid discharge and remote plasmas in the flowing afterglow configurations. However, this selective chemistry is usually restricted to low-power densities and low deposition rates (a few angstroms per second up to 10 Å/s), due to the limitations imposed by radical–radical recombinations and powder formation. Nevertheless, even when the discharge is in the γ' regime and powders tend to accumulate in the plasma, a proper control of the hydrodynamics and electrical boundary condition of the discharge allows us to take advantage of the increased resistivity of the discharge to improve the power transfer efficiency in RF discharges and the a-Si:H deposition rate while avoiding powder deposition on the substrate and energetic ion bombardment. The more appropriate configuration for large-area deposition is the capacitively coupled planar-diode RF discharge. The advantage of increasing the electrical excitation frequency from RF (13.56 MHz) to VHF (≈100 MHz) and MW (2.45 GHz) for a better power coupling is counterbalanced by the difficulty of scaling up the discharge in two dimensions for large-area deposition.

The second route is the ion-assisted deposition at low pressure in magnetically confined discharges where the Si-containing species incoming on the surface may have a low surface mobility but where a large ion flux at moderate energy (a few tens of electronvolts) enhance the surface mobility by collisional momentum transfer. These conditions are achieved both in ultra-low-pressure (≈1-mtorr range) PECVD in SiH_4 multipole or ECR discharges but also in low-pressure DC magnetron sputtering. The latter techniques can be scaled up at least along one dimension and adapted for large-area deposition on a continuously moving substrate.

The third route is a deposition assisted by electronic energy transfer rather than collisional momentum transfer. The best candidates are light species such as H atoms, H^+ or H_2^+ ions, or He* metastables, or slow He^+ ions that are present at high dilution of SiH_4 in H_2 or He. The electronic energy transfer induces chemical annealing in the a-Si:H surface growth zone, which improves the propagation of the Si network. In addition, with H atoms, the combination of chemical annealing and chemical etching and passivation of the Si surface favoring the surface mobility of SiH_3 adsorbates eventually leads to microcrystalline or epitaxial film growth. However, the growth rate is usually small.

The highest a-Si:H deposition rates (>100 Å/s) have been obtained in plasma jets, but their microstructural and optoelectronic properties are quite poor. At such fast deposition rates, the concept of surface physicochemistry developed to explain a-Si:H deposition in the 1 – 10-Å/s range may not be appropriate. However, we think that a proper tuning of the flux ratio of Si-containing species and H atoms or slow ions might help improving the a-Si:H film quality.

Acknowledgments

This analytical view of the reactor design for a-Si:H PECVD is the result of years of experience in SiH_4 plasma diagnostics and modeling, *in situ* surface diagnostics, and a-Si:H film characterizations, which I shared with all my colleagues of the Laboratoire de Physique des Interfaces et des Couches Minces at the Ecole Polytechnique and with Jacques Schmitt from the SOLEMS company. I also thank the French teams of the ARC PIRSEM-CNRS on diagnostics and modelling of a-Si:H PECVD reactors: A. Bouchoule of the GREMI in Orléans, J. P. Boeuf of the CPAT in Toulouse, and J. P. Couderc and B. Despax from ENSIGC in Toulouse. Finally I want to acknowledge many decisive contributions from several researchers of different groups with whom I had enlightening discussions and sometimes fruitful collaborations, especially A. Matsuda from the Electrotechnical Laboratory in Tsukuba, A. Gallagher from the University of Colorado at Boulder, I. Shimizu from the Tokyo Institute of Technology, and J. Abelson and M. Kushner from the University of Illinois.

References

1. K. U. Riemann, *J. Appl. Phys.* **65,** 999 (1989).
2. K. Kohler, D. E. Horne, and J. W. Coburn, *J. Appl. Phys.* **58,** 3350 (1985).
3. A. M. Pointu, *Appl. Phys. Lett.* **50,** 1047 (1987).
4. C. Böhm, J. Perrin, and P. Roca i Cabarrocas, *J. Appl. Phys.* **73,** 2578 (1993).
5. Y. Yamaguchi, A. Sumiyama, R. Hattori, Y. Morokuma, and T. Makabe, *J. Phys. D.* **22,** 505 (1989).
6. C. G. Goedde, A. J. Lichtenberg, and M. A. Lieberman, *J. Appl. Phys.* **64,** 4375 (1988).
7. M. J. Kushner, *IEEE Trans. Plasma Sci.* **PS-14,** 188 (1986).
8. V. A. Godyak and A. S. Khanneh, *IEEE Trans. Plasma Sci.* **PS-14,** 112 (1986).
9. P. Belenguer and J. P. Boeuf, *Phys. Rev. A* **41,** 447 (1990).
10. C. Böhm, J. Perrin, P. Belenguer, and J. P. Boeuf, *Proc. Int. Conf. Phenom. Ionized Gases 20th (ICPIG–XX), Pisa* **5,** 1052. Ist. Fis. At. Mol.–CNR, Pisa, 1991.
11. J. Perrin, P. Roca i Cabarrocas, B. Allain, and J. M. Friedt, *Jpn. J. Appl. Phys.* **27,** 2041 (1988).
12. C. Böhm and J. Perrin, *J. Phys. D* **24,** 865 (1991).
13. J. P. Boeuf and P. Belenguer, *J. Appl. Phys.* **71,** 4751 (1992).
14. J. Perrin, *J. Non-Cryst. Solids* **137/138,** 639 (1991).
15. F. Tochikubo, A. Suzuki, S. Kakuta, Y. Terazono, and T. Makabe, *J. Appl. Phys.* **68,** 5532 (1990); T. Makabe, F. Tochikubo, and M. Nishimura, *Phys. Rev.* **42,** 3674 (1990).
16. M. Capitelli, C. Gorse, R. Winkler, and J. Wilhelm, *Plasma Chem. Plasma Process.* **8,** 399 (1988);

R. Winkler, M. Capitelli, C. Gorse, and J. Wilhelm, *Plasma Chem. Plasma Process.* **10,** 419 (1990).
17. V. A. Godhyak and R. B. Piejack, *J. Vac. Sci. Technol., A* **8,** 3833 (1990).
18. A. Howling, C. Hollenstein, P. J. Paris, F. Finger, and U. Kroll, *Proc. Int. Conf. Phenom. Ionized Gases 20th (ICPIG–XX), Pisa* **5,** 1089. Ist. Fis. At. Mol.–CNR, Pisa, 1991.
19. M. J. Kushner, *J. Appl. Phys.* **63,** 2532 (1988).
20. J. P. Boeuf and P. Belenguer, unpublished results (1992).
21. S. Veprek and M. G. J. Veprek-Heijman, *Plasma Chem. Plasma Process.* **11,** 323 (1991).
22. J. R. Doyle, D. A. Doughty, and A. Gallagher, *J. Appl. Phys.* **68,** 4375 (1990).
23. D. A. Doughty and A. Gallagher, *J. Appl. Phys.* **67,** 139 (1990).
24. J. C. Knights, J. P. M. Schmitt, J. Perrin, and G. Guelachvili, *J. Chem. Phys.* **76,** 3414 (1982).
25. N. Hata and K. Tanaka, *J. Non-Cryst. Solids* **77/78,** 777 (1985).
26. J. Perrin, *J. Phys. D* **26,** 1662 (1993).
27. A. Gallagher, *Mater. Res. Soc. Symp. Proc.* **38,** 99 (1984).
28. P. A. Longeway, R. D. Estes, and H. A. Weakliem, *J. Phys. Chem.* **88,** 73 (1984); *J. Appl. Phys.* **57,** 5499 (1985).
29. G. Turban, Y. Catherine, and B. Grolleau, *Thin Solid Films* **60,** 147 (1979); **67,** 309 (1980); **77,** 287 (1981); *Plasma Chem. Plasma Process.* **2,** 61 (1982).
30. J. Perrin, J. P. M. Schmitt, G. De Rosny, B. Drévillon, J. Huc, and A. Lloret, *Chem. Phys.* **73,** 383 (1982).
31. L. Layeillon, B. Despax, P. Duverneuil, and J. P. Couderc, *C. R. Colloq. Int. Procedes Plasmas, 8th (CIP 91), Antibes, F., Vide Couches Minces* Suppl. No. 256, p. 32 (1991).
32. B. Drévillon, J Huc, A. Lloret, J. Perrin, G. de Rosny, and J. P. M. Schmitt, *Appl. Phys. Lett.* **37,** 646 (1980).
33. J. Perrin and J. P. M. Schmitt, *Chem. Phys.* **67,** 167 (1982).
34. J. P. M. Schmitt, *J. Non-Cryst. Solids* **59/60,** 649 (1983).
35. N. Itabashi, N. Nishiwaki, M. Magane, S. Naito, T. Goto, A. Matsuda, C. Yamada, and E. Hirota, *Jpn. J. Appl. Phys.* **29,** L-505 (1990).
36. D. A. Doughty and A. Gallagher, *Phys. Rev. A* **42,** 6166 (1990).
37. J. P. Boeuf, P. Belenguer, and J. Wang, *Mater. Res. Soc. Symp. Proc.* **165,** 17 (1990).
38. J. Perrin, A. Lloret, G. de Rosny, and J. P. M. Schmitt, *Int. J. Mass Spectrom. Ion Proc.* **57,** 249 (1984).
39. A. A. Howling, J. L. Dorier, and C. Hollenstein, *Appl. Phys. Lett.* **62,** 1341 (1993); A. A. Howling, L. Sansonnens, J. L. Dorier, and C. Hollenstein, *J. Phys. D* **26,** 1003 (1993).
40. I. Chen, *Thin Solid Films* **101,** 41 (1983); I. Chen and F. Jansen, *J. Non-Cryst. Solids* **59/60,** 695 (1983).
41. K. Tachibana, T. Okuyama, H. Harima, and Y. Urano, *Proc. Int. Symp. Plasma Chem. 7th (ISPC–7), Eindhoven, Neth.* (C. J. Timmermans, ed.), p. 588, 1985.
42. J. C. Knights, *in* "Plasma Synthesis and Etching of Electronic Materials" (R. P. H. Chang and B. Abeles, eds.), Vol. 38, p. 371. Mater. Res. Soc., Pittsburgh, Pennsylvania, 1985.
43. C. C. Tsai, J. C. Knights, G. Chang, and B. Wacker, *J. Appl. Phys.* **59,** 2998 (1986).
44. B. Drévillon, J. Perrin, J. M. Siéfert, J. Huc, A. Lloret, G. de Rosny, and J. P. M. Schmitt, *Appl. Phys. Lett.* **42,** 801 (1983); B. Drévillon, J. Huc, and N. Boussarssar, *J. Non-Cryst. Solids* **59/60,** 735 (1983); B. Drévillon and M. Toutlemonde, *J. Appl. Phys.* **58,** 535 (1985).
45. N. Blayo and B. Drévillon, *Appl. Phys. Lett.* **59,** 950 (1991).
46. G. H. Lin, J. R. Doyle, M. He, and A. Gallagher, *J. Appl. Phys.* **64,** 188 (1988).
47. M. J. Kushner, *J. Appl. Phys.* **62,** 4763 (1987); M. J. McCaughey and M. J. Kushner, *J. Appl. Phys.* **65,** 186 (1989).

48. P. Ho, W. G. Breiland, and R. J. Buss, *J. Chem. Phys.* **91,** 2627 (1991).
49. J. Perrin, Y. Takeda, N. Hirano, Y. Takeuchi, and A. Matsuda, *Surf. Sci.* **210,** 114 (1989).
50. A. Matsuda, K. Nomoto, Y. Takeuchi, A. Suzuki, A. Yuuki, and J. Perrin, *Surf. Sci.* **227,** 50 (1990).
51. J. L. Guizot, K. Nomoto, and A. Matsuda, *Surf. Sci.* **244,** 22 (1991).
52. D. A. Doughty, J. R. Doyle, G. H. Lin, and A. Gallagher, *J. Appl. Phys.* **67,** 6220 (1990).
53. P. Roca i Cabarrocas, *Mater. Res. Soc. Symp. Proc.* **149,** 33 (1989).
54. U. Kroll, F. Finger, J. Dutta, H. Keppner, A. Shah, A. Howling, J. L. Dorier, and C. Hollenstein, *Mater. Res. Soc. Symp. Proc.* **258,** 135 (1992).
55. R. Robertson and A. Gallagher, *J. Appl. Phys.* **59,** 3402 (1986).
56. D. Mataras, S. Cavadias, and D. Rapakoulias, *J. Appl. Phys.* **66,** 119 (1989).
57. A. Bouchoule, A. Plain, L. Boufendi, J. P. Blondeau, and C. Laure, *J. Appl. Phys.* **69,** 1991 (1991); L. Boufendi, A. Plain, J. P. Blondeau, A. Bouchoule, C. Laure, and M. Toogood, *Appl. Phys. Lett.* **60,** 169 (1992).
58. N. Itabashi, N. Nishiwaki, M. Magane, T. Goto, A. Matsuda, C. Yamada, and E. Hirota, *Jpn. J. Appl. Phys.* **29,** 585 (1990).
59. J. Perrin, Y. Takeda, N. Hirano, H. Matsuura, and A. Matsuda, *Jpn. J. Appl. Phys.* **28,** 5 (1989).
60. P. Roca i Cabarrocas, P. Morin, V. Chu, J. P. Conde, J. Z. Liu, H. R. Park, and S. Wagner, *J. Appl. Phys.* **69,** 2942 (1991).
61. K. Nomoto, Y. Urano, J. L. Guizot, G. Ganguly, and A. Matsuda, *Jpn. J. Appl. Phys.* **29,** L1372 (1990).
62. H. Shirai, J. Hanna, and I. Shimizu, *Appl. Phys. Lett.* **59,** 1096 (1991); *Jpn. J. Appl. Phys.* **30,** L679 (1991).
63. N. Blayo and B. Drévillon, *J. Non-Cryst. Solids* **137/138,** 775 (1991).
64. H. Shirai, J. Hanna, and I. Shimizu, *Jpn. J. Appl. Phys.* **30,** L881 (1991).
65. M. Yamanaka, Y. Hayashi, and I. Sakata, *Jpn. J. Appl. Phys.* **29,** L217 (1990).
66. A. Suzuki, Y. Toyoshima, P. McElheny, and A. Matsuda, *Jpn. J. Appl. Phys.* **30,** L790 (1991).
67. K. G. Spears, R. P. Kampf, and T. J. Robinson, *J. Chem. Phys.* **92,** 5297 (1988).
68. Y. Watanabe, M. Shiratani, Y. Kubo, I. Ogawa, and S. Ogi, *Appl. Phys. Lett.* **53,** 1263 (1988); Y. Watanabe, M. Shiratani, and H. Makino, *Appl. Phys. Lett.* **57,** 1616 (1990).
69. J. Perrin, C. Böhm, R. Etemadi, and A. Lloret, *Plasma Sources Sci. Technol.* **3,** 252 (1994).
70. L. J. Overzet, J. T. Verdeyen, R. M. Roth, and F. F. Carasco, *Mater. Res. Soc. Symp. Proc.* **98,** 321 1987); J. T. Verdeyen, J. Beberman, and L. Overzet, *J. Vac. Sci. Technol., A* **8,** 1851 (1990).
71. T. J. Sommerer, M. S. Barnes, J. H. Keller, M. McCaughey, and M. J. Kushner, *Appl. Phys. Lett.* **59,** 638 (1991).
72. G. M. Jellum, J. E. Daugherty, and D. B. Graves, *J. Appl. Phys.* **69,** 6923 (1991).
73. J. L. Dorier, C. Hollenstein, A. A. Howling, and U. Kroll, *J. Vac. Sci. Technol., A* **10,** 1048 (1992).
74. A. Lloret, E. Bertran, J. L. Andùjar, A. Canillas, and J. L. Morenza, *J. Appl. Phys.* **69,** 632 (1991).
75. J. P. M. Schmitt, *Thin Solid Films* **174,** 193 (1989).
76. P. Roca i Cabarrocas, J. B. Chevrier, J. Huc, A. Lloret, J. Y. Parey, and J. P. M. Schmitt, *J. Vac. Sci. Technol., A* **9,** 2331 (1991).
77. D. E. Carlson, *in* "Polycrystalline and Amorphous Thin Films" (L. L. Kazmerski, ed.), Academic Press, New York, 1980.
78. Y. Yamaguchi and T. Makabe, *Proc. Int. Symp. Plasma Chem., 9th (ISPC–9) (IUPAC), Pugnochiuso, Italy* (R. d'Agostino, ed.), p. 1353 (1989).
79. H. Curtins, N. Wyrsch, and A. Shah, *Electron. Lett.* **23,** 228 (1987).
80. F. Finger, U. Kroll, V. Viret, A. Shah, X.-M. Tang, J. Weber, A. Howling, and C. Hollenstein, *J. Appl. Phys.* **71,** 5665 (1992).

81. H. Chatham and P. Bath, *Mater. Res. Soc. Symp. Proc.* **149,** 447 (1989); H. Chatham, P. Bath, A. Benson, and C. Matovich, *J. Non-Cryst. Solids* **115,** 201 (1989).
82. T. Hamasaki, M. Ueda, A. Chayahara, M. Hirose, and Y. Osaka, *Appl. Phys. Lett.* **44,** 600 (1984); *Appl. Phys. Lett.* **44,** 1049 (1984).
83. J. P. Conde, K. K. Chan, J. M. Blum, M. Arienzo, and J. J. Cuomo, *J. Appl. Phys.* **71,** 3981 (1992).
84. K. Yokota, M. Takada, Y. Ohno, and S. Katayama, *J. Appl. Phys.* **72,** 1188 (1992).
85. M. Taniguchi, M. Hirose, T. Hamaski, and Y. Osaka, *Appl. Phys. Lett.* **37,** 787 (1980).
86. N. Ohnishi, H. Nishiwaki, K. Uchihashi, K. Yoshida, M. Tanaka, K. Ninomiya, M. Nishikuni, N. Nakamura, S. Tsuda, S. Nakano, T. Yazaki, and Y. Kuwano, *Jpn. J. Appl. Phys.* **27,** 40 (1988).
87. M. R. Wertheimer, M. Moisan, J. E. Klemberg-Sapieha, and R. Claude, *Proc. Int. Symp. Plasma Chem., 8th (ISPC–8) (IUPAC), Tokyo* (K. Akashi and A. Kinbara, eds.), p. 1253 (1987).
88. M. R. Wertheimer, J. E. Klemberg-Sapieha, and H. P. Schreiber, *Thin Solid Films* **115,** 109 (1984).
89. M. Kitagawa, S. Ishihara, K. Setsune, Y. Manabe, and T. Hirao, *Jpn. J. Appl. Phys.* **26,** L231 (1987).
90. Y. H. Shing, C. L. Yang, C. E. Allevato, and F. S. Pool, *Mater. Res. Soc. Symp. Proc.* **149,** 63 (1989).
91. T. Watanabe, M. Tanaka, K. Azuma, M. Nakatani, T. Sonobe, and T. Shimada, *Jpn. J. Appl. Phys.* **26,** L288 (1987).
92. A. Matsuda and K. Tanaka, *J. Appl. Phys.* **60,** 2351 (1986).
93. T. Ichimura, T. Ihara, T. Hama, M. Ohsawa, H. Sakai, and Y. Uchida, *Jpn. J. Appl. Phys.* **25,** L276 (1986).
94. A. Gallagher, *J. Appl. Phys.* **60,** 1369 (1986).
95. G. Lucovsky, P. D. Richard, D. V. Tsu, S. Y. Lin, and R. J. Markunas, *J. Vac. Sci. Technol., A* **4,** 681 (1986).
96. D. V. Tsu and G. Lucovsky, *J. Non-Cryst. Solids* **97/98,** 839 (1987).
97. L. Bardos, V. Dusek, and M. Vanecek, *J. Non-Cryst. Solids* **97/98,** 281 (1987).
98. G. J. Meeusen, E. A. Ershov-Pavlov, R. F. G. Meulenbroeks, M. C. M. van de Sanden, and D. C. Schram, *J. Appl. Phys.* **71,** 4156 (1992).
99. M. Hanabusa, *Mater. Sci. Rep.* **2,** 51 (1987).
100. J. Perrin and T. Broekhuizen, *Appl. Phys. Lett.* **50,** 433 (1987); *Mater. Res. Soc. Symp. Proc.* **75,** 201 (1987); T. Fuyuki, B. Allain, and J. Perrin, *J. Appl. Phys.* **68,** 3322 (1990).
101. Y. Matsui, A. Yuuki, N. Morita, and K. Tachibana, *Jpn. J. Appl. Phys.* **26,** 1575 (1987).
102. K. Kamisako, T. Imai, and Y. Tarui, *Jpn. J. Appl. Phys.* **27,** 1092 (1988).
103. R. E. Rocheleau, S. S. Hegedus, W. A. Buchanan, and S. C. Jackson, *Appl. Phys. Lett.* **51,** 133 (1987).
104. W. Y. Kim, M. Konagai, and K. Takahashi, *Jpn. J. Appl. Phys.* **27,** L948 (1988).
105. K. Kumata, U. Itoh, Y. Toyoshima, N. Tanaka, H. Anzai, and A. Matsuda, *Appl. Phys. Lett.* **48,** 1380 (1986).
106. T. Fuyuki, K. Y. Du, S. Okamoto, S. Yasuda, T. Kimoto, M. Yoshimoto, and H. Matsunami, *J. Appl. Phys.* **64,** 2380 (1988).
107. S. Nishida, T. Shimoto, A. Yamada, S. Karasawa, M. Konagai, and K. Takahashi, *Appl. Phys. Lett.* **49,** 79 (1986).
108. N. Gonohe, S. Shimizu, K. Tamagawa, T. Hayashi, and H. Yamakawa, *Jpn. J. Appl. Phys.* **26,** L1189 (1987).
109. Y. Toyoshima, K. Kumata, U. Itoh, and A. Matsuda, *Appl. Phys. Lett.* **51,** 1925 (1987).
110. R. Bilenchi, I. Gianinoni, and M. Musci, *J. Appl. Phys.* **53,** 6479 (1982).

111. M. Meunier, J. H. Flint, J. S. Haggerty, and D. Adler, *J. Appl. Phys.* **62,** 2812 (1987); **62,** 2822 (1987).
112. B. A. Scott, J. A. Reimer, R. M. Plecenick, E. E. Simonyi, and W. Reuter, *Appl. Phys. Lett.* **40,** 973 (1982); B. S. Meyerson, B. A. Scott, and D. J. Wolford, *J. Appl. Phys.* **54,** 1461 (1983).
113. H. Matsumura, *Appl. Phys. Lett.* **51,** 804 (1987); *J. Appl. Phys.* **65,** 4396 (1989).
114. J. Doyle, R. Robertson, G. H. Lin, M. Z. He, and A. Gallagher, *J. Appl. Phys.* **64,** 3215 (1988).
115. T. D. Moustakas, *in* "Semiconductors and Semimetals, Vol. 21: Hydrogenated Amorphous Silicon," Part A (J. Pankove, ed.), p. 55. Academic Press, New York, 1984.
116. M. J. Thompson, *Top. Appl. Phys.* **55,** 119 (1984).
117. N. Savvides, *J. Appl. Phys.* **55,** 4232 (1984).
118. M. Pinabarsi, J. Abelson, and M. J. Kushner, *J. Appl. Phys.* **68,** 2255 (1990).

5 Optoelectronic Properties of Amorphous Silicon Using the Plasma-Enhanced Chemical Vapor Deposition (PECVD) Technique

Arun Madan
MVSystems Inc.
Golden, Colorado

I. Introduction

By far the most intensively studied of the amorphous semiconductor family is amorphous silicon (a-Si), and particularly the hydrogenated variety [1–3]. Any deposition technique that does not employ a reactive environment of H (or F) leads to extremely poor materials in terms of high density of localized states (DOS) due to Si dangling bonds, such that the films cannot be electronically doped, which then precludes any electronic device possibilities. Numerous techniques [4] for depositing a-Si-based alloys have been attempted, including (1) RF glow discharge [or plasma-enhanced chemical vapor deposition (PECVD)] in

SiH_4 gas [5], (2) PECVD in SiF_4 and H_2 gas mixtures [6], (3) reactive sputtering of Si in H environments [7], (4) evaporation of Si in the presence of H [8], (5) CVD of SiH_4 gas and higher silanes [9], (6) photo-CVD [10], (7) electron cyclotron resonance (ECR) PECVD [11], and (8) remote PECVD [12].

However, by far the most widely studied and used is the PECVD technique, which typically utilizes a capacitively coupled multichamber system as shown in Fig. 1 [13]. A gas, such as SiH_4, is allowed to pass at a controlled rate between

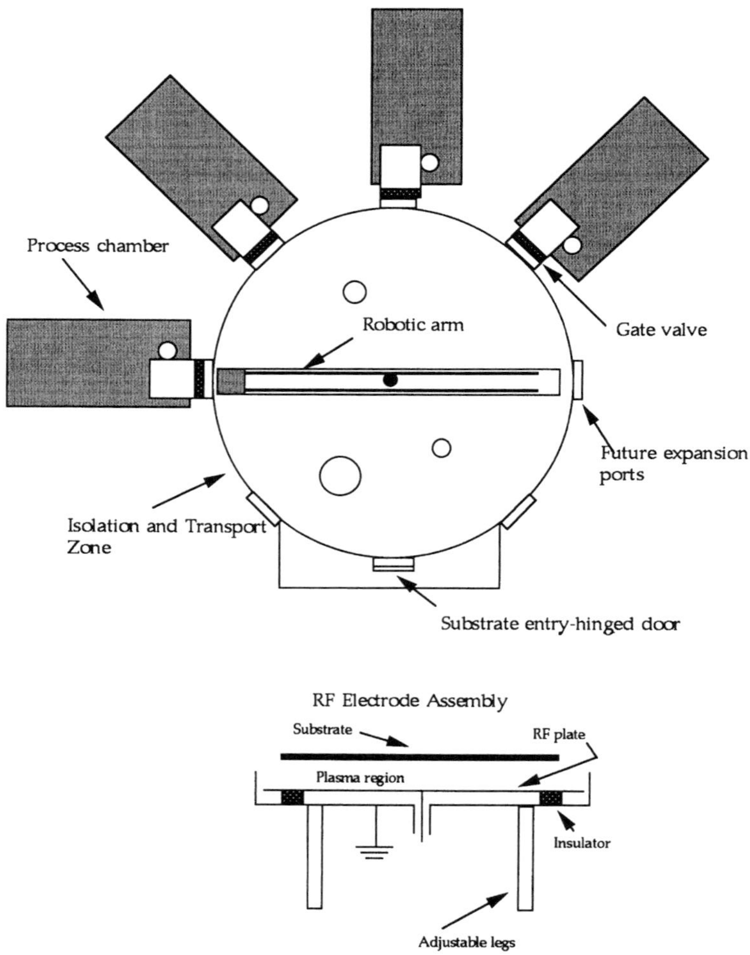

FIGURE 1. (a) A multichamber PECVD system; (b) a schematic of the RF electrode plate assembly [13].

Table 1 Typical conditions employed in the fabrication of a-SiH using the PECVD of SiH_4 gas in a research-type system

Parameters	Comments	Typical values
Gas composition		100% SiH_4
Flow rate	High	~40 sccm
Pressure	Low	200–600 m torr
Deposition temperature		250°C
RF power	Low	<25 m W cm^{-2}
Anode–cathode distance		~1–4 cm

the two plates and the plasma is generated using DC, audio-, RF, or microwave frequencies. A variety of species are produced, such as atoms, free radicals, and stable and unstable ions. The mean energy of the electron is on the order of a few electronvolts, and the electron temperature reached can be up to 100 times higher than that of the gas, and hence the electrons possess enough energy to break down the molecular bonds.

The optoelectronic properties of the a-SiH prepared by the PECVD technique are dictated by many deposition parameters, such as pressure of the gas, flow rate, substrate temperature, power dissipation in the plasma, anode–cathode distance, excitation frequency, deposition rate, diluent used, and electrode configuration [2]. Table 1 shows an example of the deposition conditions that are generally employed to produce "device"-quality amorphous hydrogenated Si (a-SiH). In Table 2, we summarize [2] some of the present state-of-the-art parameters obtained for undoped and doped a-SiH (F) material thus produced. It should be noted that the "device"-quality material exhibits semiconducting behavior since ln σ_d versus $10^3/T$ (where σ_d is the dark conductivity) exhibits a straight line with a conductivity activation energy of ~0.85 eV, which is approximately equal to half the band gap of ~1.75 eV. Typically $\sigma_d < 10^{-10}$ (Ω-cm)$^{-1}$ which changes to $>10^{-5}$ (Ω-cm)$^{-1}$ under light illumination (Global AM1.5 sun illumination of intensity 100 mW cm^{-2}). The DOS (or defects) is found to be low with a dangling-bond (DB) density, as measured by ESR (electron spin resonance), to be $\sim 10^{15}$ cm^{-3}. The inherent disorder possessed by these materials manifests itself as band tails that emanate from the conduction and valence bands and are characterized by exponential tails with energies of 25 and 45 meV, respectively; it should be noted that the broader tail from the valence band provides for dispersive transport (shallow defect controlled) for holes with a low drift mobility of 10^{-2} cm^2 s^{-1} V^{-1}, whereas the electrons exhibit nondispersive transport behavior with a higher mobility of ~1 cm^2 s^{-1} V^{-1}. Hence the material exhibits poor minority (hole) carrier transport with a diffusion length less than 0.5 μm, which puts a design limitation on electronic devices, such as solar cells, as discussed

Table 2 Typical optoelectronic parameters obtained for a-Si:H(F) alloys

Undoped		
Hydrogen content	C_H	~10%
Dark conductivity at 300 K	σ_d	$\sim 10^{-10} (\Omega\ \text{cm})^{-1}$
Activitation energy	ΔE or E_a	~0.8 – 0.9 eV
Preexponential conductivity factor	σ_0	$>10^3 (\Omega\ \text{cm})^{-1}$
Optical band gap at 300 K	E_g	1.7 – 1.8 eV
E_g variation with temperature	$E_g(T)$	$2\text{-}4.10^{-4}$ eV K^{-1}
Density of states at minimum	g_{min} or $g(E_f)$	$>10^{15}\text{–}10^{17}\ \text{cm}^{-3}\ \text{eV}^{-1}$
Density of states at conduction band edge	$g(E_c)$	$\sim 10^{21}\ \text{cm}^{-3}\ \text{eV}^{-1}$
ESR spin density	N_s	$\sim 10^{15}\ \text{cm}^{-3}$
IR spectra		$2000/640\ \text{cm}^{-1}$
Photoluminescence peak at 77 K		~1.25 eV
Extended state mobility		
Electrons	μ_n or μ_e	$>10\ \text{cm}^2\ \text{s}^{-1}\ \text{V}^{-1}$
Holes	μ_p or μ_h	$\sim 1\ \text{cm}^2\ \text{s}^{-1}\ \text{V}^{-1}$
Drift mobility		
Electrons	μ_n or μ_e	$>1\ \text{cm}^2\ \text{s}^{-1}\ \text{V}^{-1}$
Holes	μ_p or μ_h	$\sim 10^{-2}\ \text{cm}^2\ \text{s}^{-1}\ \text{V}^{-1}$
Conduction band tail slope		25 meV
Valence band tail slope		40 meV
Hole diffusion length		$< 0.5\ \mu\text{m}$
Doped amorphous		
n-Type: ~1% addition of PH3 to gas phase		$\sigma_d \sim 10^{-2} (\Omega\ \text{cm})^{-1}$ $\Delta E \sim 0.2$ eV
p-Type: ~1% addition of B_2H_4 to gas phase		$\sigma_d \sim 10^{-3} (\Omega\ \text{cm})^{-1}$ $\Delta E \sim 0.3$ eV
Doped microcrystalline		
n-Type: ~1%PH_3, added to dilute SiH_4/H_2 or 500 vppm, PH_3 added to SiF_4/H_2 (8 / 1) gas mixtures; relatively high power involved		$\sigma_d \geq 1 (\Omega\ \text{cm})^{-1}$ $\Delta E \leq 0.05$ eV
p-Type: ~1% B_2H_6 added to dilute SiH_4/H_2 gas mixtures		$\sigma_d \geq 1 (\Omega\ \text{cm})^{-1}$ $\Delta E \leq 0.05$ eV

more fully in Chapter 3. Finally, it is to be noted that the material can be doped n- and p-type with a room-temperature σ_d in excess of 10^{-4} $(\Omega\text{-cm})^{-1}$. By manipulating the plasma (use of SiF_4 and H_2) or by heavily diluting the SiH_4 plasma with H_2, microcrystalline, one can make n and p materials with σ_d in excess of 1 $(\Omega\text{-cm})^{-1}$ [6, 14].

Further improvements in the material are expected to result from a better understanding of the relationship between the processing conditions and the specific chemical reactions taking place in the plasma and at the surface that promote film growth. In previous chapters, exhaustive treatments are given on the "desirable" precursors involved in the film growth but, as noted, the understanding is far from complete. In this chapter, we shall attempt to show that an interdependence exists between some of the crucial deposition parameters and their influence on the material properties and hence on the performance of the electronic devices.

II. Effect on Properties of a-SiH Due to Parametric Variations Using the PECVD Technique

A. Deposition Temperature

One of the main controlling factors determining the optoelectronic properties of the a-SiH material is the deposition temperature T_s. The original work on a-SiH alloy showed [15–17] that the density of localized states (DOS) in the midgap position is critically controlled by T_s. From their field-effect data, the DOS was seen to decrease from $>10^{18}$ cm^{-3} eV^{-1} to $\sim 10^{16}$ cm^{-3} eV^{-1} when T_s was raised from room temperature to ~250°C. This is also consistent with the ESR data which indicates that the dangling-bond density, N_s, is $\sim 10^{15}$ cm^{-3} for a sample grown at T_s ~250°C [18], whereas $N_s > 10^{17}$ cm^{-3} for a low-T_s ~50°C-produced sample; using constant-photocurrent method (CPM) data, as shown in Fig. 2a, we note that a similar trend in the DOS with an increase of the T_s [19].

This vast change in the DOS (or defects) has been attributed to basic structural changes that occur in the film. As shown in Fig. 3 [20], films produced at low T_s (25°C) generally possess a H concentrations (>30%). Although, as shown in Figs. 2b and 4 [21], the H/Si content ratio within the film and the consequent optical band gap E_g are reduced when T_s is increased, it is the mode of incorporation of H into the Si matrix that has a major effect on the electronic behavior, as indicated by IR spectroscopy. (The absorption peaks in the IR spectra can be interpreted by recognizing that there can be either 1, 2 or 3 H atoms situated at a Si site. Since the mass of H is small compared to that of Si, the vibrational modes can be described by considering only the displacement of the H atoms.) Specifically, the Si–H environment is characterized by a bond stretching mode at 2000 cm^{-1} and a bond bending mode at 640 cm^{-1}, whereas sites with more than one H atom exhibit additional features in the bond bending frequency range, 800–950 cm^{-1}, as well as bond stretching modes between 2050 and 2190 cm^{-1}. Comparing Figs. 3 and 5, we note that for low excitation powers and at T_s = 240°C, the IR spectra are dominated by two strong absorption bands at 2000 and at 640 cm^{-1}; on the other hand, the sample deposited at 25°C show additional modes in the region of band stretching (1900–2200 cm^{-1}) and bond bending regimes, with

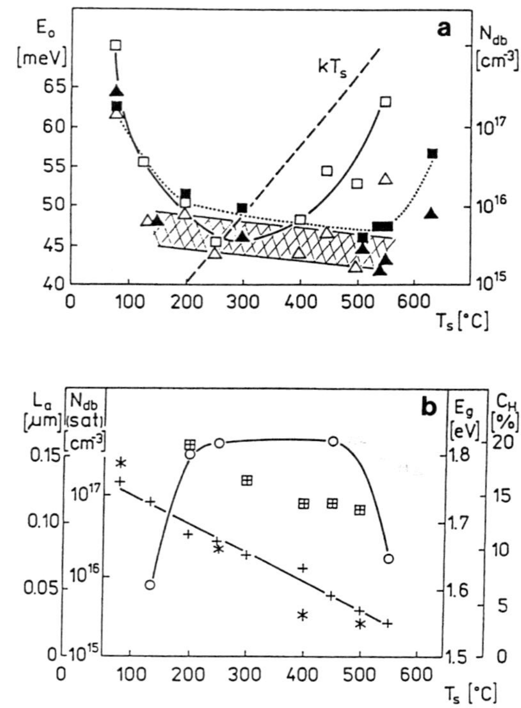

FIGURE 2. (a) Urbach slope, E_0 (triangles), and DB density (N_{db}) as a function of the deposition temperature T_s. As-grown samples from UHV PECVD system are marked by the empty symbols, from the hot-wire series by full symbols. Curves are drawn as a guide to the eye. Dashed lines represent $E_0 = kT_s$. (b) Ambipolar diffusion length, L_a (circles), Tauc optical gap, E_g (crosses), and hydrogen content C_H (asterisks) as a function of the substrate temperature, T_s for as-grown UHV PECVD samples. DB density in light-soaked, saturated state N_{db} (sat) (squares with cross) is also shown [19].

two sets of doublets in the 800–900-cm^{-1} region. For the low-T_s-produced samples, in addition to it exhibiting a 845-cm^{-1} peak, there is also an appearance of columnar morphology as observed by scanning electron microscopy [22], which indicates that the chains of SiH_2 groups $(SiH_2)n$ or polysilane exist in the interstitial regions between columns of less hydrogenated material.

Further evidence that low-T_s-produced films are structurally different from high-T_s-produced samples is provided by H effusion experiments. In this, a sample is heated in an evacuated chamber; at specific temperatures there is a sudden increase in the pressure due to effusion of H [23]. A sample deposited at low T_s produces two distinct pressure changes, one at $T = 350°C$ and the other at

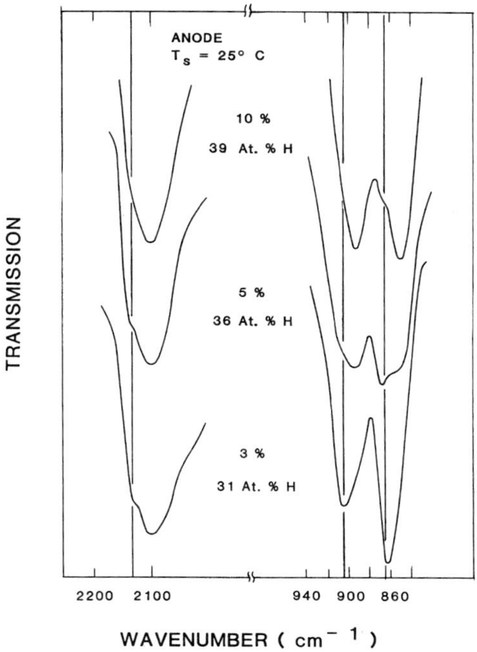

FIGURE 3. Infrared transmission of a-Si:H for T_s = 25°C. The relative dilution of SiH_4 in Ar is given as atomic percent of H [20].

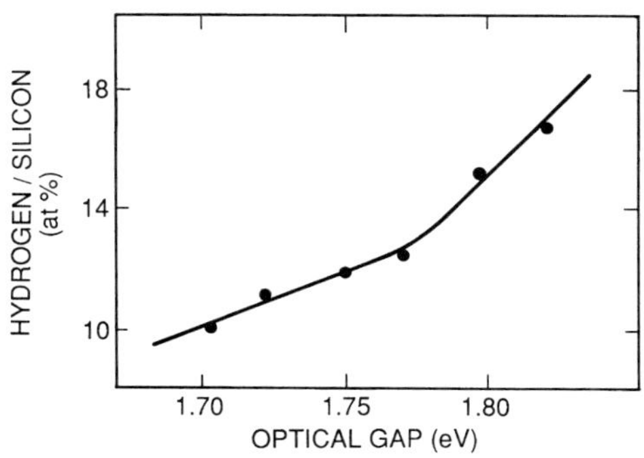

FIGURE 4. Correlation between the optical band gap E_g and the hydrogen/silicon ratio for a constant electronic and gas conditions: RF power = 70 W, SiH_4/H_2 gas ratio of 17%. The experimental points correspond to films deposited at different temperatures T_s in the range of 140–400°C [21].

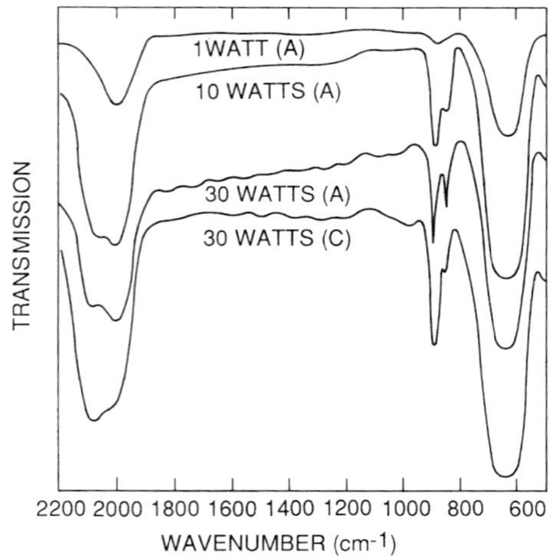

FIGURE 5. Infrared transmission for a-Si:H samples prepared at different powers with T_s held at 230°C [20].

T = 550°C, whereas high-T_s-produced samples show only the latter effect. The results are self-consistently interpreted on the basis of different H configurations within the material: T = 350°C effusion peak is associated with the H bonded in the more weakly bonded H–Si–H sites (dihydride), whereas the T = 550°C peak is related to the H bonded in a Si–H site (monohydride). This structural change is confirmed by the Raman data [24], as shown in Fig. 6, where the integrated intensity at 2100 cm^{-1} shows a minimum as a function of T_s at about 250°C, indicating primarily a reduction in the dihydride component, whereas the integration of the 625 cm^{-1} peak reinforces the data of Fig. 4, that there is a reduction in the total hydrogen concentration.

The decrease in the DOS, due to structural changes, has an effect on the optoelectronic properties as shown in Fig. 2b, where the diffusion length shows a peak in the T_s range of 250–450°C. Considering first the mobility of electrons, an interesting [25] *in situ* measurement examines the majority carrier transport using the time-resolved microwave conductivity (TRMC), in which the change of the microwave power, $\Delta P(t)/P$, from the sample, on pulsed illumination [using a 10-ns pulse width, Nd:YAG (neodymium:yttrium aluminum garnet) laser] is monitored. In this, the TRMC signal is proportional to the number of mobile excess carriers (weighted by their mobilities) and hence is a measure of the drift mobility. Figure 7 shows the mobility for samples fabricated at the different T_s as indicated. As the T_s is varied from about 125 to 250°C, the mobility slowly in-

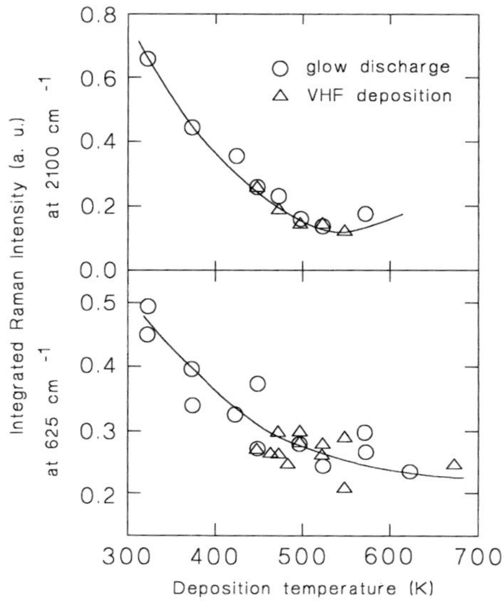

FIGURE 6. Relative measure of H content and bonding structure from Raman measurements of the SiH_x wagging (b) and stretching vibrations (a) [24].

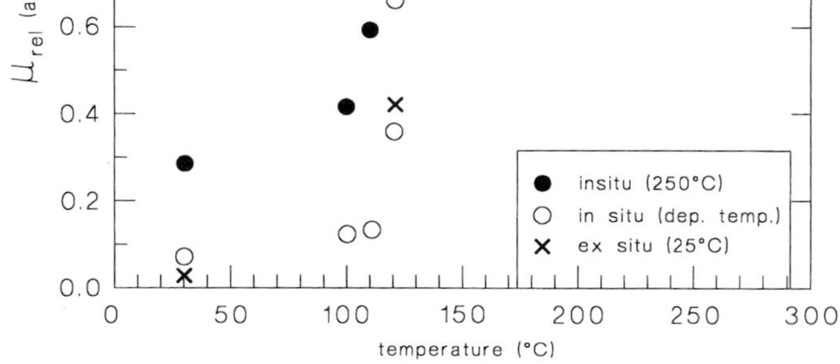

FIGURE 7. Relative effective mobility of a-Si:H as a function of T_s [25].

creases, whereas for $T_s < 125°C$, it shows a sharp decrease. In the former case the data could be interpreted on the basis of sharper band tails emanating from the conduction band; for the latter temperature case, the decrease is attributed to an increase in the DOS deep in the band gap.

Although the electron drift mobility exhibits nondispersive transport and a relatively high value of 1 $cm^2\ s^{-1}\ V^{-1}$, the hole transport shows a significantly inferior transport behavior. The transport measurements from transient-type experiments exhibit dispersive behavior with mobilities on the order of $10^{-2} cm^2\ s^{-1}\ V^{-1}$. (The mobility is related to the minority carrier diffusion length, which is a major factor in dictating the performance of solar cells, and its variation, as a function of T_s, is shown, e.g., in Fig. 2b.) From the drift mobility (for electron and holes) variation as a function of temperature one can determine that the transport is dictated by traps in the band tails emanating from valence (and conduction bands) whose energy distribution can be characterized with an exponential with a characteristic energy of 45 meV (and 25 meV). This data coupled with the DOS data result in a picture of the DOS of a-SiH as shown in Fig. 8.

One characteristic feature of the material, and hence its applicability in many product areas, is its photoconductivity, σ_{ph}. Typically under Global AM1.5 illu-

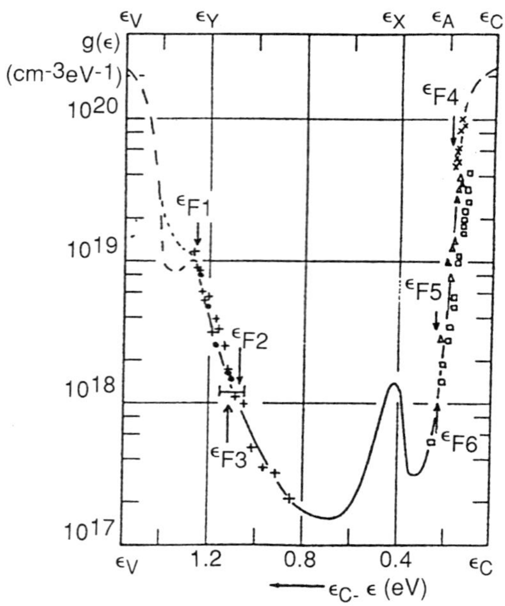

FIGURE 8. Density of states spectra.

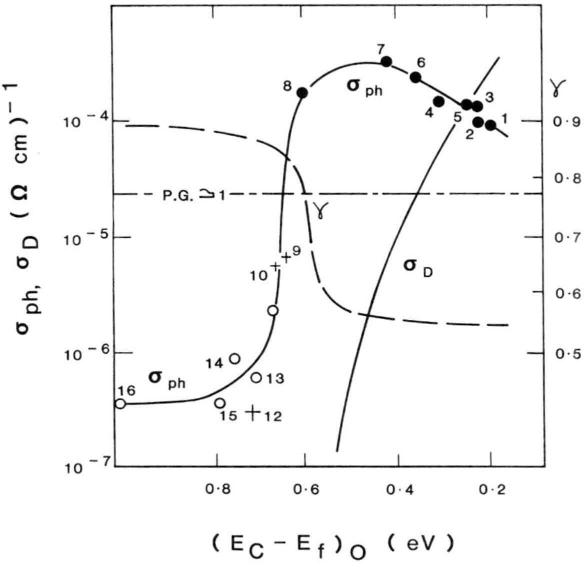

FIGURE 9. Photoconductivity at 295 K plotted against $(E_c - E_f)_0$, the position of the dark Fermi level for phosphorus-doped (solid circles), undoped (+), and boron-doped samples (open circles) samples. The broken line represents the exponent γ (see right-hand ordinate); σ_d is a typical dark conductivity curve. P.G. ~1 refers to unit photoconductive gain at a field of 3×10^3 V cm^{-1} [26].

mination (of intensity 100 mW cm^{-2}) the material exhibits σ_{ph} in excess of 10^{-5} cm^2 s^{-1} V^{-1}, which is about 10^6 larger than its dark conductivity.

A useful parameter to measure is the intensity dependence of the photoconductivity, σ_{ph} α F', where F is the intensity of the incident light. We show in Fig. 9 [26], γ as a function of $(E_c - E_f)$, where E_c is the conduction band edge and E_f is the Fermi level position; in the figure, E_f is altered in the different samples investigated by doping with PH_3. (It should be emphasized that not all the dopants atoms go into the donor configuration, since >90% act to create extra defects in the material.) Nevertheless, we note that for an intrinsic material, where $(E_c - E_f) \sim 0.8$–0.9 eV, $\gamma \sim 0.9$; this decreases to smaller values when dopants (which, in turn, increases the defect states) are added. A value of $\gamma > 0.9$ is consistent with low DOS ($<10^{16}$ cm^{-3} eV^{-1}) as calculated by McMahon and Madan [27]. Therefore, by analogy, we expect that as the deposition temperature is altered from its optimum value, γ will decrease to smaller values due to an increased DOS.

The preceding discussion suggests that an optimum T_s (~250°C) exists for producing a low DOS material. However, it may be feasible to decrease the DOS even further, since the structure may be in one of the local energy minima among many that could exist, which, in turn, is dependent on the particulars of the preparation conditions employed. It was noted initially by Ast and Brodsky [28] that rapid cooling from a temperature of $T > 200°C$ led to an increase in the dark conductivity by a factor of 2, which suggested that a basic structural change had occurred. Subsequent work has indicated that changes can occur dependent on the annealing history. For instance, Smith *et al.* [29] reported an increase in the dangling-bond density, if the material is quenched from a temperature in excess of 200°C. They reasoned that since the number of free electron–hole pairs increases with T, there is an inevitable increase in the recombination rate between the carriers, which provides sufficient energy to break the bonds; this is akin to the often quoted cause of instability known as the *Staebler–Wronski* [30] *effect.* It is argued that most of these extra defects created would be frozen in during the quenching process. Using CPM, sub-band-gap absorption was reported to increase by a factor of 3 for fast-quench (200°C/min) samples in comparison with the slow-cool samples (1°C/min). However, this was not supported by a subsequent work of Agrawal *et al.* [31], as shown in Fig. 10, where no change in sub-

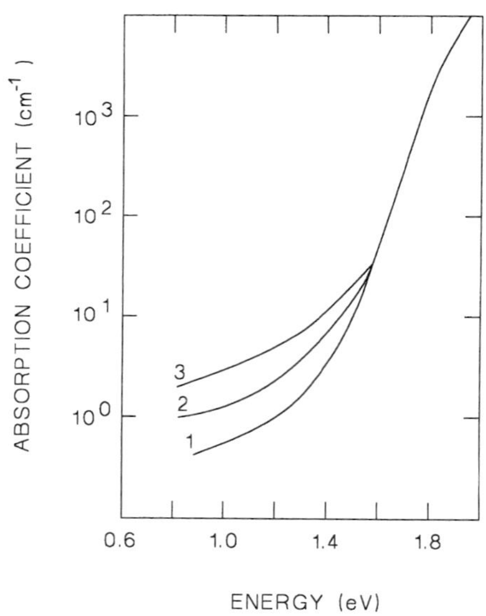

FIGURE 10. Sub-band-gap absorption as measured by the photothermal deflection spectroscopy of intrinsic a-Si:H alloy after (1) slow (1°C/min.) and fast (10°C/s) cooling from 220°C, (2) subsequent fast cooling from 260°C, and (3) subsequent slow cooling from 260°C [31].

Table 3 Fill factor of solar cell under different conditions of cooling rate from 220C

	Zero bias		Reverse bias	
	Fast cool	Slow cool	Slow cool	Fast cool
AM1	0.67	0.67	0.72	0.69
Red	0.64	0.65	0.7	0.67
Blue	0.73	0.72	0.74	0.74

From Agrawal *et al.* [31].

band-gap absorption is evident for slow or fast cooling from a $T > 220°$C; further, the slow and fast anneals from a higher T of 260°C show the opposite trend to the work of Smith *et al.* [29]. By comparing the solar cell device performance, shown in Table 3, we observe virtually no change in the fill factor (FF) (at the red or the blue end of the illuminated spectra) is found irrespective of slow or fast cooling. (It should be noted that the physical interpretation of the FF is complex as it depends on many factors, such as the DOS at the p^+/i interface, the bulk DOS spectra, the mobility, and the lifetime of the carriers.) Nevertheless, the FF for the blue wavelength not only remains unchanged but is also of a high value; this implies that the interface DOS remains at a low value. The same could be said of the bulk DOS, as the FF in the red also does not change.

Street [32] noted that the $\sigma_d(T)$ data for n-type (Fig. 11) samples and for p-type samples exhibited different branches dependent on the magnitude of the cooling rates. Figure 11 shows the data for the n-type samples warmed beyond 130°C, with the different branches coalesced, and suggests that the sample is now in thermal equilibrium (TE); in this, the TE temperature is defined as TE = 130C since the relaxation time is approximately equal to the experimental time scale. This is qualitatively similar to the behavior of glasses, where the V–T (volume–temperature) curve generally exhibit a departure from linearity at the glass transition temperature T_g: further, T_g was found to increase with faster cooling. The analogy with a glass is made complete by attributing the "viscous" behavior to the H, which can move within the material; it is speculated that this motion is responsible for the structural changes as it is a catalyst or is intermediate between the shallow defects (weak Si — Si bonds) and deep defects (DB).

However, as discussed in Chapters 1 and 4, the growth and the quality of a-SiH is dictated by various reactions, all of which have a temperature dependence. We mentioned in Chapter 4 that SiH_3 could be the major precursor for growth, and the data indicate that approximately 30% of these species arriving on the surface

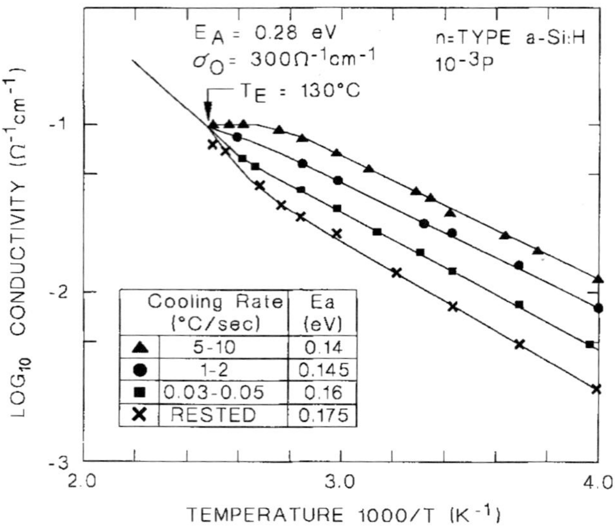

FIGURE 11. The temperature dependence of the DC conductivity immediately after annealing when cooled at different rates, and after being stored at room temperature for 2 months. Note the change in activation energy above the equilibration temperature [32].

are adsorbed and diffuse by a temperature-activated process given by $D = D_h \exp(-E_h/kT)$, where $E_h = 0.2$ eV. Thereafter, the adsorbed precursor competes with other temperature-activated processes occurring at the surface, such as (1) reacting with the surface bonded H and desorbing SiH_4, leaving a DB on the surface (abstraction); (2) chemisorbing on to the surface DB created by another precursor in (1); and (3) reacting with another precursor and desorbing as Si_2H_6. It is expected that for $T < 350°C$, the dominant mechanism for creation of DBs is (1), but for $T > 350°C$, thermal desorption of H from the surface becomes an important source of growth sites. This reaction implies that for temperatures beyond the normal "optimal" T_s, the diffusion coefficient of SiH_3 will increase and lead to a decrease in the effective number of surface DBs and thereby lower the bulk DOS. A test could be to increase the flux of SiH_3 at the surface via manipulation of the deposition parameters (by increasing the RF power, adjusting the flow rates, and simultaneously using low deposition pressure). The resulting data [33] are shown in Fig. 12, where we note that (1) the deposition rate is increased and (2) the DOS decrease as the SiH_3 flux is increased. The fact that for $T_s = 400°C$, the DOS is lower than with respect to 250°C argues against the TE model discussed above.

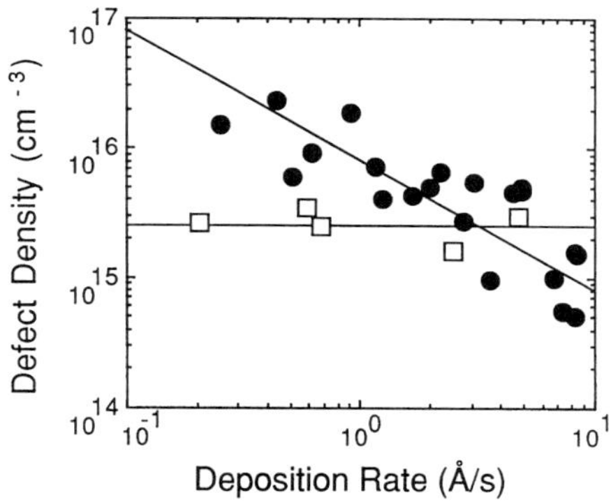

FIGURE 12. Defect density of a-Si:H films deposited at 250°C (open squares) and 400°C (closed circles) at different deposition rates [33].

B. Influence of Ion Bombardment, Area of Electrodes, and Bias Voltage

As discussed previously, the normal mode of deposition is to use a capacitively coupled planar-type PECVD system. Figure 13 shows [34] the time-averaged potential distribution between the RF plate (cathode) and the grounded plate. As there is a large difference between the mobility of ions and the electrons, the cath-

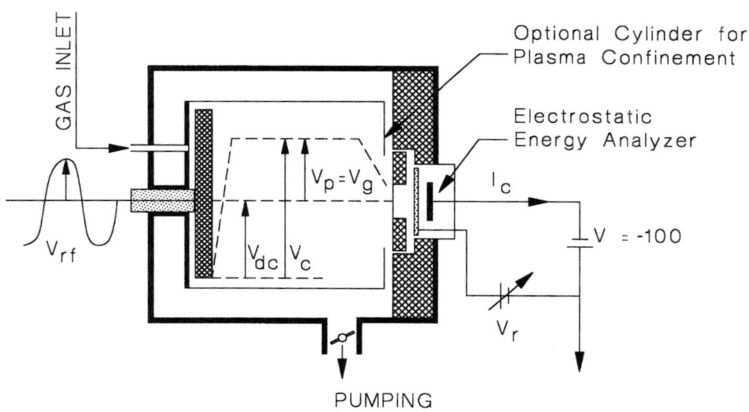

FIGURE 13. Schematic view of a capacitively coupled PECVD reactor [34].

ode self-biases to a negative voltage of V_{DC}. The plasma region has a net positive potential of V_p. There are two sheath regions, one near the cathode and the other near the anode. For perfect symmetry between the anode and the cathode, the plasma potential is given by $V_p = (V_{DC} + V_{RF})/2$, where V_{RF} is the peak RF voltage. However, in a realistic situation, the electrodes are asymmetrical, and

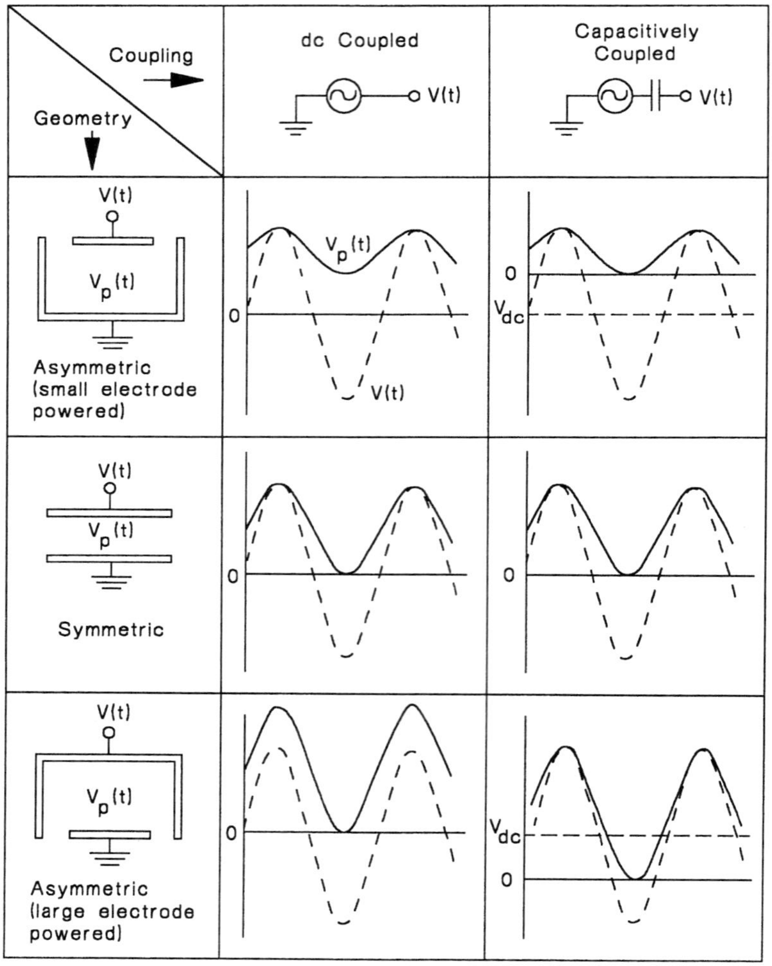

FIGURE 14. Illustrative plasma potentials $V_p(t)$ (solid curves) and excitation electrode voltages $V(t)$ (dashed curves), assuming purely capacitive sheath behavior, for three system geometries and for DC-coupled and capacitively coupled excitation electrodes [35].

hence the plasma potential can be expressed as $V_p = V_{RF}[1 + (A_a/A_c)]^{-1}$, where A_a and A_c are the areas of the anode and cathode, respectively. Therefore, as the ratio A_a/A_c increases, the plasma potential decreases and vice versa. This is illustrated [35] in Fig. 14 for different electrode symmetries.

Generally, it is found that "device"-quality films are produced with the substrate situated at the anode, heated to a $T_s \sim 250°C$. However, as shown in Fig. 13, there is potential drop at the anode, and hence it is possible that the growing surface can experience ion bombardment. In a study carried out by Kasper *et al.* [36], and shown in Fig. 15 (for H_2 plasma), that as the electrode configuration becomes more asymmetrical (i.e., $A_a/A_c > 1$), the plasma potential decreases in relation to the less symmetrical configuration. Therefore, it is conceivable that the film properties could be dictated by these changes in the potential drop, which influence the extent of the ion bombardment. Whether this is beneficial is unclear, as this is intertwined with other issues such as the different deposition conditions and system configuration employed by the researchers and not always stated. In principle, there are many effects due to ion–surface interactions during film growth formation, including (1) enhancement of adatom (adsorbed atom) migration, (2) desorption of impurity atoms on the substrate, (3) displacement of surface atoms, (4) trapping or sticking of incident ions, (5) sputtering, and (6) implantation.

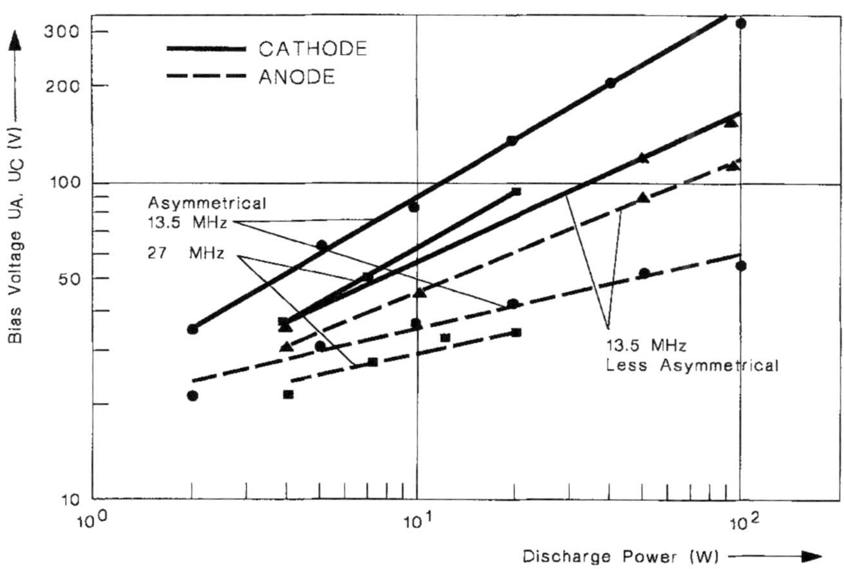

FIGURE 15. Bias voltage versus discharge power for a H_2 plasma [36].

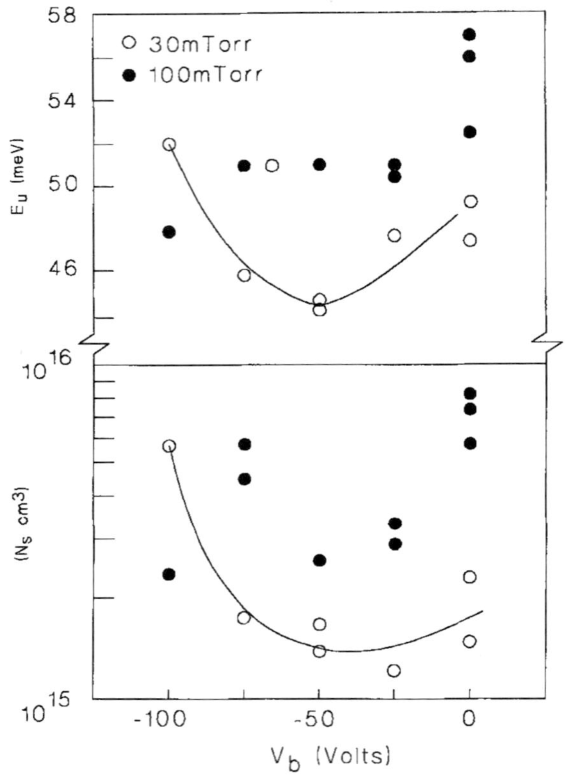

FIGURE 16. Bulk density of states and Urbach energy measured by the CPM as a function of the substrate bias. The solid lines are guides [38].

There have been many studies of imposing a bias on the substrate in an attempt to understand the ion bombardment issue. For instance, Hattori *et al.* [37] do not see any changes in the spin density (10^{16} cm^{-3}) irrespective of whether there is a positive or negative bias of 50 V, although the deposition rate is seen to decrease with increasing negative bias. On the other hand, Roca i Cabarrocas *et al.* [38] see a minimum in the DOS (as determined by the CPM) for samples prepared at 30 mtorr and shown in Fig. 16; the situation was less clear at a higher pressure of 100 mtorr. By comparing the data from the Photothermal Deflection Spectroscopy (PDS) technique and the CPM, they find that the surface state density is not affected by the negative bias. However, these results are not conclusive, as the σ_d remained almost constant, as did the photoconductivity and γ at 0.85. They [38] note that with increasing negative bias, E_g decreases due to a reduction in the H concentration, whereas Ando *et al.* [39], using a different electrode configuration note an increase in the H concentration. It should be noted that most systems do not employ a bias imposed onto the anode.

C. Effect of Deposition Pressure and Electrode Spacing

The pressure of the gasses during the deposition affects the properties of the film, since gas-phase polymerization is encouraged at high pressure [40] with the consequence that the SiH_2 component within the film becomes more pronounced. The different bonding arrangements within the film can be understood from Paschen's law, which states that the voltage required to sustain the plasma between the two electrodes is a two-branched function of the product of plasma pressure (p) and the distance between the two electrodes (d); this is represented schematically in Fig. 17. In region I (as p is lowered or as d is decreased), the electron energy is limited by collisions with the electrodes. When p or d is made to decrease, more

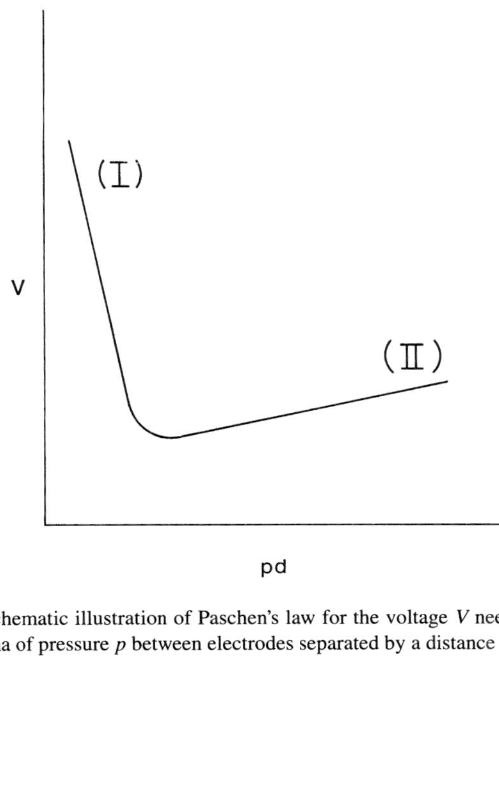

FIGURE 17. A schematic illustration of Paschen's law for the voltage V needed to sustain a glow discharge in a plasma of pressure p between electrodes separated by a distance d.

voltage is required to make up for the energy lost to the walls of the system. On the other hand, in region II (increase of p or d), the electrons are more likely to collide with the plasma constituents than the electrodes. The voltage required to sustain the discharge also increases, this time to make up for the energy lost to collisions within the plasma, which, in turn, promotes polymerization and can lead to the inclusion of SiH_2 chains within the film.

The deposition rate, the uniformity, and the properties of the films are dictated by the pd product. Ishihara *et al.* [41] note that for deposition at 2 torr (which is considered high as the normal deposition pressure used is <600 mtorr), the deposition rate increased from 3 to 18 Å/s when d was changed from 18 to 10 mm. With $d = 5$ mm, the deposition decreased to less than 1 Å/s, and the deposit occurred primarily at the edge of the substrate while uniform deposit occurred at an approximately 10 mm spacing. The IR spectra, shown in Fig. 18, indicate that as the pd product increases (curves 1 and 3), the films tend to exhibit a pronounced peak at 2090 cm^{-1}, which is a characteristic of SiH_2 inclusion whereas the lower-pd-produced film (curve 2) shows a predominant peak at 2000 cm^{-1}, characteristic of singly bonded SiH alloy.

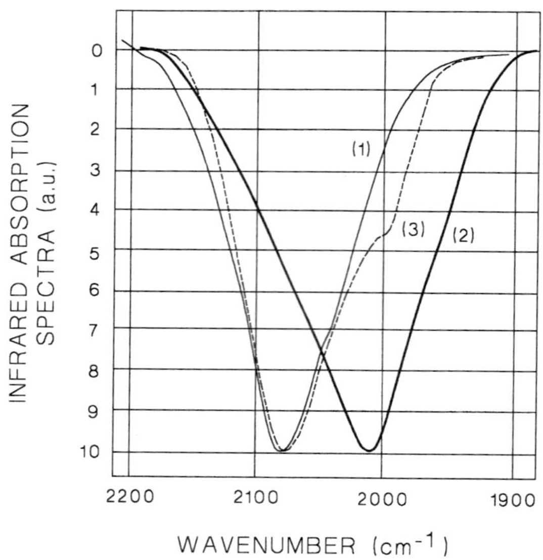

FIGURE 18. Normalized infrared absorption spectra centered about 2000 cm^{-1}: (1) electrode spacing $d = 30$ mm and pressure, $p = 0.5$ torr; (2) electrode spacing $d = 10$ mm and pressure $p = 1$ torr; (3) electrode spacing $d = 4$ mm and pressure $p = 4$ torr [41].

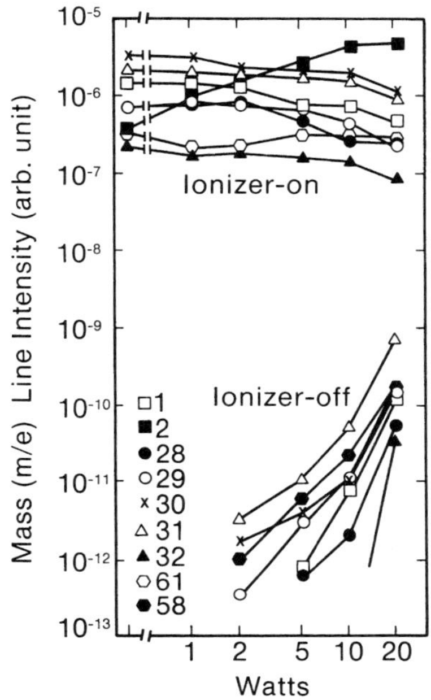

FIGURE 19. Line intensities of the observed mass numbers (mass/charge ratio *m*/*e*) from a SiH_4 plasma as functions of RF power (5 sccm, 50 m torr) for ionizer on/off operation [42].

D. Effect of Power and Gas Flow Rates

Matsuda and Tanaka [42] have identified the relative concentration of ionic and neutral species contained within the plasma by mass spectroscopy. Figure 19 shows the comparison of mass spectra with ionizer on and off. It is clearly evident that for low excitation power, the density of ionic species is lower by 10^4 compared to the density of the neutral species. Ions such as SiH_3^+, SiH_2^+, SiH^+, H^+, and $Si_2H_2^+$ increase by more than two orders of magnitude as the RF power is increased from 2 to 20 W. A rapid rise in the population of ions with power implies an increase in the electron density. Further, the predominant ionic species appears to be SiH_3^+, in agreement with the results of Drévillon *et al.* [43] and discussed in Chapter 2.

The results of Turban *et al.* [44] shown in Fig. 20 illustrate that for low power levels (e.g., 5 W) up to 50% of the SiH_4 remains undissociated and this percentage decreases to 20% when the films are prepared at a power above 50 W. The figure

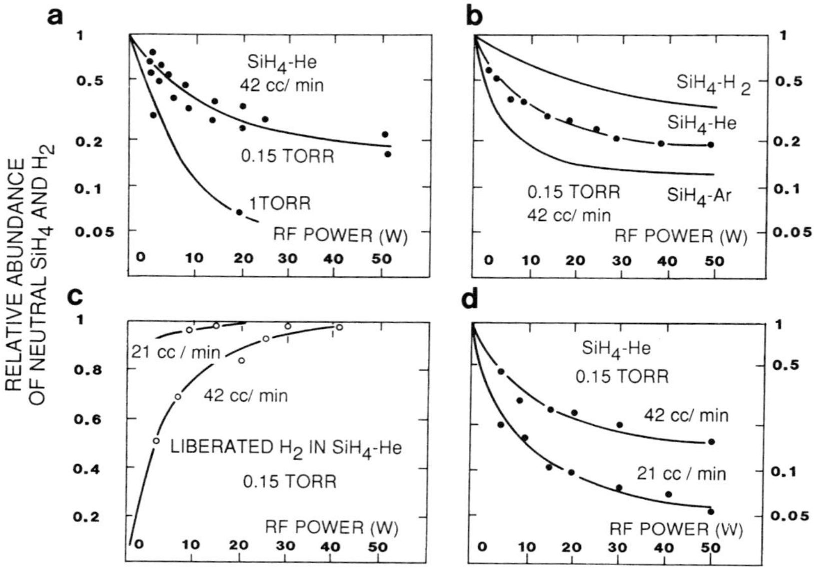

FIGURE 20. Relative concentration of neutral molecules in a silane discharge as a function of power: (a), (b), and (d) show the concentrations of SiH_4: (c) concentration of H_2 released by silane decomposition [44].

also shows the formation of H_2 due to the decomposition of SiH_4 gas, and that the decomposition proceeds more readily with lower flow rates. These observations, coupled with the corresponding IR measurements performed on the resultant films, suggest that for conditions under which the silane is not entirely decomposed, the films contain a majority of SiH units. Those films in which the silane is strongly dissociated contain a majority of SiH_2 units. This is in agreement with the work of Knights and Lujan [22] and Street *et al.* [18], who have obtained a correlation between the spin density of the a-SiH films and the deposition condition employed. Similar conclusions were also arrived at by Hirose [45] using optical emission spectroscopy (OES). By studying the emission lines of Si (288 nm), SiH (414 nm), and H (656 nm) from the plasma, it was noted that the emission corresponding to atomic H increased faster than that for SiH when the flow rate of SiH_4 was decreased. The importance of the reactive SiH and H species is shown by the correlation between the concentration ratio of SiH to H in the plasma and the content of SiH_2 units in the resulting film, as shown in Fig. 21. The incorporation of SiH_2 units is significant at low flow rates where the SiH_4 gas depletes rapidly.

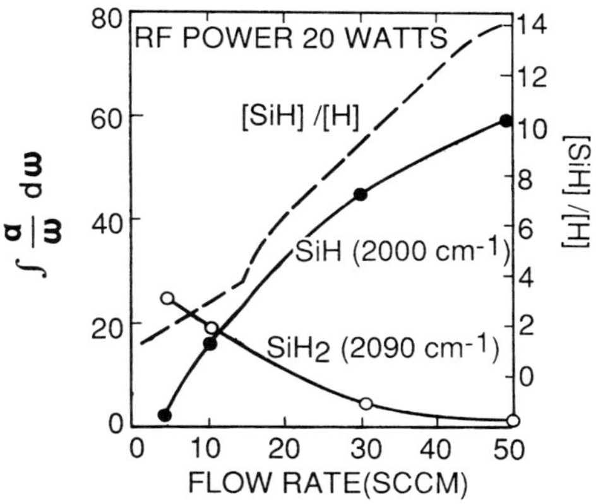

FIGURE 21. Integrated absorption of SiH and SiH_2 stretching modes as a function of silane flow rate. The dashed line refers to the concentration ratio of the reactive SiH to H [45].

It is interesting to note that by changing the electrode configuration, a result exactly opposite that discussed above can be found. Conde *et al.* [46] used a concentric electrode confinement instead of a planar configuration; in this the reactor consists of a quartz tube, lined on the inside by a grounded cylindrical (anode), which is the counterelectrode for a 13.56-MHz RF-powered paddle-shaped electrode placed at the center of the reactor (cathode). In this arrangement, electrons are confined and moderate ion bombardment is expected since the ratio $A_a/A_c > 1$. In contrast to the conventional studies, the substrate was situated on the cathode. Figure 22 shows that as the flow rate is increased, the ratio of the IR peaks 2000/2090 cm^{-1} decreases by nearly two orders of magnitude, with the result that at a high flow rate of 100 sccm, the film exhibits primarily SiH_2 peak, which is representative of poor material, as discussed previously.

E. Effect of Excitation Frequency

For many applications, such as electrophotography, where a-Si:H thickness in excess of 20 μm are required, microwave frequencies (instead of 13.56 MHz) are employed to obtain high deposition rates (>1 μm/min). However, the optoelectronic properties degrade especially in the hole transport characteristics; this is less of a concern as the electrophotographic application relies primarily on electron transport. However, to obtain rates higher than those normally used (<5 Å/s) could be important in other applications, such as solar cells, even though

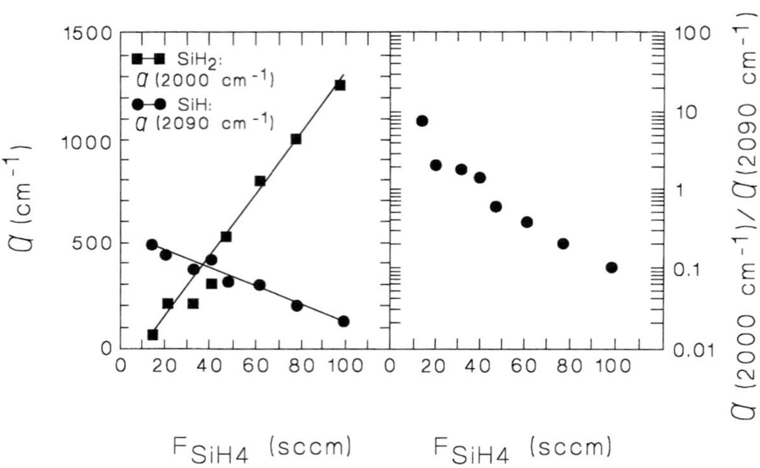

FIGURE 22. Infrared adsorption coefficient at 2000 (full circles) and 2090 cm^{-1} (open squares), plotted as a function of the flow of silane. The straight lines are least-square fits to the experimental data. In the inset, the ratio of alpha at 2000 and 2090 cm^{-1}, corresponding to the relative proportion of SiH vs. SiH_2 bonds in the a-SiH film, is shown as a function of the flow of silane [46].

the thickness is on the order of only 2500–5000 Å, as it has implications for throughput and cost of production.

The crucial aspect of the high-deposition-rate technique is that the optoelectronic properties of a-Si:H must be maintained. Numerous approaches have been attempted, such as the following: (1) increase of RF excitation power in a conventional PECVD reactor leads to an increase in the dihydride (SiH_2) component, increase in DOS, decrease in the (μt)h etc.; importantly, it also leads to an increase in the dust, which affects the yield of the product; and (2) the plasma confinement method. A more successful approach is the use of higher frequencies in the range of 70–110 MHz, where encouraging results of materials [47], solar cell efficiencies in the range of 10% [48], and very good thin-film transistor (TFT) characteristics [49] have been reported.

Figure 23 shows [50] the IR spectra for films produced at different deposition (R) rates (curve A, 3.6 Å/s); curve B, 9.5 Å/s; curve C, 15.7 Å/s), and Fig. 24 shows the H concentration (C_H) of the corresponding films. According to the data, the dihydride component (IR peak at 2090 cm^{-1} and its scissors mode at 890 cm^{-1}) seems to decrease with R, C_H is constant for $R > 8$ Å/s, and the H is bonded primarily in a monohydride configuration. As shown in Table 4, the Urbach tail parameter (E_0) seems to increase slightly and the DOS remains virtually constant with σ_d and σ_{ph} both slightly decreasing. In order to understand

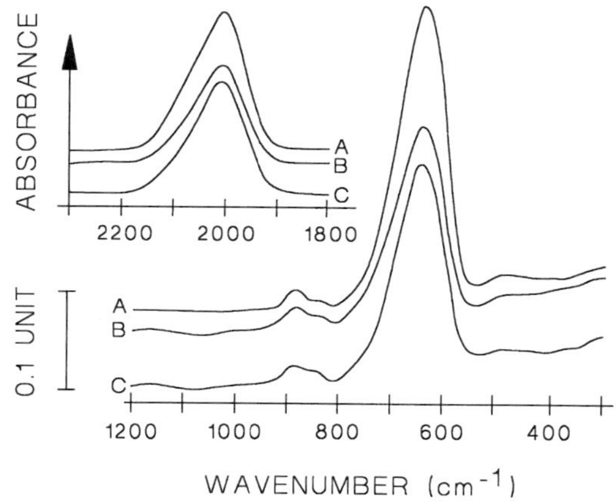

FIGURE 23. Infrared absorption spectra for samples prepared at different deposition rates: A = 3.6, B = 9.5, and C = 15.7 Å/s. The other parameters are as listed in Table 4 [50].

this result, we recapitulate an earlier analogous work of He discharges. As shown [51] in Fig. 25, the electron plasma density can be increased substantially from 10^9 to 10^{11} cm^{-3} when the frequency of excitation is altered from 10^9 to 10^{11} cm^{-3} when the frequency of excitation is altered from 13.56 to 110 MHz at a pressure of 250 mtorr. In this, the sheath thickness was essentially independent of voltage for constant pressure and electrode spacing. However, the electron density and ion current were proportioned to w^2, where w is the angular frequency of

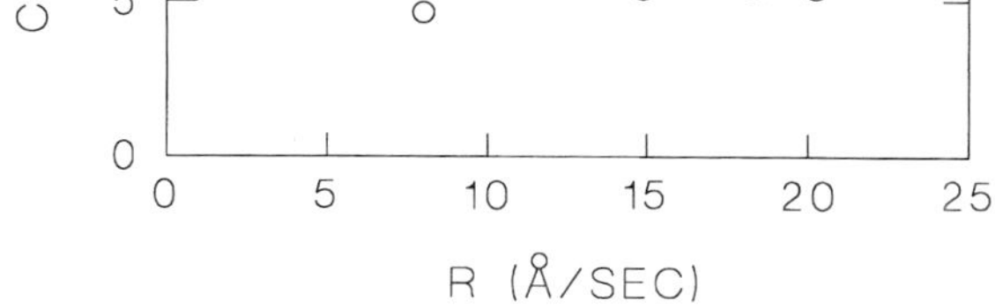

FIGURE 24. Bonded H content C_H as a function of the deposition rate R [50].

Table 4 Deposition parameters and thin-film properties for 3 samples A, B, and C deposited at different rates R at a plasma excitation frequency of 70 MHz

Parameter	A	B	C
T_s (°C)	300	300	300
P (W/cm^3)	0.015	0.09	0.10
p (mbar)	0.14	0.28	0.26
F_{SiH_4} (sccm)	6.6	9.5	18.3
f (MHz)	70	70	70
d (μm)	1.95	1.87	2.16
R (Å/s)	3.6	9.5	15.7
C_H (atom %)	8.9	6.3	5.4
E_{opt} (eV)	≈1.72	≈1.70	≈1.68
N_D (cm^{-3})	1.5×10^{16}	1.8×10^{16}	1.7×10^{16}
E_0 (meV)	55	56	59
σ_{dark} (1/Ω cm)	8.5×10^{-10}	1.5×10^{-10}	1.7×10^{-10}
σ_0 (1 Ω cm)	2200	8400	3200
E_a (eV)	0.73	0.81	0.79
σ_{AM} (1/Ω cm)	6.5×10^{-5}	3.5×10^{-5}	1.8×10^{-5}

From Curtins *et al.* [50].

the RF excitation. As shown in Fig. 26 [51], keeping w constant (at 120 MHz) and changing the pressure can alter the sheath thickness. This manipulation of the plasma, in terms of sheath thickness and electron/ion densities and energies, has also been demonstrated in the preparation of a-Si:H layer, as shown in Fig. 27 [52], where this, we plot σ_{ph} and σ_d as functions of power and pressure. We note that as the pressure is increased, σ_d decreases from 10^{-9} (Ω-cm)$^{-1}$ to about 10^{-11} (Ω-cm)$^{-1}$ whereas σ_{ph} remains constant at $>10^{-5}$ (Ω-cm)$^{-1}$. The decrease of $\sigma_d \sim 10^{-11}$ (Ω-cm)$^{-1}$ is indicative of a low DOS. During these experiments, the authors note that as the pressure increases, the visible plasma region shrinks toward the powered electrode. They surmise that the contribution of the short-lifetime radicals to the deposition of a-Si:H layer decreases. However, we note that the average ion energy is reduced, and hence there could be a decrease in the ion bombardment of the growing surface.

Dutta *et al.* [53] have performed systematic work focusing on stress and bombardment energy as the excitation frequency is altered from 13.56 to 70 MHz, with $T_s = 200$°C, flow rate at 30 sccm, and gas pressure at 0.3 mbar with the total power density in the plasma kept constant at 19 mW cm^{-3}. They note that the stress, which is compressive, reduces from 10^8 to 10^7 N m^{-2} when the frequency changes from 13.56 to 70 MHz. This decrease is associated with a decreased ion bombardment at the higher frequencies. The ion energy is estimated

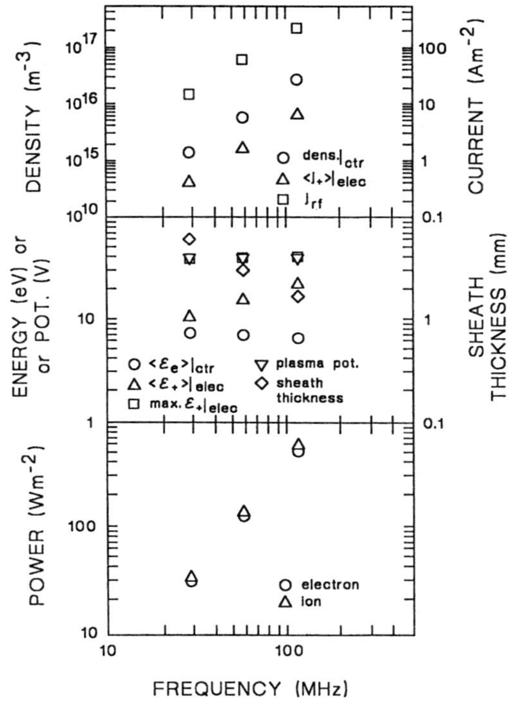

FIGURE 25. (a) Plasma density (center of discharge), ion current at the electrode, and RF current as a function of frequency. (b) Mean electron energy in the center ($<e_q>$/ctr), mean ion energy at the electrode ($<e_+>$/elec.), maximum ion energy at the electrode (max e_+/elec.), plasma potential, and sheath thickness as a function of frequency. (c) Electron power and ion power as a function of frequency. Conditions 250 mtorr, 50 V_{RF}, and 4-cm electrode gap.

from the peak-to-peak voltage, V_{pp} and given by $eV_{\mathrm{pp}}/4$ for a symmetrical discharge to be 46 eV at 13.56 MHz and 14 eV at 70 MHz. Dutta *et al.* further demonstrate the viability of this approach, by fabricating a simple TFT structure with $I_{\mathrm{on}}/I_{\mathrm{off}} > 10^7$ with $\mu_{\mathrm{fe}} \sim 1.5\ \mathrm{cm}^2\ \mathrm{s}^{-1}\ \mathrm{V}^{-1}$, which is one of the highest values reported.

In order to understand the role of *w,* we discuss the work of Wertheimer [54], who considers the ratio of the two characteristic frequency, ν/w, where ν is the electron–neutral collision frequency for momentum transfer. This ratio dictates the characteristics of the electron energy distribution function (EEDF) as modeled by Ferreira and Loureiro [55]. Their calculations show that in an Ar plasma, there is an increase in the population of high-energy electrons as the frequency is in-

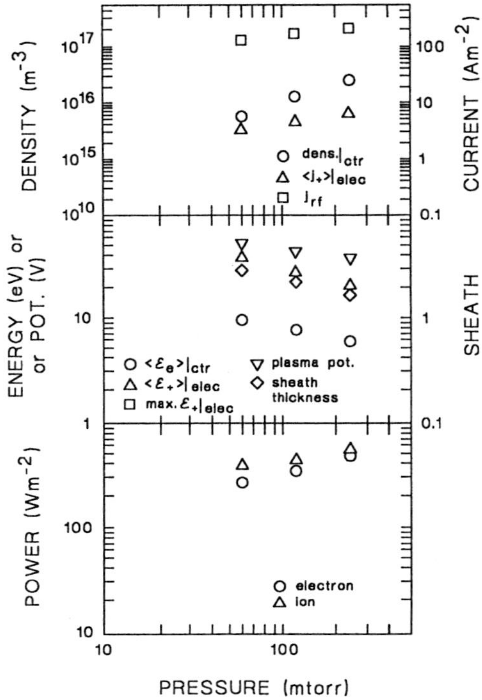

FIGURE 26. (a) Plasma density (center of discharge), ion current at the electrode, and RF current as a function of pressure. (b) Mean electron energy in the center, mean ion energy at the electrode, maximum ion energy at the electrode, plasma potential, and sheath thickness as a function of pressure. (c) Electron power and ion power as a function of pressure. Conditions 250 mtorr, 50 V_{RF}, and 4-cm electrode gap.

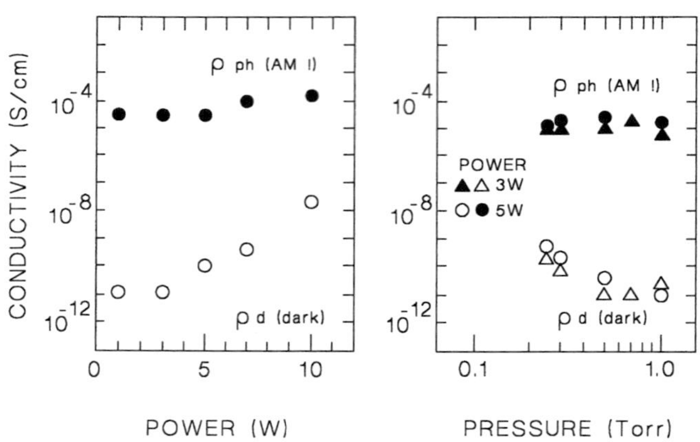

FIGURE 27. Photoconductivity (global AM1.5) (a) and dark conductivity (b) for different power and pressure at 144 MHz.

creased. At microwave frequency, $\nu/w \sim 0$, whereas Curtins *et al.* [47] have estimated that $\nu/w \sim 1.5$ at 25 MHz and 0.2 at 150 MHz; hence at a frequency of ~70 MHz, there is an increase in the population of high-energy electrons (>11–15 eV), with the effect that there is a more efficient dissociation of the SiH_4 gas. For $\nu/w > 1$, applicable in the low frequency ranges, then, in effect the plasma "does not know" that it is being sustained by alternating RF field.

III. Alternative Deposition Techniques

As mentioned previously, many different approaches have been attempted to produce a-SiH material other than the conventional PECVD technique, which is by far the most studied and used in industry. One of the major impediments for the deployment of amorphous silicon solar cells has been the inherent degradation or the so-called Staebler–Wronski (SW) effect [30], in that the application of light can affect the electronic properties of the material. Two types of states can be distinguished: a fully annealed state attained by slowly cooling the material from approximately 500 K to room temperature in the dark (state A); and a light-soaked state, attained after the application of about 10^{21} photons/cm^2 to the material (state B). Changes have been observed in the carrier diffusion length, unpaired spin density, DOS in the gap, etc. (see, e.g., ref. [56]). The transition from state A to state B seems to be induced by any process that creates free carriers that recombine with an energy less than that of the band gap. This transition is relatively inefficient since the increase in the spin density takes place at a rate of 10^{-8} spins per absorbed photon. The recovery from state B to state A seems to be complete on annealing, suggesting that (1) state B is metastable and (2) the photoinduced defects are distinct from the defects present in state A. Various explanations for the SW effect have been put forth, but no clear consensus has emerged as yet.

One link to the SW effect has been the diffusion of H, which creates extra metastable defects within the semiconductor. Therefore, in this context, alternative deposition techniques are being sought to produce device-quality material with lower C_H and that could lead to lower instability effects. One of those techniques is the chemical annealing technique [57]. In this, the PECVD approach is combined with a source of H atoms from a 2.45-GHz microwave source. The film growth involves alternating the growth (of 20-Å thickness); the film is then subjected to chemical annealing (or exposure) by the H for a certain duration of time to control the C_H in the film. For $T_s \sim 300$C, Shirai *et al.* [57] were able to produce films with $C_H \sim 5\%$, $E_g \sim 1.6$ eV and with dark/light conductivity ratio of $>10^4$. However, the Urbach edge parameter E_0 was found to be shallower than in conventional PECVD-grown a-SiH film. Importantly, the drift mobility of holes was found to be nondispersive with a high mobility of 0.2 $cm^2\ s^{-1}\ V^{-1}$. This surprising result indicates that the valence band tail is sharper than in the conventional PECVD a-SiH and is a result of a lower extent of disorder in the material; however, they [57] observe a high $\sigma_d > 10^{-8}$ (Ω-cm)$^{-1}$ with a corresponding high activation energy, $E_a \sim 0.8$ eV. (It should be noted that, generally, the high dark conductivity is associated with a defective material, but in this case it is due to a high σ_0, the preexponent in the conductivity expression.) They surmise that this approach produces a much more rigid structure and is linked to the efficient abstraction of H from the surface; they suspect that the reconstruction of the surface is promoted by the diffusion of the DBs via the H. In any case, the films exhibited improved characteristics in that the photoconductivity did not show marked deterioration with continuous illumination, at least in the first 20 h or so.

IV. Surface States, Interface States, and Their Effect on Device Performance

The detailed structure and its distribution of defect states in the bulk are known only approximately, which has also complicated the understanding of surface states. In contrast, the surface of crystalline Si and its respective reconstructions are well known. Because the number and energy of the bulk states is known, the amount of band bending at the crystalline Si/vacuum interface, as measured by the photoemission technique, correlates well with the bulk and surface state densities calculated or measured experimentally using various techniques. On unoxidized surfaces the surface state density is high ($>10^{14}$ $cm^{-2}\ eV^{-1}$), and E_f in nondegenerate crystalline Si is controlled (pinned) by these surface states. Himpsel *et al.* [58] report that when the crystalline Si(111) surface is oxidized, the number of surface states is reduced significantly. Similarly, in a-SiH it might be expected that many surface states would be eliminated by H passivation, and to some extent this seems to be the case. Surface state density ranging from $4{\cdot}10^9$ $cm^{-2}\ eV^{-1}$ to as high as $1.5{\cdot}10^{13}$ $cm^{-2}\ eV^{-1}$ have been reported [59]. For instance, we discussed above that the films deposited at low T_s possess columnar structure and hence are porous [23]. Street and Knights [60] have used these types of films to study the a-SiH/oxide interface and concluded that the increase in ESR spin density with time is associated with dangling bonds at the surface of the columns. The spin density was measured over a period of time for samples that were exposed to ambient air. Samples with columnar structure exhibited an increase in the spin density, but homogeneous samples did not show any change. They deduced that the minimum increase in the spin density corresponded to a bulk density of $5{\cdot}10^{17}$ cm^{-3} for a sample with an internal surface per unit volume of $1.6{\cdot}10^6$ cm^{-1}, leading to a surface state density of $3{\cdot}10^{11}$ cm^{-2}.

Table 5 Amorphous silicon

Type of device	Product
Products available commercially	
Photovoltaic cell	Calculators, watches, battery chargers, power modules
Photoreceptor	Electrophotography (photocopying machines), LED printer
Photoconductor	Color sensors, light sensors, etc.
Image sensor	Contact-type image sensor [facsimile (FAX) machines], electronic write boards
Solar control layer	High-reflecting float glass
Thin-film field-effect transistors (FETs)	Displays, television monitors
High-voltage thin-film transistors	Printers
Proposed applications of amorphous silicon	
Image-pickup tubes	Fast detectors and modulators
Optical waveguides	Position sensors
Optical recording	LEDs
Spatial light modulators	Neural networks
FETs for logic circuits	Diodes
Passive layers	FETs for ambient sensors
DiFETs	Charge-coupled devices
Strain gauges	Bipolar transistors
Photolithographic masks	Nuclear particle detectors
Memories	

As discussed previously, a-Si is being used in many applications and is summarized in Table 5 [2]. There is an underlying commonality in these applications as it involves cojoining of layers (undoped, doped, and related alloys) into device configurations as shown in Fig. 28. The most important structures are the p–i–n and SiN_x/i/n as they are used in applications such as thin-film transistors (TFT) for displays, sensors, and solar cells. As most of these layers are thin, the interface state densities and their manipulations become important from a device point of view.

An a-Si:H TFT is generally constructed in a sequence of layers: SiN_x/i (a-SiH)/n (a-SiPH) of approximate thicknesses, 2000, 1500, and 300 Å, respectively. Deposition of the dielectric and the i-layer in the same deposition chamber can have a deleterious effect on the performance of the device. For instance, when consecutive layers are deposited (SiN_x layer followed by an a-SiH layer), and as the persistence of NH_3 is long, then on reinitiating the plasma, the NH_3 species desorb from the walls. The residual impurity (N) can act as a n-type dopant and will increase the off-current of the TFT and lower the field effect mobility, μ_{fe}. It has been shown that despite extensive pumping and purging between the dielectric

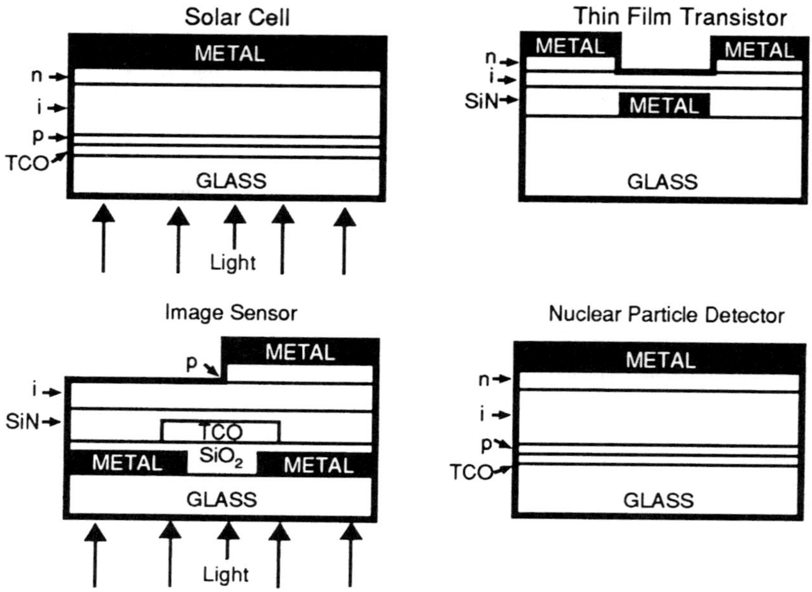

FIGURE 28. Amorphous silicon device configurations.

layer and the i-layer depositions, secondary-ion mass spectroscopy (SIMS) data revealed that there was still a large concentration of N in the i-layer close to the interface [61].

Normally the TFTs are constructed in the configuration SiN_x/i/n and passivated with a-SiN_x layers. The sequence of deposition of i-layer on SiN_x or vice versa has an effect on the performance of the device [62] where it was found, using the Rutherford backscattering technique, that a less sharp interface existed between the a-Si : H and a-SiN_x layers when the sample was deposited in a sequence of Si:H on SiN_x in comparison with the other way. This result has been confirmed [63] using the *in situ* spectroellipsometry (SE) technique for samples grown in a single-chamber system. SE measured for a-SiH on SiN_x ("top") revealed that the data could be fitted well to a sharp interface; on the other hand, data for samples deposited the other way ("bottom") could be fitted only with a progressive shift from SiN_x to a-SiH, i.e., a smeared interface. Although the "bottom" configuration leads to an atomically abrupt interface, TFTs are not fabricated in this fashion as their performance is inferior in comparison with Si on SiN_x.

To study the interface effect in more detail, Matsumuto [64] instead used a

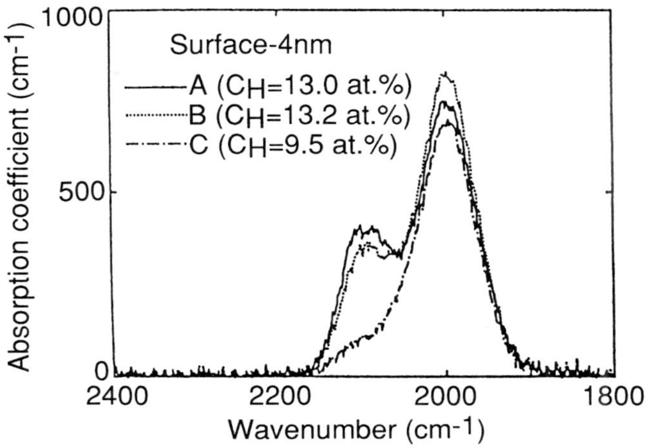

FIGURE 29. Subtracted ATR spectra by the first etching for three types of a-Si:H samples. Sample A underwent no additional processing. Sample B was kept for 5 min in a SiN chamber after the deposition of a-SiH. In sample C, SiN (50 nm) was deposited on a-SiH after the same processing as for sample B [64].

multichamber load-locked system and the resultant films studied using the Fourier transform IR (FTIR) technique. Figure 29 shows the results of how the H in a-SiH/SiN interface is affected by the deposition of a SiN overlayer; in this, curve A represents the data for a-SiH layer only, curve B represents a sample kept in the SiN chamber for 5 mins. after the deposition a-SiH to evaluate any thermal effects, and curve C represents a sample in which SiN overlayer was deposited onto the a-SiH semiconductor. The spectrum for C differs fundamentally from the A and B spectra in that the latter possess large dihydride components as exemplified by the 2100-cm^{-1} IR peak, which is reduced drastically for the curve C sample. In Fig. 30, we show the data for the sample configuration, crystalline Si/SiN/SiH (5 nm)/SiD (5 nm). We note that the data show a large SiH_2 (IR peak at 2100 cm^{-1}) component as well as an IR peak at 2000 cm^{-1}; as the a-Si:H is devoid of the SiH_2 IR component, any residual signal (i.e., the SiH_2 IR peak) is to be attributed to the interface of SiN_x a-SiH. The fact that a-SiH on SiN_x has thus far resulted in better TFTs could be related to SiH_2 (residing on the surface), which may provide a mechanism to relax the strained bonds between the a-SiH and the SiN_x layers.

Similar considerations apply to the solar cells if manufactured in a single-chamber system. A solar cell is typically fabricated in the sequence p (a-SiCBH)/i (a-SiH)/n (a-SiPH) on a textured tin-oxide-coated glass substrate. The p-type ma-

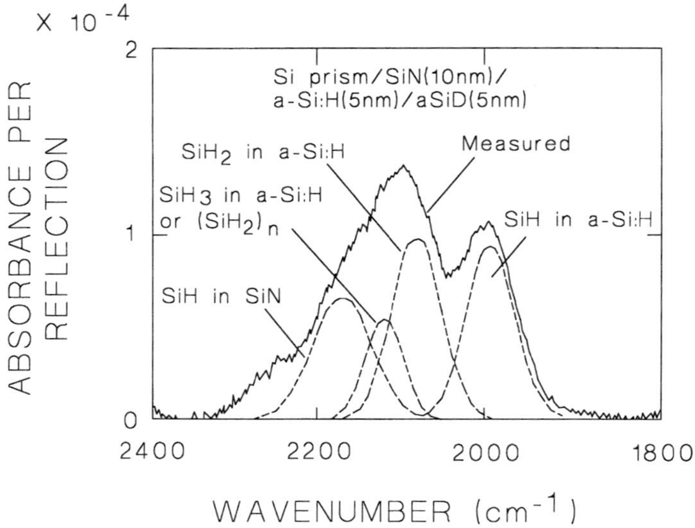

FIGURE 30. ATR spectra for Si prism/SiN (10 nm)/a-SiH (5 nm)/a-SiD (5 nm). An ATR spectrum for Si prism/SiN (10 nm) is subtracted from the original spectrum. The solid line indicates the measured spectrum. Dashed lines show deconvoluted spectra [64].

terial is fabricated, in most instances, from a gas mixture of SiH_4, CH_4, and B_2H_6, whereas the n-layer is fabricated from SiH_4 and PH_3 gas mixtures. It is well known that even 1 ppm of B (or P) decreases the $\mu\tau$ (where μ is the mobility and τ is the recombination lifetime) product of electrons and holes. The consequence is that the minority carrier diffusion length decreases and the solar cell efficiency drops significantly in comparison with the data obtained from multichamber systems, which can routinely fabricate devices in excess of 10% [13].

A significant improvement can be made in solar cell performance by manipulating the p^+/i interface. As an example, we show in Table 6 [65] data for two types of cells that differ only with respect to an insertion of a graded layer (GL) between the p^+- and the i-layers. The GL is usually inserted at the end of the p^+ by decreasing the concentration of the CH_4 and B_2H_6 in the plasma to zero for a few seconds. The improvement occurs in all three solar cell characteristics defining the conversion efficiency. In a separate study by Tran *et al.* [66], who used much thicker graded carbon layers (see Fig. 31), the improvement in the open-circuit voltage V_{oc} could be as much as 100 meV to values in excess of 0.85 V. In a similar study, Arya *et al.* [67], also note that the short-circuit density J_{sc} can be improved especially at the blue end of the spectrum, as shown in Fig. 32. The

Table 6 Dark and light parameters for devices made with and without the graded layer (GL)[a]

GL (s)	Direct 1.5 illumination V_{oc} (V)	J_{sc}	FF	Blue illumination V_{oc} (V)	J_{sc}	FF	Red illumination V_{oc} (V)	J_{sc}	FF	Dark n
0	0.759	13.62	0.69	0.556	0.073	0.71	0.616	0.347	0.73	1.4
10	0.817	15.21	0.70	0.610	0.094	0.73	0.668	0.392	0.74	1.5

[a]The light parameters use direct AM1.5 illumination (100 mW cm^{-2}), which underestimates the performance of the device. The GL thickness is related to the deposition time, and n refers to the diode quality factor (J_{sc} is in mA cm-$^{-2}$).

reason for this could be (1) the built-in field is increased, thereby preventing back recombination at the p$^+$/i interface; (2) there are relaxations of bonds at the heterojunction, which, in turn, reduces the density of recombination centers, as in the case of SiN_x/a-Si:H interface discussed above. In a theoretical study by Tasaki *et al.* [68] the p$^+$/i interface region was modeled as a 25-Å-thick section with short lifetimes. Figure 33 shows that as the defect states in this particular region increase, all three parameters defining the solar cell conversion efficiency are affected.

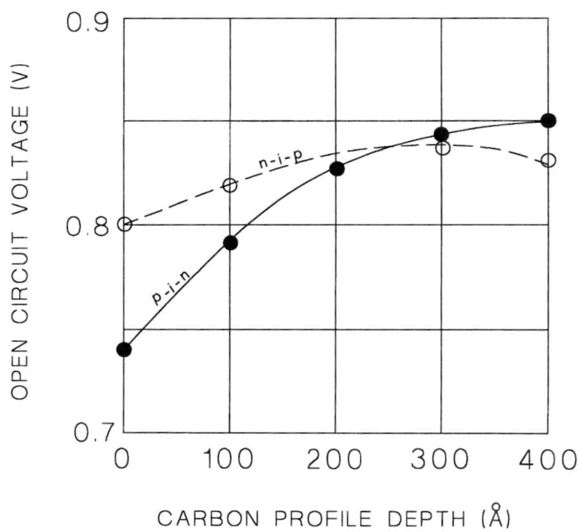

FIGURE 31. Dependence of V_{oc} on the carbon profile at the buffer layer of p–i–n and n–i–p amorphous silicon solar cells [66].

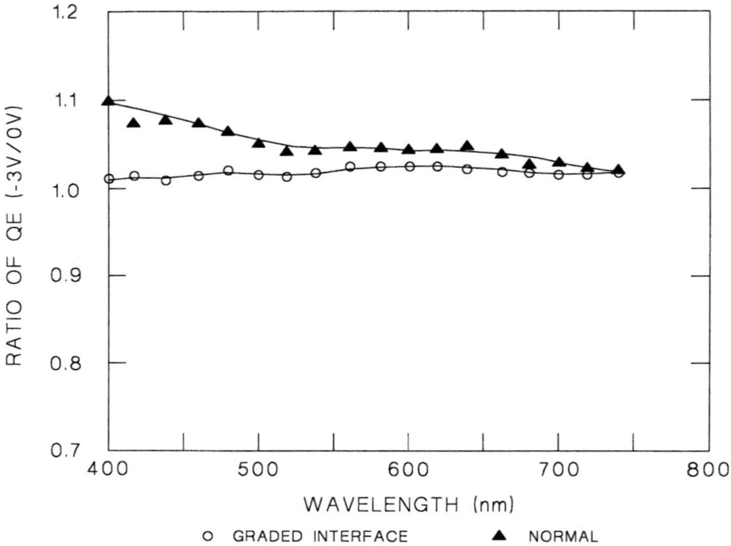

FIGURE 32. Ratio of quantum efficiency (− 3 V/0 V) as a function of wavelength [67].

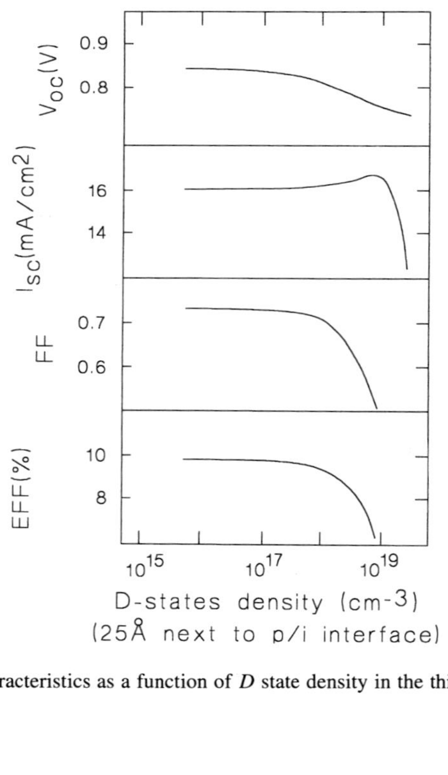

FIGURE 33. Cell characteristics as a function of D state density in the thin i-layer (25 Å) next to the p–i interface [68].

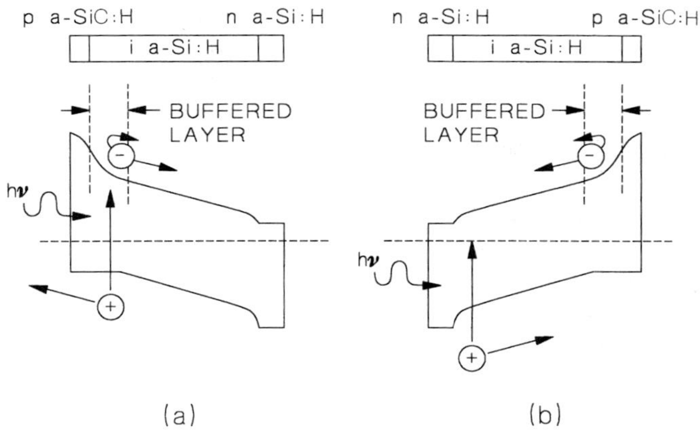

FIGURE 34. Band diagrams with carbon buffer layer for (a) p–i–n and (b) n–i–p amorphous silicon solar cells [69].

In order to understand these results, we note from Evangelisti *et al.* [69] that a valence-band discontinuity does not exist at the a-Si:C:H/a-Si:H interface and hence the energy-band diagram could be as shown in Fig. 34. To ascertain whether the interface states are reduced, we turn to the work of Asano *et al.* [70], who have characterized these surfaces (a-SiC/a-SiH) using the PDS technique with films fabricated in a multichamber system. By comparing a-SiC/a-SiH and a-SiC/GL/a-SiH, they note that no fundamental difference in sub-band-gap absorption had occurred and thus concluded that the graded layers do not decrease the interface defect density. In contrast to this work, Petrauskas *et al.* [71] studied a-Si:H/a-SiCH quantum-well structures, and measured the lateral ambipolar motion using the transient grating effect technique; they found that the lateral mobility of optically free-carrier decreases with the reduced thickness of the quantum wells and attributed this to an interface roughness as the dominating scattering process. There is evidence from TEM and SEM (tunneling and scanning electron-microscopic) techniques, that the heterojunction interfaces cause this roughness; TEM data have revealed that the multilayers are flat near the substrate but become increasingly rough after several layers are deposited [72]. Using similar quantum-well structures Muschik *et al.* [73] suspect that the recombination lifetimes (using transient photoluminescence technique) are affected only when the well thickness is less than 30 Å, and hence the interface state density plays a role.

In these observations there is no consensus as to the increase of the conversion efficiency parameters when the graded layer is inserted between the p- and the i-layers. However, it is feasible that with the GL present, the device could be more appropriately modeled as p–insulator–i–n. In an analogous situation when comparing a Schottky barrier (SB)-type device with a MIS (metal–insulator–semiconductor) using a-Si:H, the insertion of the insulator has resulted in major improvements in every device parameter [2, 74]; namely, the quantum efficiency at the blue end of the illuminated spectra increases substantially (and hence J_{sc} increases), the V_{oc} increases by 250 mV, and the FF improves. In comparing the MIS with a SB-type device, we note that primarily the dark current is suppressed as a result of tunneling and that the diode quality factor increases because the applied voltage is dropped across the insulator. It is the suppression of the dark current that provides for an improvement in all the parameters that dictate the conversion efficiency. We therefore suspect that with different types of graded layer inserted between the p^+- and the i-layers, and paying attention to the electron affinity, the performance of the device could be improved substantially over and above the currently obtained efficiency of approximately 14% [75].

V. Summary

Amorphous silicon technology has now matured to the point where many products are now commercially available. Further expansion into the marketplace can be expected as some of the remaining issues, such as the instability under illumination, are either circumvented or solved entirely.

References

1. N. F. Mott and E. A. Davis, "Electronic Processes in Non Crystalline Materials." Oxford Univ. Press, London, 1979.
2. A. Madan and M. Shaw, "The Physics and Applications of Amorphous Semiconductors." Academic Press, San Diego, 1988.
3. R. A. Street, "Hydrogenated Amorphous Silicon." Cambridge Univ. Press, New York, 1991.
4. W. Luft and Y. S. Tsuo, "Hydrogenated Amorphous Silicon Alloy Deposition Process." Dekker, New York, 1993.
5. R. C. Chittick, J. H. Alexander, and H. F. Sterling, *J. Electrochem. Soc.* **116,** 77 (1969).
6. A. Madan, S. R. Ovshinsky, and E. Benn, *Philos. Mag.* **40,** 259 (1979).
7. T. D. Moustakas and W. Paul, *Phys. Rev. B* **16,** 1564 (1977).
8. R. D. Andas and D. E. Brodie, *Can. J. Phys.* **67,** 195 (1989).
9. M. Hirose, *in* "Semiconductors and Semimetals, Vol. 21: Hydrogenated Amorphous Silicon," (J. Pankove, ed.), p. 109. Academic Press, New York, 1984.
10. M. Konogai, *Mater. Res. Soc. Symp. Proc.* **70,** 257 (1986).
11. F. S. Pool, J. M. Essick, Y. H. Shing, and R. T. Mather, *Mater. Res. Soc. Symp. Proc.* **258,** 173 (1992).
12. N. M. Johnson, J. Walker, C. M. Doland, K. Winer, and R. A. Street, *Appl. Phys. Lett.* **54,** 1872 (1989).
13. A. Madan, P. Rava, R. E. I. Schropp, and B. Von Roedern, *Appl. Surf. Sci.* **70/71,** 716 (1993).

14. S. Usui and M. Kukuchi, *J. Non-Cryst. Solids* **34,** 1 (1979).
15. W. E. Spear and P. G. LeComber, *J. Non-Cryst. Solids* **8–10,** 727 (1972).
16. A. Madan, Ph.D. Thesis, Univ. of Dundee, Dundee, Scotland, 1973.
17. A. Madan, P. G. LeComber, and W. E. Spear, *J. Non-Cryst. Solids* **20,** 239 (1976).
18. R. A. Street, J. C. Knights, and D. K. Biegelsen, *Phys. Rev. B* **19,** 3027 (1978).
19. M. Venecek, B. P. Nelsen, A. H. Mahan, and R. S. Crandall, *J. Non-Cryst. Solids* **137/138,** 191 (1991).
20. G. Lucovsky, R. J. Nemanich, and J. C. Knights, *Phys. Rev. B* **19,** 2064 (1979).
21. J. Perrin, J. Solomon, G. Bourdon, J. Fontenille, and E. Ligeon, *Thin Solid Films* **62,** 327 (1979).
22. J. Knights and R. A. Lujan, *Appl. Phys. Lett.* **35,** 244 (1979).
23. H. Fritzche and C. C. Tsai, *Solid State Technol.* **21,** 55 (1978).
24. M. B. Schubert, H. D. Mohring, R. Zedlitz, and G. H. Bauer, *J. Non-Cryst. Solids* **137/138,** 195 (1991).
25. C. Swiatkowski, W. Hirsch, and M. Kunst, *Mater. Res. Soc. Symp. Proc.* **258,** 75 (1992).
26. D. A. Anderson and W. E. Spear, *Philos. Mag.* **36,** 695 (1977).
27. T. McMahon and A. Madan, *Appl. Phys. Lett.* **57,** 5302 (1985).
28. D. G. Ast and M. Brodsky, *Philos. Mag., Part B* **41,** 273 (1980).
29. Z. Smith, S. Aljishi, D. Slobodin, V. Chu, S. Wagner, P. M. Lenahan, R. R. Arya, and M. S. Bennett, *Phys. Rev. Lett.* **57,** 2450 (1986).
30. D. L. Staebler and C. R. Wronski, *J. Appl. Phys.* **51,** 3262 (1980).
31. S. C. Agrawal, J. C. Payson, and S. Guha, *Phys. Rev. B* **36,** 9348 (1987).
32. R. A. Street, *Sol. Cells* **24,** 211 (1988).
33. G. Ganguly and A. Matsuda, *Mater. Res. Soc. Symp. Proc.* **258,** 39 (1992).
34. P. Roca i Cabarrocas, *Mater. Res. Soc. Symp. Proc.* **149,** 33 (1989).
35. K. Kohler, J. W. Coburn, D. E. Horne, and E. Kay, *J. Appl. Phys.* **57,** 59 (1985).
36. W. Kasper, H. Böhm, and B. Hirschauer, *J. Appl. Phys.* **71,** 4168 (1992).
37. T. Hattori, S. Mizuki, K. Mackawa, H. Okamoto, and Y. Hamakawa, *Tech. Dig. Int. Photovoltaic Sci. Eng. Conf., 3rd, Tokyo* p. 731 (1987).
38. P. Roca i Cabarrocas, P. Morin, V. Chu, J. P. Conde, J. Z. Liu, H. R. Park, and S. Wagner, *J. Appl. Phys.* **69,** 2942 (1991).
39. K. Ando, M. Aozasa, and R. G. Pyon, *Appl. Phys. Lett.* **44,** 413 (1984).
40. M. H. Brodsky, M. Cardona, and J. J. Cuomo, *Phys. Rev. B* **16,** 3556 (1977).
41. S. Ishihara, M. Kitagawa, and T. Hirao, *J. Appl. Phys.* **62,** 488 (1987).
42. A. Matsuda and K. Tanaka, *Thin Solid Films* **92,** 171 (1982).
43. B. Drévillon, J. Huc, A. Lloret, J. Perrin, G. De Rosny, and J. P. M. Schmitt, *Proc. Symp. Plasma Chem., 5th, Edinburgh, 1981.*
44. G. Turban, Y. Catherine, and B. Grolleau, *Thin Solid Films* **60,** 147 (1979).
45. M. Hirose, *J. Phys. (Paris), Colloq.* **42,** C4-705 (1981).
46. J. P. Conde, K. K. Chan, J. M. Blum, and M. Arienzo, *J. Appl. Phys.* **71,** 3990 (1992).
47. H. Curtins, N. Wyrsch, and A. V. Shah, *Electron. Lett.* **23,** 228 (1987).
48. P. K. Bhat, C. Marshall, J. Sandwich, H. Chatham, R. E. I. Schropp, and A. Madan, *IEEE Proc. Photovoltaics, 20th* p. 91 (1988).
49. Y. Watanabe, M. Shiritani, Y. Kubo, I. Ogawa, and S. Ogi, *Appl. Phys. Lett.* **53,** 1263 (1988).
50. H. Curtins, N. Wyrsch, M. Farre, K. Prasad, M. Brechet, and A. V. Shah, *Mater. Res. Soc. Symp. Proc.* **95,** 249 (1987).
51. M. Surendra and D. B. Graves, *Appl. Phys. Lett.* **59,** 2091 (1991).
52. S. Oda and M. Yasukawa, *J. Non-Cryst. Solids* **137/138,** 677 (1991).
53. J. Dutta, U. Kroll, Chabloz, A. V. Shah, A. A. Howling, J. L. Dorier, and C. Hollenstein, *J. Appl. Phys.* **72,** 3220 (1992).

54. N. R. Wertheimer, *J. Vac. Sci. Technol., A* **3,** 2643 (1985).
55. C. M. Ferreira and J. Loureiro, *J. Phys. D* **17,** 1175 (1984).
56. B. L. Stafford, ed., *AIP Conf. Proc.* No. 234 (1991).
57. H. Shirai, J. Hanna, and I. Shimizu, *Jpn. J. Appl. Phys.* **30,** L881 (1991).
58. F. J. Himpsel, G. Hollinger, and R. A. Pollak, *Phys. Rev. B* **28,** (1983).
59. M. Hirose, T. Sususki, and G. H. Dohler, *Appl. Phys. Lett.* **34,** 234 (1979); P. Nielsen and R. Greden, *J. Vac. Sci. Technol., A* **1,** 583 (1983).
60. R. A. Street and J. C. Knights, *Philos. Mag.* **43,** 1091 (1981).
61. M. J. Thompson, *Top. Appl. Phys.* **56,** 119 (1984).
62. J. R. Abelson, C. C. Tsai, and T. W. Sigmon, *Appl. Phys. Lett.* **49,** 850 (1986).
63. M. Stchakovsky, B. Drévillon, and P. Roca i Cabarrocas, *J. Appl. Phys.* **70,** 2132 (1991).
64. T. Matsumuto, *Appl. Phys. Lett.* **60,** 1940 (1992).
65. E. Terzini and A. Madan, unpublished results (1991).
66. N. T. Tran, F. R. Jeffrey, and D. J. Olsen, *Mater. Res. Soc. Symp. Proc.* **95,** 545 (1987).
67. R. R. Arya, M. S. Bennett, A. Catalano, and K. Rajan, *Tech. Dig. Int. Photovoltaic Sci. Eng. Conf., 3rd, Tokyo* (1987).
68. H. Tasaki, W. Y. Kim, M. Hallerdt, M. Konogai, and K. Takahashi, *J. Appl. Phys.* **63,** 550 (1988).
69. F. Evangelisti, P. Fiorini, C. Giovannela, F. Patella, P. Perfetti, Quaresima, and M. Capozi, *Appl. Phys. Lett.* **48,** 1538 (1984).
70. A. Asano, T. Ichimura, Y. Uchida, and H. Sakai, *J. Appl. Phys.* **63,** 2346 (1988).
71. M. Petrauskas, J. Kolenda, A. Galeckas, R. Scwarz, F. Wang, T. Muschik, T. Fischer, and H. Wienert, *Mater. Res. Soc. Symp. Proc.* **258,** 553 (1992).
72. H. Itoh, S. Matsubara, S. Muramatsu, N. Nakamura, and T. Shimada, *Tech. Dig. Int. Photovoltaic Sci. Eng. Conf., 3rd, Tokyo* p. 37 (1987).
73. T. Muschik, T. Fischer, and R. Scwarz, *Mater. Res. Soc. Symp. Proc.* **258,** 565 (1992).
74. A. Madan, J. McGill, J. Yang, and S. R. Ovshinsky, *Appl. Phys. Lett.* **37,** 826 (1980).
75. J. Yang, *Proc. IEEE Photovoltaic Spec. Conf., 20th, Las Vegas, Nev., 1988* p. 241. IEEE, New York, 1989.

6 Amorphous-Silicon-Based Devices

Yoshihiro Hamakawa, Wen Ma, and Hiroaki Okamoto
Faculty of Engineering Science
Osaka University
Toyonaka, Osaka, Japan

Recent progress in tetrahedrally bonded amorphous semiconductors and their technological applications to optoelectronic devices are reviewed in this chapter. First, some significant advantages of these materials are pointed out, and tangible instances are demonstrated from current live topics. The present state of the art in optoelectronic device development with this new kind of thin film is then reviewed, and the technical data are summarized and discussed.

I. Introduction

In the past two decades, remarkable progress has been seen in the field of disordered materials in both theoretical and experimental aspects. One major reason for this progress is the great advances made in material preparation technologies. These advances are supported by ultra-high-vacuum techniques, ultrapurification of inorganic elements and precisely synthesized heat-treatment technologies, including the rapid quenching of thin-film materials. Another reason for this progress might be development of a series of new material characterization methods with their computerized measurement systems.

The recent discovery of an existence of valence controllability in hydrogen passivated amorphous silicon (a-Si) strongly promotes the evaluation of amorphous semiconductor as a new electronic material. This new amorphous material is able to form both p–n and p–i–n junctions and has excellent photoconductivity with a considerably high absorption coefficient. These characteristics, coupled with mass-production capabilities of large-area nonepitaxial growth on any substrate material, strongly satisfy the current need for the development of a low-cost solar cell as a new energy resource. With the aid of the national project for renewable energy development, substantial progress in the amorphous silicon field has been seen in recent years in both basic physics and technology. These integrated new knowledges have opened some other new application fields such as thin-film transistor (TFT), electrophotography, three-dimensional integrated devices, and quantum-well devices, and has triggered many related research and development (R&D) efforts in the broad areas of optoelectronics and electronics.

In this chapter, recent advances in the a-Si device applications are reviewed. First, some unique advantages of these tetrahedrally bonded amorphous semiconductors are enumerated and explained with some concrete evidence from current technologies. Progress of some typical devices is introduced, together with the individual R&D efforts to improve the performance of each device.

II. Significant Advantages of a-Si and Its Alloys as a New Optoelectronic Material

First, some unique physical properties and remarkable advantages of a-Si alloys as new optoelectronic materials are enumerated as shown in Table 1 in view of both basic physics and the applied technology [1].

As has been reported elsewhere, a-Si and microcrystalline Si have more than one order of magnitude larger optical absorption coefficients compared to those of single-crystal silicon at the maximum solar photon energy region near 5000 Å as shown in Fig. 1. Moreover, this material system has an excellent photoconductivity in this visible photon energy region. Figure 2 shows the dark conductivity and photoconductivity versus the optical energy gap in nondoped a-Si alloy films deposited under high hydrogen dilution conditions of the plasma CVD [2]. As can be seen from the figure, a-$Si_{1-x}C_x$ alloy has a considerably high photoconductivity with σ_{ph}/σ_d ratio $>10^4$. Another notable property of these hydrogenated tetrahedrally bonded amorphous semiconductors is the existence of valence electron control by doping with substitutional impurity atoms. Effects of impurity doping on the conductivity, optical absorption coefficient, and optical energy gap have been intensively investigated on a-Si by Okamoto *et al.* [3], fluorinated

Table 1 Significances of a-Si alloys as a new electronic material that opens up a technological innovation from bulk crystalline age to multilayered thin-film age.

Physical properties	Fabrication technology
Excellent photoconductivities with high absorption coefficient for sunlight	Wide-area thin film (plasma CVD, ECR CVD, etc.)
Low dark conductivity	Low temperature deposition $100°C < T_s < 380°C$
High structural sensitivity (valency electron controllability)	p–n control can be accomplished only by mixture gas regulations
Wide range of energy controllability $1.0\ eV < E_{g(opt)} < 3.6\ eV$ a-SiGe–a-Si–a-SiC	Can be deposited on any inexpensive substrates
Easy to make heterojunction (relatively low interface states)	Easy to apply integration technology (heterojunction, superlattice, tandem cells)
Mechanically strong (because of amorphous network)	Low cost and good mass-production capability

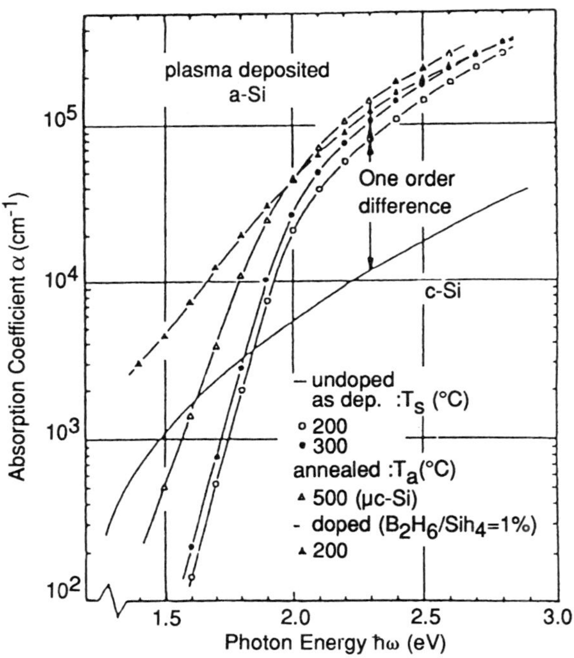

FIGURE 1. Comparison of absorption coefficient spectra of a-Si:H, microcrystalline Si, and single-crystalline Si with the solar radiation spectrum.

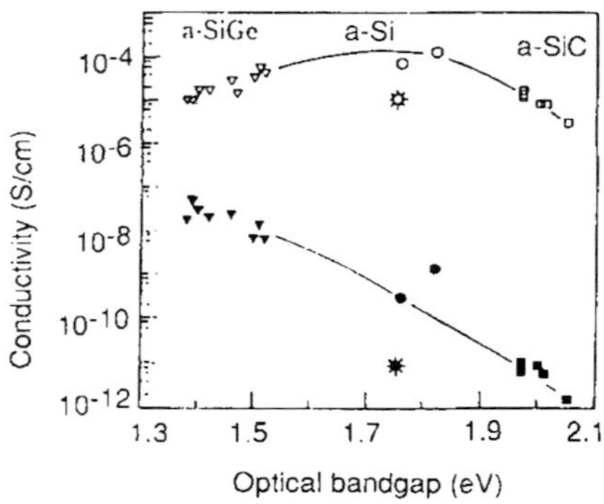

FIGURE 2. Dark and photoconductivities of a-SiGe:H (triangles), a-Si:H (circles), a-SiC:H (squares) vs. optical gap. Star marks ☆ and ★ show the property of IBD-produced film.

amorphous silicon (a-Si:F:H) by Madan and Ovshinsky [4], hydrogenated amorphous silicon carbide (a-SiC:H) by Tawada *et al.* [5], (still in progress) on amorphous silicon germanium (a-SiGe:H) by Yukimoto [6], and amorphous silicon tin (a-SiSn:H) by Kuwano and Tsuda [7]. Figure 3 summarizes the valence control for amorphous silicon carbide (a-$Si_{1-x}C_x$) fabricated from the plasma decomposition of a gaseous mixture of $(SiH_4)_{1-y} + (CH_4)_y$ [8].

Since a-Si film is deposited by a kind of vapor growth technology called *plasma CVD,* junction formation can be easily made in the same reaction chamber by mixing substitutional impurity gases into SiH_4 or SiF_4. Moreover, the interconnection of cells can be made in the process of a-Si film deposition with conventional integrated-circuit photomask processing and also by laser-beam lithography. Combining the mass-production lines could be easily accomplished in an all-dry process. Figure 4 shows the mass-production sequence of the glass substrate integrated a-Si solar cells.

As the preparation technology for a-SiC alloy, the plasma CVD is now widely utilized everywhere, whereas electron cyclotron resonance (ECR), CVD and ion-beam CVD (IB-CVD) have been intensively investigated in the last few years. Figure 5a shows a schematic illustration of the ECR CVD apparatus. Microwave power at 2.45 GHz is introduced into the ECR plasma excitation chamber through a rectangular waveguide through a fused-quartz-plate window. The ECR excitation chamber forms a cylindrical resonator of TE_{113} mode of the introduced microwave. In the system, the magnetic flux required for satisfying the ECR condition is about 875 G at the center of the magnetic coil. The generated ECR plasma is extracted from the ECR excitation chamber into the deposition chamber along with the gradient of dispersed magnetic field as shown in Fig. 5b. The extracted

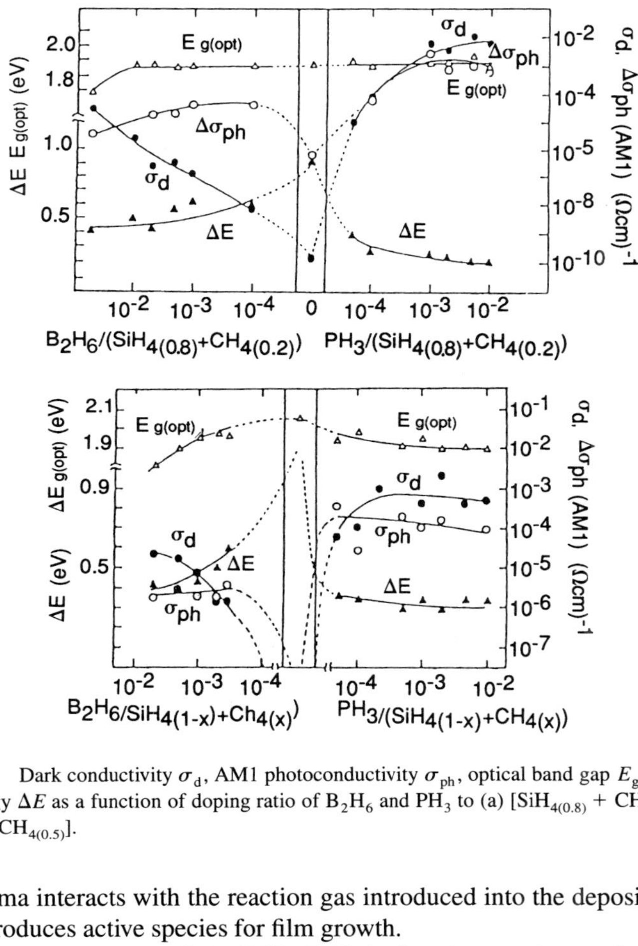

FIGURE 3. Dark conductivity σ_d, AM1 photoconductivity σ_{ph}, optical band gap $E_{g(opt)}$, and activation energy ΔE as a function of doping ratio of B_2H_6 and PH_3 to (a) $[SiH_{4(0.8)} + CH_{4(0.2)}]$ and (b) $[SiH_{4(0.5)} + CH_{4(0.5)}]$.

ECR plasma interacts with the reaction gas introduced into the deposition chamber and produces active species for film growth.

The unique advantage of the ECR CVD is that the growing surface does not experience bombardment damage by electrons and/or other heavy species since there is an effect of a soft landing with an energy of several tens of electronvolts [9]. This effect might result in the prevention of weak bonds from being introduced into the network, and furthermore there could be a suppression of the diffusion of long-lifetime radical species due to the raised surface temperature. It is expected that films with dense network and low defect density are formed.

For the deposition of a-SiC and μc-SiC, hydrogen is used as an ECR plasma excitation gas, and a mixture of SiH_4, CH_4, and B_2H_6 or PH_3 is usually employed as a reaction gas for the growth of p- and n-type SiC:H. Details of the preparation

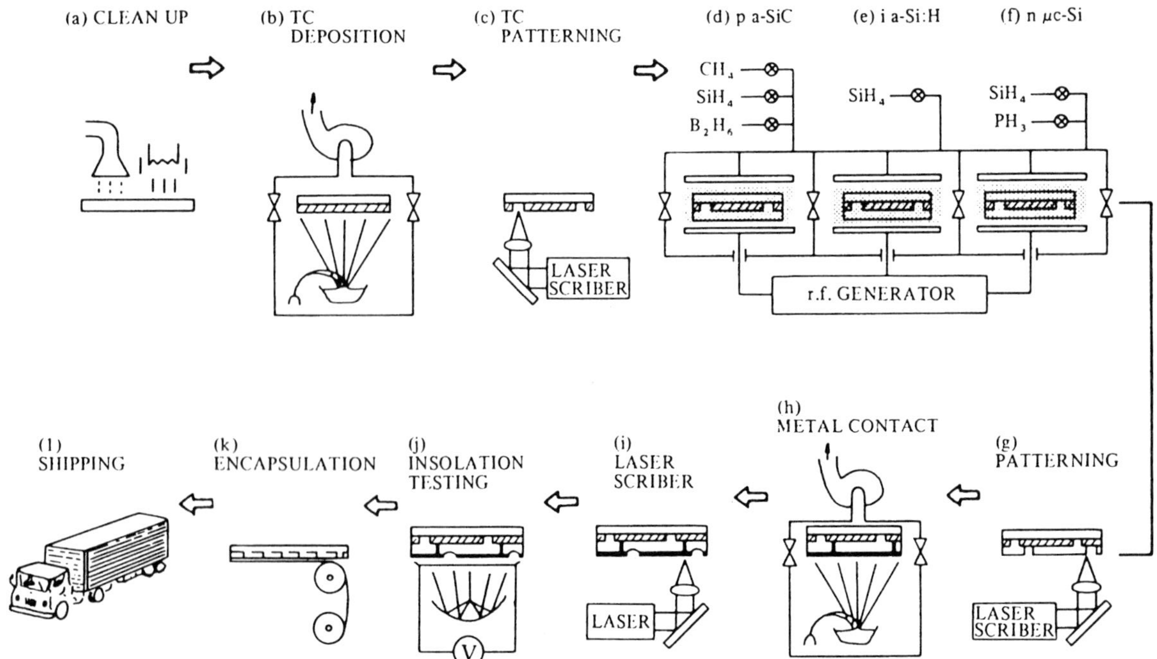

FIGURE 4. A production sequence of a-Si solar cells having the advantages of (1) large-area nonepitaxial growth for any foreign substrate material at low temperature, (2) all-dry process, and (3) high mass producibility.

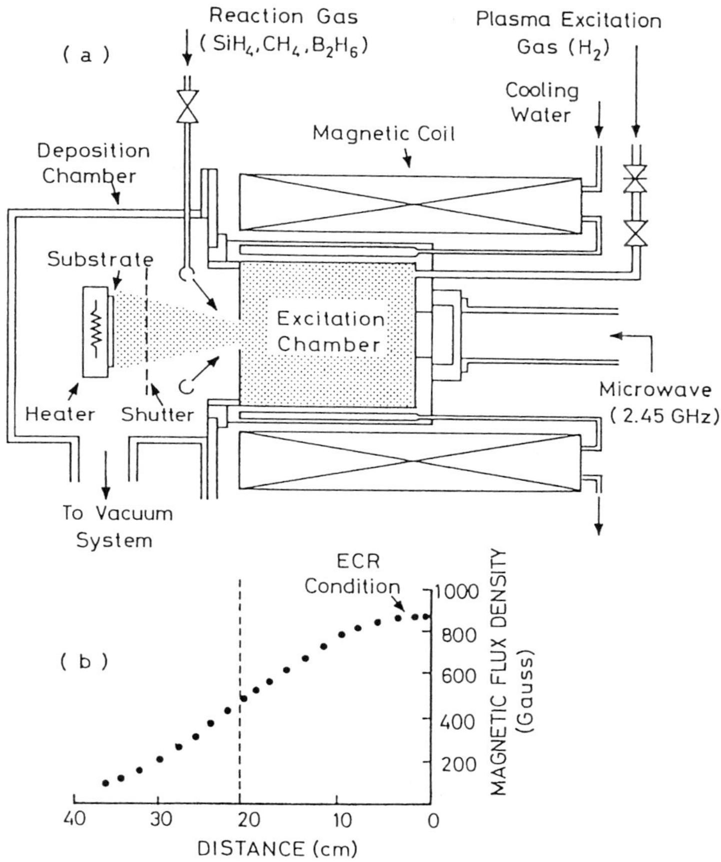

FIGURE 5. A schematic diagram of electron cyclotron resonance (ECR) plasma CVD system (a) and the profile of the magnetic field for the extraction of the plasma from the excitation chamber into the deposition chamber (b).

conditions are summarized in Table 2. Since the operating pressure is in the range of 10^{-3}–10^{-4} torr, and the lifetime of chemically active hydrogen radicals is quite long, a large amount of hydrogen radicals will reach the growing surface and play an important role in determining the properties of the growing films. Therefore, the dependence of the material properties on the hydrogen dilution ratio in the reaction gas has been investigated. Figure 6 shows the dependence of the optical energy gap and dark conductivity of the samples on the H_2 dilution ratio increases; it should be noted that the optical gap (E_0) and also the dark con-

Table 2 Preparation conditions of p-type amorphous and microcrystalline SiC in ECR plasma CVD.

Substrate temperature	RT–400°C
Microwave power	150–400W
Microwave frequency	2.45 GHz
Magnetic flux density	0.0875 T (tesla)
Total gas pressure	10^{-3}–10^{-4} torr
Plasma excitation gas (flow rate)	H_2 (10–100 sccm)
Reaction gas (flow rate)	SiH_4 (10–50 sccm)
	CH_4 (10–50 sccm)
	B_2H_6 (40–100 sccm)

ductivity (σ_d) of both p- and n-type films increase with H_2 dilution. As can be seen from Fig. 6, there are two main factors that determine the optical energy gap; one is the composition ratio of Si:C:H corresponding to the source gas ratio CH_4/SiH_4, and the other is the extent of H_2 dilution that might be related to the details of the network structure. The film properties are strongly dependent not only on the substrate temperature and microwave power but also on the ratio of hydrogen

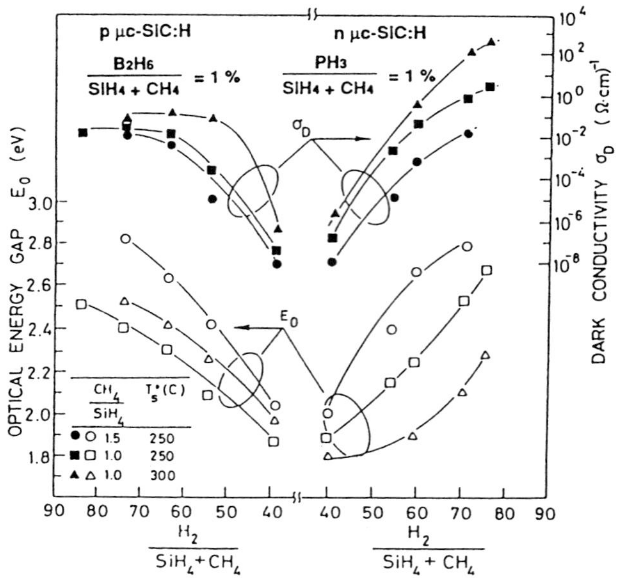

FIGURE 6. Dependences of the optical energy gap and dark conductivities of p- and n-type μc-SiC:H on the hydrogen dilution ratio in the reaction gas in ECR CVD.

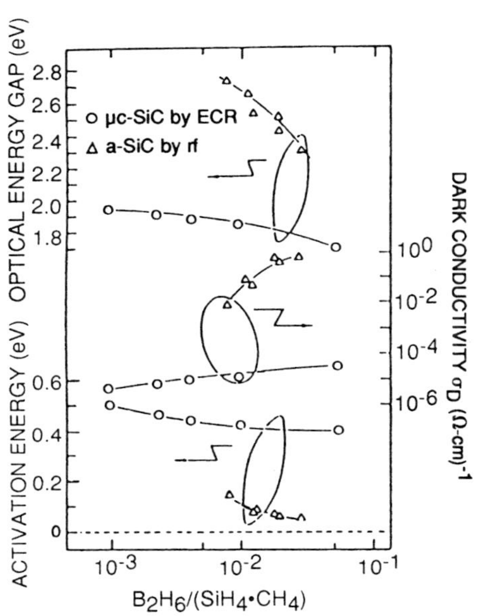

FIGURE 7. Dependence of dark conductivity, activation energy, and optical energy gap on the dopant gas ratio in the reaction gases for p-type μc-SiC:H.

to reaction gases $H_2/(CH_4 + SiH_4)$ during deposition. Although the optical energy gap increases with the flow rate of CH_4, the effect is not as remarkable as the dependence of hydrogen dilution. Hydrogen dilution has the effect of reducing the hydrogen content in the film, and also of enhancing the degree of microcrystallinity.

The formation of Si and SiC microcrystallites is confirmed by Raman spectra as mentioned in the original work [9]. The Raman spectrum of the films prepared at microwave powers is >250 W and exhibits distinct structures at around 520 and 740 cm^{-1}, which correspond to TO phonon modes of crystalline Si and SiC clusters.

The conductivity of μc-SiC can also be controlled by adjusting the dopant gas/host reaction gas flow ratio. Figure 7 shows the dependence of the optical energy gap and dark conductivity of μc-SiC:H prepared by ECR CVD on the flow ratio of B_2H_6 p-type doping gas. Here, the hydrogen dilution ratio is kept constant at 74. The data for p-type a-SiC:H prepared by RF plasma CVD are also shown for comparison. The carbon content x in both the cases is about 0.3. It is clear that the total doping efficiency in p-type μc-SiC:H is higher than that in p-type a-SiC:H by several orders of magnitude.

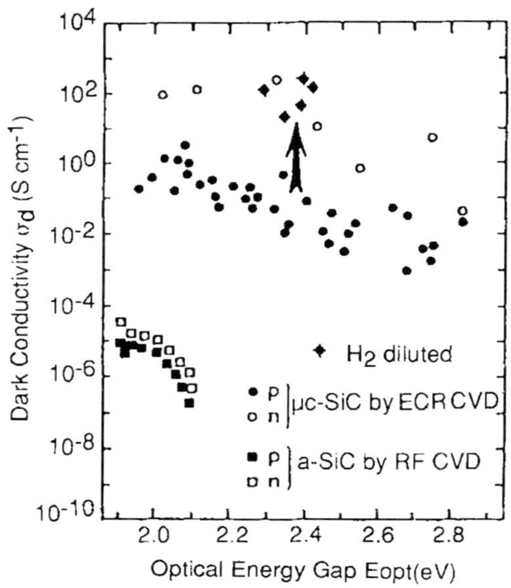

FIGURE 8. Relationship between dark conductivity and optical energy gap of amorphous and microcrystalline SiC:H prepared by RF and ECR plasma CVD.

Figure 8 summarizes the relation between the dark conductivity and the optical energy gap of p- and n-type a-SiC:H prepared by conventional RF plasma CVD and p- and n-type μc-SiC:H prepared by ECR plasma CVD. As the optical energy gap increases, the dark conductivity of the films prepared by the RF plasma CVD rapidly decreases, while that of the films prepared by the ECR plasma CVD remains higher than 10^{-3} S/cm even when the optical energy gap exceeds 2.5 eV. In Fig. 8, the star marked "H_2 diluted p-type μc-SiC:H" film has more than 3×10^{19} cm^{-3} hole concentrations with a mobility of 24 cm^2 V^{-1} s^{-1} as measured by the Hall effect [9].

Another topic in the material preparation technology field is the ion-beam deposition (IBD) method [10]. It should be emphasized that a stable film against light induced degradation with a high deposition rate is still an important key issue in the field. To conquer these problems, tremendous R&D efforts have been in progress using a wide variety of approaches, e.g., the chemical annealing treatment (CAT) [11], intense xenon light-pulse-assisted plasma CVD [12], IBD, and so on. Among these, IBD method is a unique challenge since it offers a wide range of control of both hydrogen content and the ion energy of the decomposed species.

Figure 9 shows a schematic illustration of the IBD system. Typical conditions for a-Si film growth are ion acceleration voltage $V_{acc} = 100–300$ V, ionic current

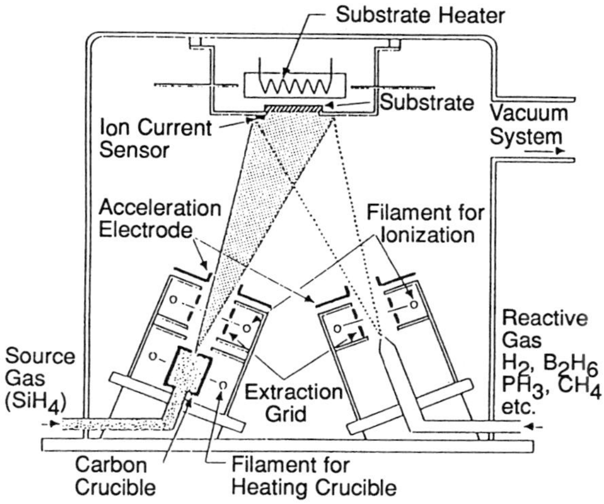

FIGURE 9. A schematic representation of ion-beam deposition (IBD) system.

$I_e = 0.8$A, with voltage V_e at 400 V, and substrate temperature $T_S = 100-300°$C, with chamber base pressure $\leq 10^{-4}$ torr.

A systematic investigation on the undoped a-Si film deposition has been made by a series of deposition parameters. The result shows a good film quality with 10^4-10^6 photo-/dark conductivity ratio (σ_{ph}/σ_d) with optical energy gap in the range of 1.7–1.8 eV. A noticeable feature of the IBD-produced film is improved stability against light exposure. Figure 10 shows a comparison of changes in the photoconductivity σ_{ph} with AM1 light illumination for IBD produced film and

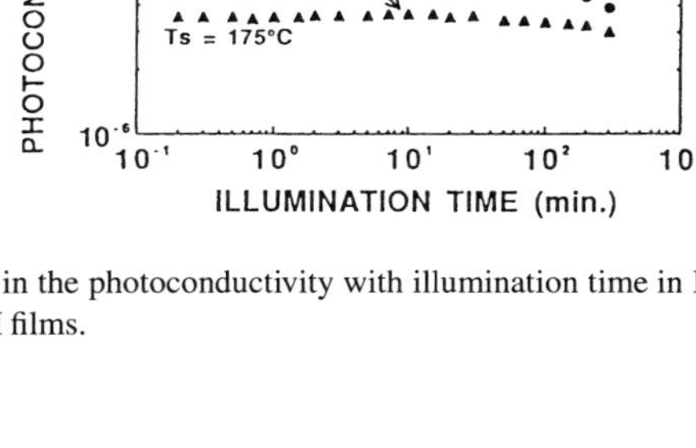

FIGURE 10. Change in the photoconductivity with illumination time in IBD produced a-Si:H and PCVD produced a-Si:H films.

Table 3 Physics and applications of a-Si alloy thin film.

Basic process	Energy perturbation	Controllable property	Practical application
Low-field conductivity	Temperature	Conductivity	Thermometer
Double injection; impact ionization; filamentary breakdown	Electric field	Conductivity	Threshold switch; triode analog device (4 dots)
Thermal instability; negative resistance	Electrical energy ($V \times I \times t$)	Conductivity	Thermistor; bistable switch
Interface effect; valence control	Bias voltage	Potential, interface potential, and current across the barrier	Polarized switch; amorphous transistor; FET
	Light (optical image)	Photoconductivity (surface charge)	Electrostatic printing; electrophotographic printing (xerox)
		Photoconductivity; photovoltaic effect	Photosensor; solar cell (a-Si:H:F); image converter; image pickup (Saticon)
Photoelectron emission	Light	Secondary-electron emission rate	Electron-beam memory
Photochemical effect	Light (optical image)	Etching rate	Mask processor; photolithography
Photostopping effect	Light (laser beam)	Absorption coefficient	Light switch; optical modulator
Electrooptical effect	Acoustic wave	Refractive index	Light switch; beam deflector; optical modulator

conventional plasma CVD produce film [10]. Table 3 summarizes various physical effects observed in this material system and their potential device applications.

III. Progress in Amorphous Silicon Solar Cell Technology

A basic difference in the photovoltaic process of a-Si solar cell to that of single-crystal p–n junction is an existence of a high electric field in the photocarrier generation region. Moreover, the internal electric field in the i-layer varies very sharply as functions of both the mid-band-gap state density, g_{min}, and the induced space charge distributions. This is called the *drift-type photovoltaic effect* in a-Si solar cells [13]. Obviously, observable J–V characteristics of the a-Si p–i–n junc-

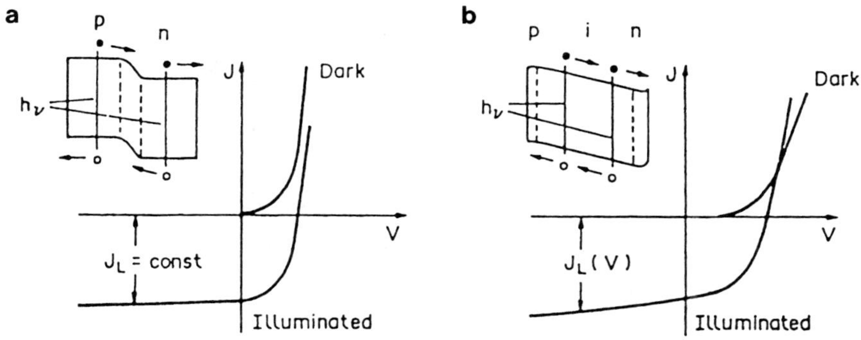

FIGURE 11. Comparison of *J–V* characteristics of crystal and amorphous solar cells: (a) crystal p–n junction and (b) a-Si p–i–n junction for dark and light illuminated conditions.

tion are quite different from those of single-crystal p–n junction as shown in Fig. 11.

On the basis of the optoelectric properties of these materials, systematic calculations have been made on the optimum thickness of the solar photovoltaic active region for various kinds of solar cell materials [14]. These results indicate that the ratio of the optimum thickness for the active layer in a-Si solar cells is less than 1/500 than in a c-Si solar cell. It is evident from this fact that a-Si represents both an energy-saving and a resource-saving solar cell material for anticipated future widespread demands in solar photovoltaic applications. Because of its amorphous structure, a-Si can be deposited onto any inexpensive substrate, which needs to be treated to only a relatively low temperature, viz., $<200-300°C$. Moreover, it is possible to form a very wide-area solar cell, because it can be deposited directly by vapor-phase growth on to flexible substrates like stainless-steel sheets and polymer films.

Utilizing the concept of nonepitaxial deposition technology, it could be possible to reduce the balance-of-system (BOS) costs in photovoltaic arrays by the hybridization of already-built units. Solar tile and sticker-form (like an adhesive label) solar cell might be useful to realize this concept. Figure 12 shows an example of various a-Si solar cells deposited on glass [15], stainless steel [16], Kapton, and polymer films [17, 18]. Al_2O_3 ceramic substrate has also been attempted for the building tile [19]. As pointed out in the early stage of the work, junction formation can be easily achieved in the same reaction chamber by mixing substitutional impurity gases into the SiH_4 or SiF_4 gases. Moreover, the interconnection of cells can be introduced into the process of a-Si film deposition with conventional integrated-circuit mask technology. Combining these unique advantages in the physical properties with some merit of fabrication in a-Si technology, automati-

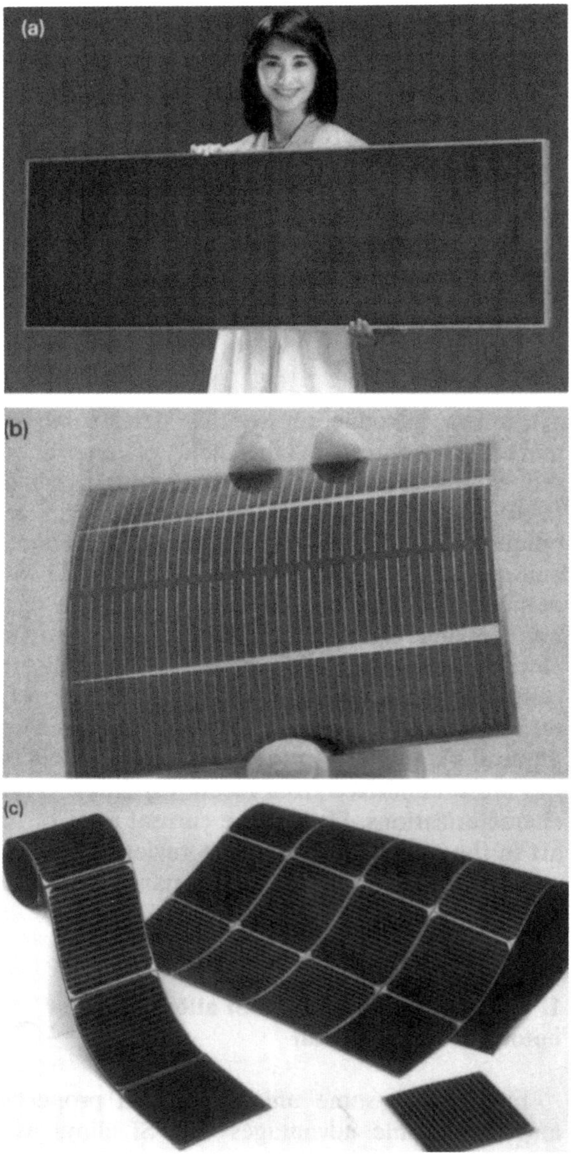

FIGURE 12. Various types of a-Si solar cells: (a) NEDO—standard large-area cell deposited on glass (presented by Sanyo); (b) deposited on stainless-steel substrate (presented by Kaneka); (c) the flexible a-Si solar cell deposited on a polymer film (presented by Sanyo).

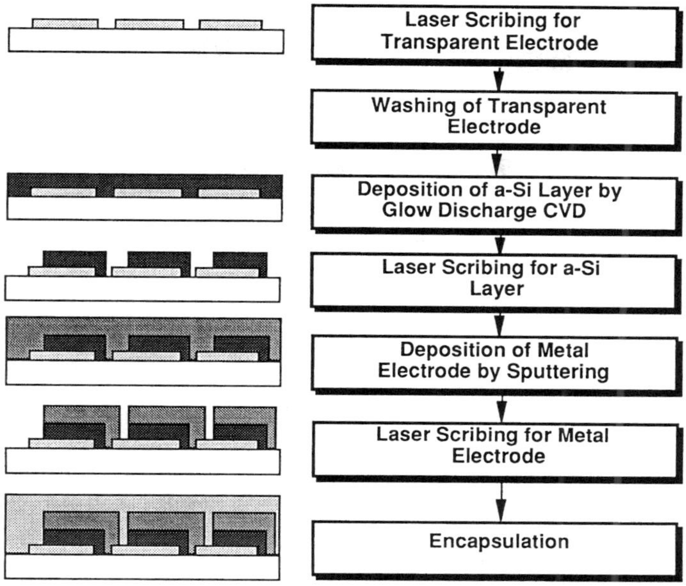

FIGURE 13. A mass-production sequence of the glass substrate integrated a-Si solar cell.

zation of a mass-production line could be easily accomplished in an all-dry process. Figure 13 shows an example of production sequence of the integrated a-Si solar cells.

Another noticeable advance in the R&D effort in recent years has been the development of new materials, e.g., a-SiC, a-SiGe, and a-SiSn [20]. The energy gap of this material system can be controlled from 1.0 to 3.6 eV by the fractional composites x and y in a-$Si_{(1-x)}Ge_x$ to a-$Si_{(1-y)}C_y$, for example. All these materials have considerably good valence controllability by the hydrogen passivation of dangling bonds and doping of substitutional impurity with proper gas mixture technique in the plasma CVD process. The existence of energy-gap controllability with the valence electron controllability means this material system as a synthetic material. Since 1978, a systematic investigation on the valence electron control of amorphous mixed alloy have been made by the Hamakawa group at Osaka University [5]. As an application of their results, they developed a-SiC/a-Si heterojunction solar cell having an efficiency of more than 8% in 1980 [8]. The best record of the a-SiC/a-Si heterojunction solar cells is 13.2% efficiency as of December 1993 [21] as shown in Fig. 14. Table 4 shows a summary of unique advantages of a-Si alloys for the low-cost solar cells.

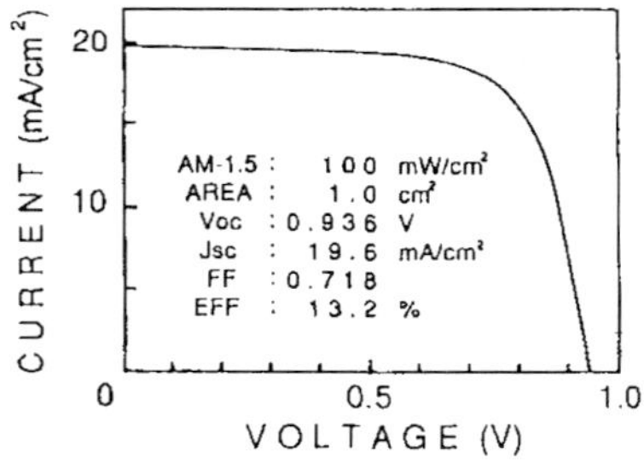

FIGURE 14. The *I–V* characteristic using a-SiC:H film (ADHT) in the p–i buffer layer.

Figure 15 shows the transitions of cell efficiency for various types of a-Si solar cells since 1976. As can be seen from this figure, there is a step-like increase in the cell efficiency in the region around 1981, and slope A (before 1981) corresponds to the improvement of the film quality and routine cell fabrication progresses. The key technologies that led to the steep slope change from A to B at 1981 were due to development of heterojunction solar cells with a-SiC:H [22] and a-SiGe:H [23]. One important remaining area for further improvement of a-Si solar cell efficiency is the more efficient collection of low-energy photons just above the band edge of a-Si, because the penetration depth of the 1.8 eV photon, for example, is on the order of 5 μm, whereas the thickness of a-Si solar cell is only 0.6 μm. This concept has been extended to more efficient utilization of optical and carrier confinement in the multilayered heterostructure junction [24]. Recently, Fujimoto *et al.* [25] have developed a practical technology with the cell structure of ITO/n μc-Si/i–p a-Si/TiO_2/Ag-plated semitextured stainless steel having an efficiency of 9.17%. Quite recently, Taiyo-Yuden/ETL groups [26] have reported 10.26% efficiency with the optical confinement effect employed by milky transparent glass (MTG).

Another way to collect the longer-wavelength photons is the absorption with the stacked junction of the lower-energy-gap semiconductor [27]. It has been reported elsewhere [28] that the percentage of light-induced degradation increases with increasing i-layer thickness. The reason is that the volume recombination of

Table 4 Unique advantages of a-Si alloy as a promised candidate of low-cost solar cell.

Synthetic material	Low cost
Energy-saving material processing	Abundant resources
Short energy payback time	Nonpolluting material

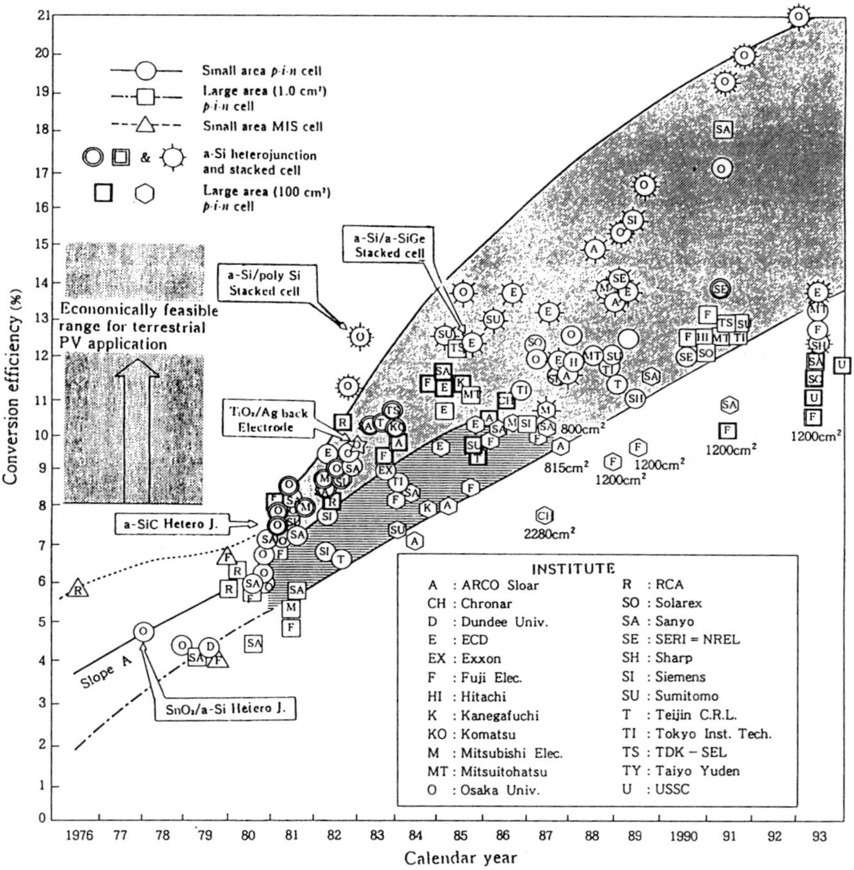

FIGURE 15. Progress of a-Si solar cell efficiencies for various types of junction structures as of August 1992. A steep slope change is seen with the appearance of a-Si alloys such as a-SiC, μc-Si, or a-SiGe.

the photogenerated carriers becomes relatively large with decreasing electric field in the i-layer. For the purpose of suppressing this effect, the tandem-type solar cell has been recommended as a more reliable a-Si solar cell. Figure 16 shows light-induced change in the conversion efficiencies of a-Si single, a-Si/a-Si tandem and a-SiC/a-SiGe triple tandem cells [2].

Recently, a concept of the band profiling design has been initiated as an optimum design of an ambipolar carrier transport in i-layer of the multi-band-gap junction. As an example of junction structure, the band profile of the a-SiC/a-Si/

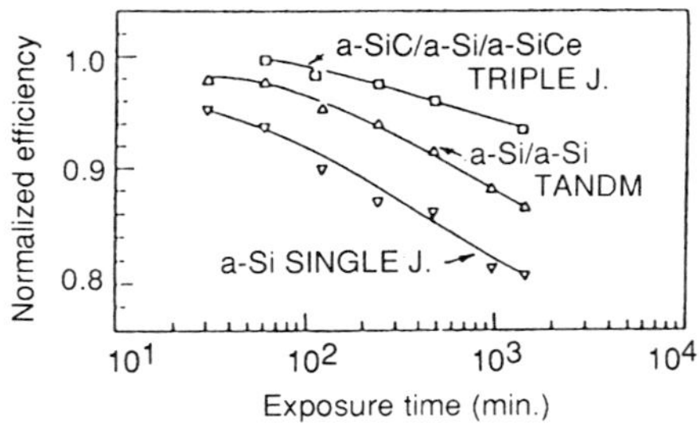

FIGURE 16. Light-induced change in conversion efficiencies of single, tandem, and triple a-Si solar cells, which is normalized to initial values (inverted triangles—a-Si single cell, triangles—a-Si/a-Si tandem cell, squares—a-Si/a-Si/a-SiGe triple cell).

a-SiGe triple-band-gap tandem solar cell is illustrated in Fig. 17, with its photovoltaic performance [2]. Figure 18 shows another example of a module structure and voltage–current V–I characteristics of a large area a-Si/a-Si stacked solar cell having more than 10% efficiency by Fuji Electric Co. Ltd. [29]. Quite recently, the USSC group has reported 11.8% initial and 10.2% stable efficiencies with 1 × 1-ft module [30], which is the best record on the practical size of a-Si submodule. However, in the R&D level investigation on a-Si/poly-Si tandem junctions, a conversion efficiency of 15.04% with V_{oc} = 1.478 V, J_{sc} = 16.17 mA/cm^2, and FF = 63% has been obtained on a two-terminal cell of sensitive area 5 × 6 mm^2 under AM1 illumination. Quite recently, by the same combination four-terminal cell 21.0% efficiency has been obtained by Ma Wen *et al.* as shown in Fig. 19 [31].

The light-induced degradation in a-Si solar cell is still a big headache. Continuous efforts to clarify the mechanism as the basic physics on one hand, and technological solution, on the other hand have been in progress. The light induced degradation seems to be mostly based upon the Staebler–Wronski (SW) effect and its recovery with thermal annealing. The apparent photovoltaic performance changes with various operation parameters such as incident-light intensity, with its spectrum, operating current density, and with the duration of light exposure. The amount of degradation also depends on the film quality, and the initial conversion efficiency. Therefore, tremendous R&D efforts have been made on this bottleneck technology.

The key questions related to the SW effects are

1. What is the source of metastable defect?
2. What promotes the creation of metastable defect?
3. What stabilizes the metastable defects, or modification of the defect structure?

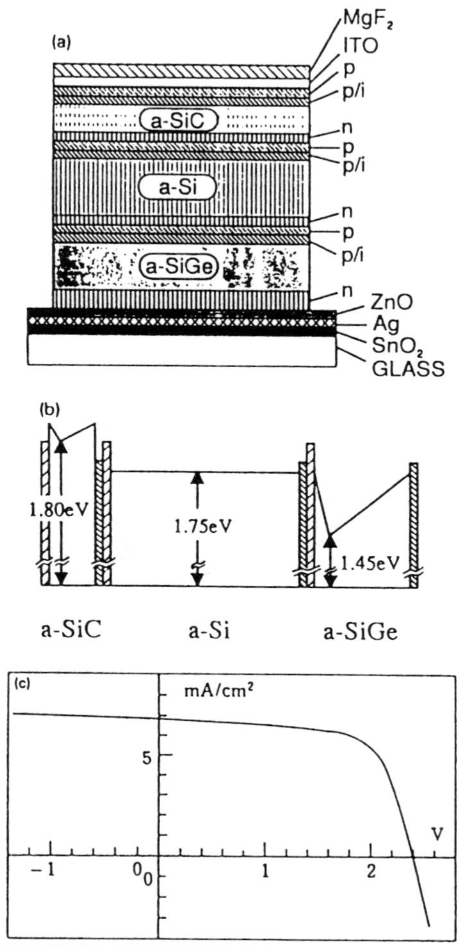

I_{sc} (mA/cm²)	V_{oc} (V)	F.F.	Efficiency (%)
6.81	2.39	0.68	11.0

FIGURE 17. Cell structure (a), band-gap profile of a-SiC/a-Si/a-SiGe multi-band-gap stacked solar cell (b), and voltage–current characteristics of a-SiC/a-Si/a-SiGe multi-band-gap stacked solar cell (c).

Various possible mechanisms have been proposed; the best-known model is the weak-bond-breaking model, first proposed by Pankove and Berkeyheiser [32] and extended by Stutzmann [33]. In the a-Si, there exist some weak Si — Si bonds that

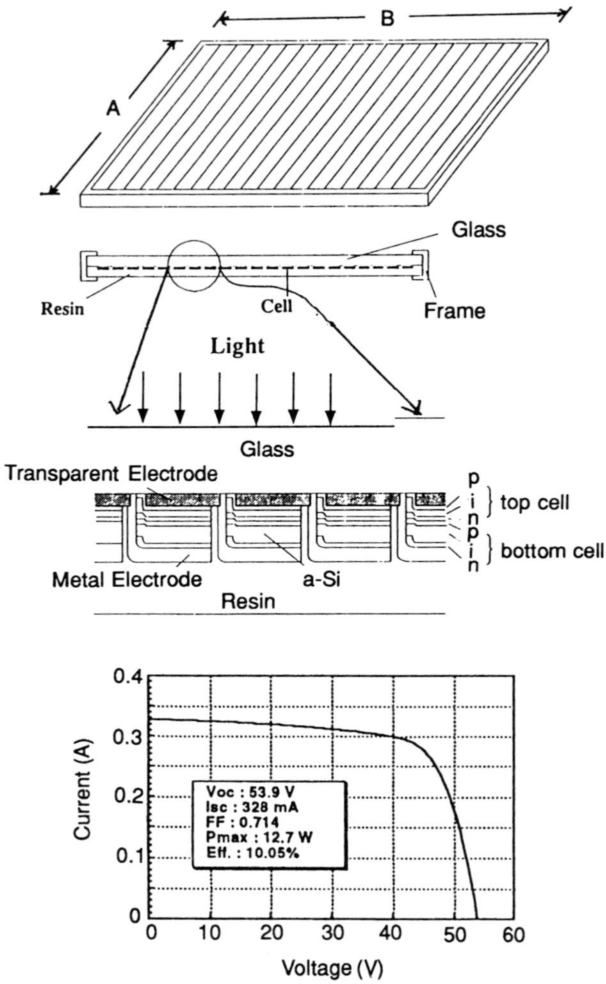

FIGURE 18. Module structure and V–I characteristics of large-area ($A \times B = 30 \times 40\ \text{cm}^2$) a-Si tandem solar cell.

consist of bonding states near the top of the valence-band edge and antibonding states near the bottom of the conduction-band edge. The total system energy at a particular weak-bond site will be increased by placing an electron–hole pair of either two electrons or two holes at the site. If these states are sufficiently local-

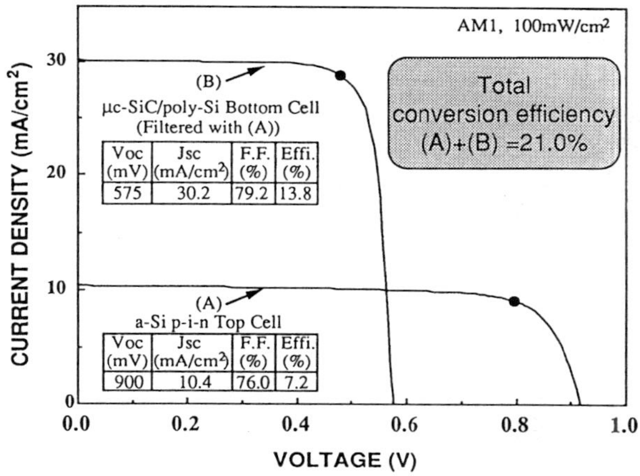

FIGURE 19. Output performances of a-Si(A)/poly-Si(B) four-terminal tandem solar cell.

ized, energy may be gained by the formation of a dangling-bond pair [34]. There is a finite chance that a backbonded hydrogen will switch its position and thereby hinder the bond healing after the excitation. This process promotes the dangling-bond pair to be located at the metastable position. Further stabilization of the metastable dangling bonds may be possible through successive bond switching, mediated by hydrogen motion [35], which promotes a larger separation of two metastable dangling bonds.

A wide variety of application systems have been developed in recent years, particularly consumer electronic applications, which are expanding very rapidly, as shown in Fig. 20. For instance, about 5 million sets per month of a Si-driven pocket calculator are fabricated in Japan as of 1984. On the other hand, the field of semipower and bulk power applications are still in the experimental phase. For example, an a-Si photovoltaic tiled building by Sanyo Electric Co. is shown in Fig. 21.

IV. Integrated Photosensor and Color Sensor

Full use of advantages such as a large-area, uniform film growth in the glow discharge produced a-Si:H is applied to the development of integrated photosensors. Pioneer work has been done by the Musashino ECL group for a non-storage-type linear sensor in 1980 [36]. Cross-sectional views of three types basic structure of photosensors, and its combination of driving thin-film transistors (TFTs), are il-

FIGURE 20. Expanding a wide variety of application systems in consumer products.

lustrated in Fig. 22. In 1982, a-Si storage mode linear photosensors were developed by Fuji-Xerox, Fuji Electric, Sony, and Fujitsu. A 512-element linear photosensor for the facsimile readout element was developed by Kanoh *et al.* [37] of the Sony group. The a-Si linear sensor consists of 1728 bits of a-Si integrated elements. This means that a 2048-element a-Si sensor can be operated with a frame time of <5 ms.

The Fuji-Xerox group has also reported 1056 bits (8 bits/cm), and 210 mm 1728 bits 98 bits/mm) linear sensors [38]. In this device, complementary metal oxide semiconductor (CMOS) analog switches, shift resistors, and interconnecting circuits are mounted on a single ceramic substrate having dimensions of 240 × 62 mm. With merit of low cost and performance uniformity, more than 45% of market share of the facsimile readout are occupied by a-Si sensors. Figure 23 shows the outlook of these developed facsimile readout devices.

FIGURE 21. Solar photovoltaic roofing tile made of amorphous silicon equipped on Japanese house (presented by Sanyo) (a), and 2KWp residential roofing tile solar houses field experiment in Rokko Island (presented by Kansai Electric Power Co. Ltd.) (b).

A
SANYO
morton
コン

B

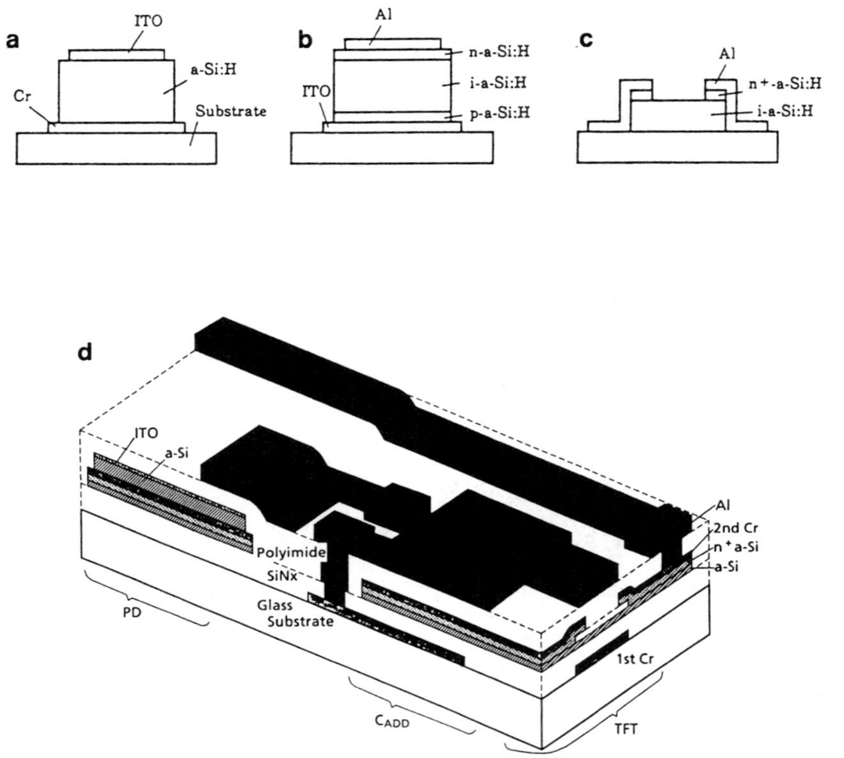

FIGURE 22. Cross-sectional views of a-Si:H photodiode integrated with switching TFT on a glass substrate: (a) Schottky structure; (b) p–i–n junction structure; (c) planar structure; (d) composite view of the device.

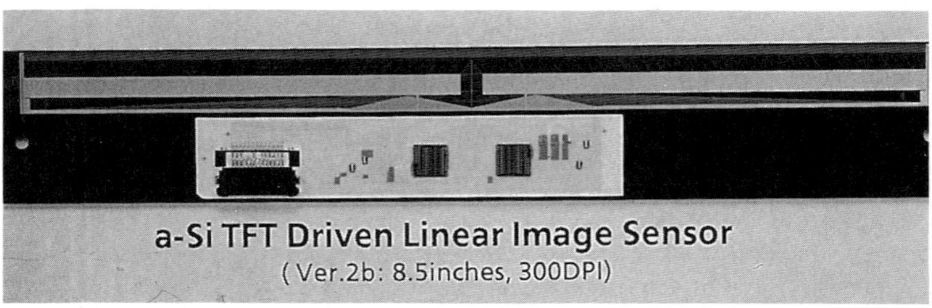

FIGURE 23. Amorphous-silicon TFT-driven linear image sensor for facsimile readout element, vertical 2 lines/mm, 8 dots/mm for B4-size paper.

FIGURE 24. Amorphous-silicon integrated color sensors.

In early 1982, the Kuwano group of Sanyo developed an integrated a-Si color sensor [39]. Recently, color sensors of many kinds, sizes, and shapes have been developed, as shown in Fig. 24. Recently, the potential needs of this device have been expanded to a wide range of applications such as automatic color recognition and color identification in the field of medical science, industrial processes, food science, agriculture, and architecture.

V. Aspect of a-Si Imaging Device Applications

The development of "Saticon," a vidicon tube using Se–As–Te chalcogenide, is famous as the first practical application of amorphous semiconductors in 1973 invented by Maruyama's group [40], and they extended their activities to a-Si, and developed a-Si vidicon in 1979 [41]. Recently this group has developed a supersensitive vidicon called HARPICON by utilization of avalanche multiplication of Se target layer [42]. A remarkable merit of these new vidicon is lower power dissipation of peripheral circuit, which enable us to employ a LSI (large-scale integrated circuit) driver circuit with size reduction.

A solid-state color imaging device using a-Si photosensitive layer has also been developed by the Hitachi group in 1981 [41]. A cross-sectional view of the picture element is shown in Fig. 25. As can be seen in the figure, a-Si photosensitive layer is directly stacked on the MOS signal processing part. A 485-vertical $\times$ 384-horizontal picture-element (pixel) device has been fabricated. The signal charge stored in the p–n junction is read through a n-MOS transistor, and on the thin n^+ side of the a-Si layer. With this stacked structure, the effective aperture reaches up to 73% of the raster area, while an ordinary charge-coupled device (CCD) imaging

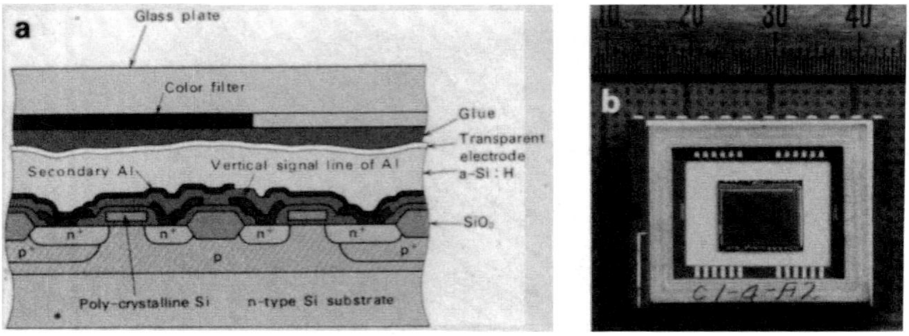

FIGURE 25. Monolithic tricolor a-Si image sensor stacked on Si MOS (a) and its external appearance (b) (presented by Hitachi Ltd.).

device has less than 47% aperture [41]. The highlight exposure was about 250 times as intense as the saturation level of the photocurrent. Because of the high photoconductivity in a-Si, the developed imaging device has much higher sensitivity with nonblooming performance as compared with ordinary devices.

VI. a-Si Electrophotographic Applications

With the aim of a full use of wide-area uniform deposition capability in a-Si, the development of electrophotographic receptor with a-Si has been initiated in 1979, and more than 10 groups have worked in this field. In principle, the photoreceptor consists of four separated functional layers as shown in Fig. 26: (1) a surface

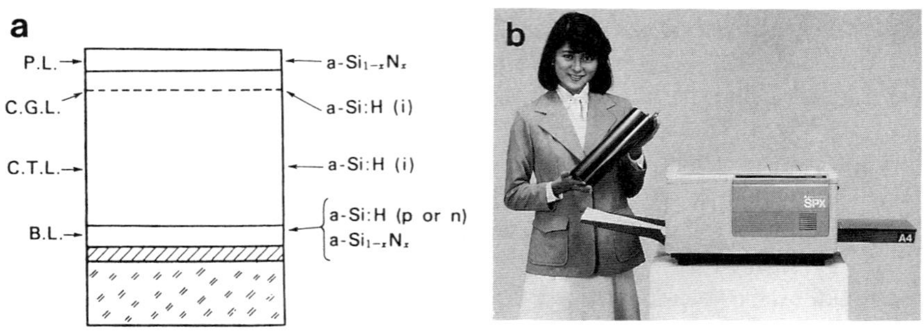

FIGURE 26. Structure of a multilayered photoreceptor.

passivation layer (SPL), (2) a charge generation layer (CGL), (3) a charge transport layer (CTL), and (4) a bottom-blocking layer (BL). To obtain full use of electrical and optoelectronic property controllability in a-Si and its alloys with controlling impurity doping and compositional regulation, a wide variety of material combinations have recently been proposed [43]. For example, Shimizu *et al.* have developed a-Si (SPL)/a-Si (CGL-CTL)/a-SiN(BL) multilayer, and Nishikawa *et al.* have substituted a-SiC layers for SPL–BL combinations. For the purpose of small-size laser printers and also a new type intelligent copier, there exists a considerable potential need to develop an infrared sensitive high-speed photoreceptor. On this objective, Kawamura [43] and his group at the University of Osaka Prefecture have recently developed the boron-doped a-Si SPL with a-SiGe alloy CTL and CGL. Through these basic investigations in the university institute, Kyocera Co. announced the successful mass production of an "amorphous silicon drum" for electrophotography in November 1981, developed with the cooperation of Kawamura's group. Stanley Electric Co. also announced the same kind of machine in May 1982. Remarkable advantages of this a-Si drum are (1) long life due to the high surface hardness; (2) high sensitivity even for the infrared region; (3) high thermal stability, which makes possible high-speed operation with nontoxicity; and (4) its bichargeable property, which can be applied to new functional devices. Quite recently, Sanyo has also commercialized an a-SiCa/a-Si drum-driven copy machine. Efforts have also been made to develop an intelligent copier using an a-Si photoreceptor in combination with a semiconductor laser diode and a microcomputer (see review in Hamakawa [44]).

VII. Visible-Light Thin-Film Light-Emitting Diode (TFLED)

Since the first observation of an infrared emission in the forward-biased a-Si junction in 1976 by Pankove and Carlson [45], electroluminescence studies on a-Si and its alloys have been made by several groups. Recently, Kruangam *et al.* succeeded in producing a visible-light emission in an a-SiC p–i–n junction [46], and developed a multicolor LED (red to green) by adjusting the carbon content. Figure 27 shows typical *V–I* characteristics for these a-SiC LED devices. Some examples of areal emission display of a-SiC LEDs are shown in Fig. 28. Multicolor emissions (red, orange, yellow, and green) have been easily obtained by controlling the optical energy gap in a-SiC, while a blue emission having a dragon pattern, is obtained in the TFEL-mode operation using a cell structure of glass/TCO/ Y_2O_3/a-C:H/Y_2O_3/Al. Although the brightness of this diode is still low

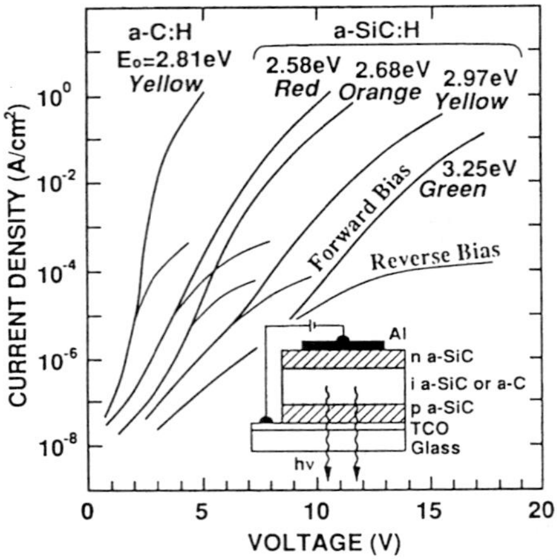

FIGURE 27. Current–voltage (J–V) characteristics of LEDs consisting of a-SiC and a-C luminescent i-layers.

[10-candela (cd) m^2 level], this TFLED has some significant advantages over the conventional crystal LEDs, such as large area, color emission in any spectrum of the visible region, and tunable coloring with color mixing of an integrated multilayered structure.

A new type of OEIC (optoelectronic integrated diode) combining a-Si integrated photosensors with this TFLED has also been postulated [47]. A series of experimental trials to improve emission efficiency are in progress; these involve the use of wide-gap carrier injectors, superlattice electrodes, and so on [48]. Figure 29 shows a cross-sectional view of the thin-film imaging storage device stacked with a-SiC n–i–n photoconductive layer and a-SiC thin-film light-emitting diode arrays [49]. The light-emitting diode has about a 500-Å active layer of i-layer a-SiC having an optical energy gap of 2.5–3.0 eV, whereas an active layer of photoconductive layer has 1.9-eV energy gap. Therefore, this device has an image converter function with storage of that image with input light of infrared to visible light. Quite recently, an improvement of spectral sensitivity has been accomplished by substituting the photoconductive layer to a high-gain photodiode with gain G corresponding to the multiplication factor M in the avalanche photodiode (APD). With thin-film electroluminescence mode employing amorphous carbon (a-C) as the light-emission layer, image conversion from red light to blue light has also been succeeded as shown in Fig. 29 [49].

New developments that augur the technological evolution leading to "macro-

FIGURE 28. Typical emission colors of a-SiC LED and a-C TFEL. Red-to-blue color emission can be obtained by controlling the fraction x in a-$Si_{1-x}C_x$. A large-area display is also available for any pattern, e.g., tiger, horse, or dragon.

electronics" in the 1990s include thin-film EL elements with new functions, such as tunable color by controlling applied voltage, which is a promising candidate for solid-state display panels for HDTV, and "intelligent" glass using electrochromic devices. As precursors of technological innovations in the twenty-first century, new "waves" in these and other fields are well worth watching. What kinds of new flowers and fruits will bloom on new trees, and how big will they be? It will be great fun to find out.

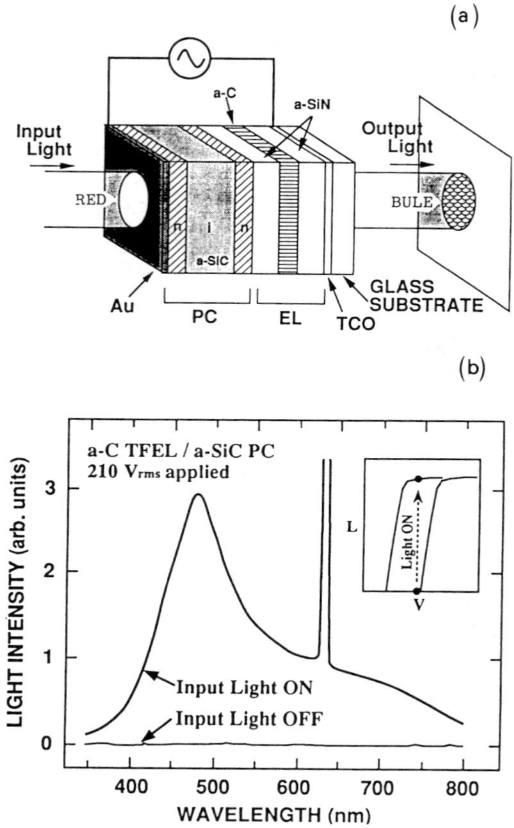

FIGURE 29. Schematic illustration of image converter device utilized by a-C EL display (a) and its output light spectrum (b).

References

1. Y. Hamakawa, *in* "Current Topics in Amorphous Materials—Physics and Technology" (Y. Sakurai *et al.*, eds.), Part 4, pp. 301–421. Elsevier, Amsterdam, 1993.
2. Y. Nakata, H. Sannomiya, S. Moriuchi, Y. Inoue, K. Nomoto, A. Yokota, M. Itoh, and T. Tsuji, *Optoelectron.—Devices Technol.* **5,** 209 (1990).
3. H. Okamoto, Y. Nitta, T. Yamaguchi, and Y. Hamakawa, *Sol. Energy Mater.* **3,** 313 (1980).
4. A. Madan and S. R. Ovshinsky, *Philos. Mag., Part B* **40,** 259 (1979).
5. Y. Tawada, H. Okamoto, and Y. Hamakawa, *Appl. Phys. Lett.* **39,** 237 (1982).
6. Y. Yukimoto, *in* "Amorphous Semiconductor—Technologies and Devices, JARECT" (Y. Hamakawa, ed.), Vol. 16, pp. 136–147. Ohm-Sha/North-Holland, Amsterdam, 1984.

7. Y. Kuwano and S. Tsuda, *in* "Amorphous Semiconductor—Technologies and Devices, JARECT" (Y. Hamakawa, ed.), Vol. 16, pp. 108–118. Ohm-Sha/North-Holland, Amsterdam, 1984.
8. Y. Tawada, K. Tsuge, M. Kondo, H. Okamoto, and Y. Hamakawa, *J. Appl. Phys.* **31,** 5273 (1982).
9. Y. Hamakawa, Y. Matsumoto, and G. Hirata, *Mater. Res. Soc. Symp. Proc.* **164,** 291 (1990).
10. Y. Hamakawa, K. Hattori, and H. Okamoto, *a-Si Sol. Cell Contractors Meet., Sunshine Proj. MIT, 1990.*
11. H. Shirai, D. Das, J. Hanna, and I. Shimizu, *Tech. Dig. Photovoltaic Sci. Eng. Conf., 5th, Kyoto* p. 59 (1990).
12. Y. Tawada and H. Yamagishi, *a-Si Sol. Cell Contractors Meet., Sunshine Proj., MIT, 1990.*
13. H. Okamoto, H. Kida, S. Nonomura, and Y. Hamakawa, *Sol. Cells* **9,** 53 (1983).
14. H. Okamoto, Y. Nitta, T. Yamaguchi, and Y. Hamakawa, *Sol. Energy Mater.* **3,** 313 (1980).
15. M. Ohnishi, H. Nishiwaki, S. Tsuda, and Y. Kuwano, *Proc. IEEE Photovoltaic Spec. Conf., 17th* p. 1121 (1984).
16. Y. Uchida and H. Haruki, *New Energy Ind. Symp., 3rd, NEF* p. 24 (1983).
17. H. Okaniwa, M. Asano, K. Nakatani, M. Yano, and K. Suzuki, *Jpn. J. Appl. Phys.* **21,** Suppl. 21-2, 239 (1982).
18. Y. Koshi, H. Tanaka, H. Nishiwaki, and Y. Kuwano, *Proc. IEEE Photovoltaic Spec. Conf., 17th* p. 1213 (1984).
19. K. Ishibitsu, Y. Nitta, and K. Kimiura, *Spring Meet. Jpn. Soc. Appl. Phys,,* 1984.
20. Y. Hamakawa and H. Okamoto, *in* "Advances in Solar Energy" (K. W. Boer, ed.), Vol. 5, pp. 1–98. Plenum, New York, 1980.
21. K. Miyachi, N. Ishigo, and N. Fukuda, *Proc. E. C. Photovoltaic Sol. Energy Conf., 11th* p. 88 (1993).
22. Y. Hamakawa, *Int. J. Sol. Energy* **1,** 125 (1982).
23. G. Nakamura, M. Kato, H. Kondo, Y. Yukimoto, and K. Shirahara, *J. Phys. (Paris), Colloq.* **42,** C-483 (1981).
24. Y. Hamakawa, Y. Tawada, K. Nishimura, K. Tsuge, M. Kondo, K. Fujimoto, S. Nonomura, and H. Okamoto, *Proc. IEEE Photovoltaic Spec. Conf., 16th* p. 679 (1982).
25. K. Fujimoto, H. Kawai, H. Okamoto, and Y. Hamakawa, *Sol. Cells* **11,** 357 (1984).
26. H. Iida, T. Miyado, and Y. Hayashi, *Fall Meet. Jpn. Soc. Appl. Phys., 1983.*
27. K. W. Mitchel, *Optoelectron.—Devices Technol.* **5,** 275 (1990).
28. Y. Hamakawa, *Proc. Euroforum New Energy Cong.* **1,** 194 (1988).
29. Y. Ichikawa, *Proc. E. C. Photovoltaic Sol. Energy Conf., 11th* p. 203 (1992).
30. S. Guha, personal communication (1994).
31. W. Ma, T. Horiuchi, C. C. Lim, M. Yoshimi, H. Okamoto, and Y. Hamakawa, *Proc. IEEE Photovoltaic Spec. Conf., 23rd* p. 338 (1993).
32. J. L. Pankove and J. E. Berkeyheiser, *Appl. Phys. Lett.* **37,** 705 (1980).
33. M. Stutzmann, *Appl. Phys. Lett.* **47,** 21 (1985).
34. M. Stutzmann, *Philos. Mag., Part B* **56,** 63 (1987).
35. K. Morigaki and F. Yonezawa, *J. Non-Cryst. Solids* **164–166,** 215 (1993).
36. T. Kagawa, N. Matsumoto, and K. Kumabe, *Proc. Conf. Solid State Devices, 13th* p. 311 (1981).
37. Y. Kanoh, S. Usui, S. Sawada, and M. Kikuchi, *Tech. Dig. 1981 IEDM* p. 313 (1981).
38. S. Tomiyama, *in* "Current Topics in Amorphous Materials" (Y. Sakaurai, Y. Hamakawa, T. Masumoto, K. Shirae, and K. Suzuki, eds.), p. 403. Elsevier, Amsterdam, 1993.
39. M. Matsumura, H. Hayama, Y. Nara, and K. Ishibashi, *Proc. Conf. Solid State Devices, 13th* p. 311 (1980).
40. E. Maruyama, T. Hirai, T. Fujita, N. Goto, Y. Isozaki, and K. Sidara, *Proc. Conf. Solid State Devices, 6th* p. 97 (1974).

41. T. Baji, Y. Shimomoto, H. Matsumura, N. Koike, T. Akiyama, S. Sasano, and T. Tsukada, *Proc. Conf. Solid State Devices, 14th* p. 269 (1981).
42. S. Ishioka, Y. Imamura, Y. Takasaki, C. Kusano, T. Hirai, and S. Nobutoki, *Proc. Conf. Solid State Devices, 14th* p. 461 (1981).
43. T. Kawamura, *in* "Amorphous Semiconductors—Technologies and Devices, JARECT" (Y. Hamakawa, ed.), Vol. 6, pp. 245–254. Ohm-Sha/North-Holland, Amsterdam, 1983.
44. Y. Hamakawa, ed., "Amorphous Semiconductors—Technologies and Devices, JARECT," Vol. 6. Ohm-Sha/North-Holland, Amsterdam, 1983.
45. J. L. Pankove and D. E. Carlson, *Appl. Phys. Lett.* **29,** 620 (1976).
46. D. Kruangam, M. Deguchi, H. Okamoto, and Y. Hamakawa, *Optoelectron.—Devices Technol.* **1,** 76 (1986).
47. Y. Hamakawa, D. Kruangam, H. Okamoto, and H. Takakura, *Proc. ESSDERC 87* p. 1025 (1987).
48. D. Kruangam, M. Deguchi, H. Okamoto, and Y. Hamakawa, *Proc. Mater. Res. Soc. Spring Meet.* p. 64 (1987).
49. T. Toyama, M. Yoshimi, H. Okamoto, and Y. Hamakawa, *Optoelectron.—Devices Technol.* **9,** 401 (1994).

Index

T

U

Plasma–Materials Interactions

Orlando Auciello and Daniel L. Flamm, *Plasma Diagnostics:* Volume 1, *Discharge Parameters and Chemistry;* Volume 2, *Surface Analysis and Interactions*

Dennis M. Manos and Daniel L. Flamm, *Plasma Etching: An Introduction*

Riccardo d'Agostino, *Plasma Deposition, Treatment, and Etching of Polymers*

Giovanni Bruno, Pio Capezzuto, and Arun Madan, *Plasma Deposition of Amorphous Silicon-Based Materials*

ISBN 0-12-137940-X

90038

9 780121 379407